STUDY GUIDE

TO ACCOMPANY

CHEMISTRY

The Study of Matter and Its Changes

JAMES E. BRADY
St. John's University, New York

JOHN R. HOLUM
Augsburg College, Minnesota

JOHN WILEY & SONS, INC.
NEW YORK / CHICHESTER / BRISBANE / TORONTO / SINGAPORE

ISBN 0-471-57876-2

Printed in the United States of America

10 9 8 7 6 5 4 3 2 1

PREFACE

Our goal in preparing this Study Guide was to provide the student with a structured review of important concepts and problem solving approaches. We begin with a preliminary chapter that introduces students to the text and to the study guide. Here we explain how to use the text and study guide together most effectively, and we explain the importance of regular class attendance and how to develop proper study habits. One of the principal features of the textbook is the organized approach to problem solving employing the chemical toolbox analogy. Because this approach is likely to be new to the student, we discuss in some detail how this analogy can help students expand their problem solving skills.

Each of the remaining chapters in the Study Guide begins with a brief overview of the chapter contents, followed by a list of Learning Objectives. Because students tend to study one section at a time, we divide each chapter in the Study Guide into sections that match one-for-one the sections in the text. Each section provides a review of the topics covered in the text. Here we call to students' attention key concepts and important facts. In many places, additional explanations of difficult topics are provided, and where students often find particular difficulty, additional worked examples are given.

In keeping with our aim of providing students with frequent opportunities to hone their skills and test their knowledge, almost all sections of the Study Guide include a brief Self-Test that consists of questions and problems that supplement those in the text. The answers to all the Self-Test exercises appear at the ends of the chapters. Many sections also contain a Thinking It Through question of the type found in the textbook, and for each there is a worked-out answer at the end of the chapter.

Following the Self-Test there is a list of new terms introduced in the section. As an exercise, the student is encouraged to write out the definitions of these terms in their notebook.

As an additional aid in problem solving, tables listing the Chemical Tools and their functions as well as summaries of important equations and other useful information are found on separate tear-out pages at the ends of chapters. The aim is to provide the student with another means to reinforce the key problem solving concepts.

James E. Brady

John R. Holum

CONTENTS

Before You Begin...

Before you begin your general chemistry course, read the next several pages. They're designed to tell you how to use this study guide and to give you a few tips on improving your study habits.

How to Use the Study Guide

This book has been written to parallel the topics covered in your text, *Chemistry: The Study of Matter and Its Changes*. Each chapter begins with a very brief overview of the chapter contents followed by a list of learning objectives. Read these before beginning a chapter, and then read them again after you've finished to be sure you have met the goals described. For each section in the textbook, you will find a corresponding section in the Study Guide. In the Study Guide, the sections are divided into **Review, Thinking It Through, Self-Test** and **New Terms**.

After you've read a section in the text, turn to the study guide and read the **Review**. This will point out specific ideas that you should be sure you have learned. Sometimes you will be referred back to the text to review topics there. Sometimes there will be additional worked-out sample problems. Work with the Review and the text together to be sure you have mastered the material before going on.

In some sections you will find questions titled **Thinking It Through**. The goal of these questions is to allow you to test you ability in figuring out *how* to solve problems. The emphasis is on the *method*, not the *answer*. (We will have more to say about this later.) In most sections you will also find a short **Self-Test** to enable you to test your knowledge and problem-solving ability. The answers to all of the Thinking It Through and Self-Test questions are located at the ends of the chapters in the Study Guide. However, you should try to answer the Thinking It Through and Self-Test questions without looking up the answers. A space is left after each Self-Test question so that you can write in your answers and then check them all after you've finished.

Chemical Vocabulary

An important aspect of learning chemistry is becoming familiar with the language. There are many cases where lack of understanding can be traced to a lack of familiarity with some of the terms used in a discussion or a problem. A great deal of effort was made in your textbook to adequately define terms before using them in discussions. Once a term has been defined, however, it is normally used with the assumption that you've learned its meaning. It's important, therefore, to learn new terms as they appear, and for that reason, most of them are set in boldface type in the text. At the end of each section of the study guide there is a list of these **New Terms**. To test your knowledge of them, you are asked to write out their meanings. This will help you review them later when you prepare for quizzes or examinations. At the end of the textbook there is a Glossary which you can use to be sure you understand the meanings of the new terms.

Study Habits

You say you want to get an A in chemistry? That's not as impossible as you may have been led to believe, but it's going to take some work. Chemistry is not an easy subject—it involves a mix of memorizing facts, understanding theory, and solving problems. There is a lot of material to be covered, but it won't overwhelm you if you ***stay up to date***. Don't fall behind, because if you do, you are likely to find that you can't catch up. Your key to success, then, is *efficient* study, so your precious study time isn't wasted.

Efficient study requires a regular routine, not hard study one night and nothing the next. At first, it's difficult to train yourself, but after a short time you will be surprised to find that your study routine has become a study habit, and your chances of success in chemistry, or any other subject, will be greatly improved.

To help you get more out of class, try to devote a few minutes the evening before to reading, in the text, the topics that you will cover the next day. Read the material quickly just to get a feel for what the topics are about. Don't worry if you don't understand everything; the idea at this stage is to be aware of what your teacher will be talking about.

Your lecture instructor and your textbook serve to complement one another; they provide you with two views of the same subject. Try to attend lecture regularly and take notes during class. These should include not only those things your teacher writes on the blackboard, but also the important points he or she makes verbally. If you pay attention carefully to what your teacher is saying in class, your notes will probably be somewhat sketchy. They should, however, give an indication of the major ideas. After class, when you have a few minutes, look over your notes and try to fill in the bare spots while the lecture is still fresh in your mind. This will save you a lot time later when you finally get around to studying your notes in detail.

In the evening (or whatever part of the day you close yourself off from the rest of the world to really study intensely) review your class notes once again. Use the text and study guide as directed above and really try to learn the material presented to you that day. If you have prepared before class and briefly reviewed the notes afterward, you'll be surprised at how quickly and how well your concentrated study time will progress. You may even find yourself enjoying chemistry!

As you study, continue to fill in the bare spots in your class notes. Write out the definitions of new terms in your notebook. In this way, when it comes time for an exam you should be able to review for it simply from your notes.

At this point you're probably thinking that there isn't enough time to do all the things described above. Actually, the preparation before class and brief review of the notes shortly after class takes very little time and will probably save more time than they consume.

Well, you're on your way to an A. There are a few other things that can help you get there. If you possibly can, spend about 30 minutes to an hour at the end of a week to review the week's work. Psychologists have found that a few brief exposures to a subject are more effective at fixing them in the mind than a "cram" session before an exam. The brief time spent at the end of a week can save you hours just before an exam (efficiency!). Try it (you'll like it); it works.

There are some people (you may be one of them) who still have difficulty with chemistry even though they do follow good study habits. Often this is because of weaknesses in their earlier education. If, after following intensive study, you are still fuzzy about something, speak to your teacher about it. Try to clear up these problems before they get worse. Sometimes, by having study sessions with fellow classmates you can help each other over stumbling blocks. Group study is very effective, because if you find you can explain something to someone else, you really know the subject. But if you can't explain a topic, then it requires more study.

Problem Solving—Using Chemical Tools

Your course in chemistry provides a unique opportunity for you to develop and sharpen your problem solving skills. Just as in life outside the classroom, the problems you will encounter in chemistry are not only numerical ones. In chemistry, you will also find problems related to theory and the application of concepts. The techniques that we apply to these various kinds of problems do not differ much, and one of the goals of your textbook and this Study Guide is to provide a framework within which you can learn to solve all sorts of problems effectively.

If you've read the "To The Student" message at the beginning of the textbook, you learned that we view solving a chemistry problem as not much different than solving a problem in auto repair. Both involve the application of specific tools that accomplish specific tasks. A mechanic uses tools such as screwdrivers and wrenches; you will learn to use a different set of tools—ones that we might call *chemical tools*.

C*hemical tools* are the simple one-step tasks that you will learn how to do, such as changing units from feet to meters, or degrees Fahrenheit to degrees Celsius. Solving more complex problems just involves combining simple tools in various ways. The secret to solving complex problems, therefore, is learning how to choose the chemical tools that must be used.

Building a Chemical Toolbox

Our first goal is to clearly identify the tools you will have at your disposal. As you study the text, the concepts you will need to solve problems are marked by an icon in the margin when they are introduced. (To see what the icon looks like, refer to the "To The Student" message in the text.) The chemical tools are summarized at the end of a textbook chapter in a section titled *Tools You Have Learned* and they are also collected in table form at the end of each of the chapters in this Study Guide.

In both the text and the Study Guide there are worked examples that illustrated a wide variety of problems and their solutions. You will notice that in many of them there is a section titled *Analysis*. The Analysis section describes the thinking that goes into solving the problem and identifies the tools needed to do the job. Be sure to study the Examples thoroughly, and also be sure to work on the Practice Exercises that follow the Examples.

Solving Problems

You should always think of solving a problem as a two-step process. The first step is figuring out *how* to solve it. The second step is obtaining the answer. Of course, once you know how to solve the problem, obtaining the answer is easy. Therefore, let's look at a method you can use when working on a problem you haven't seen before—one for which the solution is not immediately obvious. To do this, we will look at a problem of the type you will encounter in Chapter 3. If you've had a previous course in chemistry, you will recognize much of the concepts presented. If they are unfamiliar, don't be concerned. The goal at this time is to illustrate how the chemical tools approach can be used to help find a solution to a problem.

Problem

Assemble all the information needed to determine the number of grams of Al that will react with 900 molecules of O_2 to form Al_2O_3, and then describe how the information can be used to find the answer.

The first step in solving the problem is determining what kind of problem it is. In this case, it is a problem dealing with a subject we call *stoichiometry*. (Don't worry, you will learn about all this later.)

Now that we have identified the *kind* of problem, we look over the tools that apply to stoichiometry problems. Here is a table that lists the tools.

Tools that apply to stoichiometry:

Tool	Function
Atomic mass	convert between grams and moles for element
Formula mass (molecular mass)	convert between grams and moles for compound
Chemical formula	gives atom ratio in a compound gives mole ratio in a compound
Chemical equation	gives mole ratios in a reaction
Avogadro's number	converts between number of particles and moles
Molarity	converts between moles and volume for a solution

Next, we examine the problem to identify the quantities that relate to the tools we have at hand. Notice that we've drawn boxes around the quantities.

> Assemble all the information needed to determine the number of [grams of Al] that will react with [900 molecules] of [O_2] to form [Al_2O_3] and then describe how the information can be used to find the answer.

Now we begin to assign specific numbers to quantities as we assemble the final set of tools we will use to solve the problem. We've collected the information in a table just to make it easier for you to follow. Notice that we have not used all the tools related to stoichiometry. Instead, we have selected just the tools that apply the the quantities in the problem.

Quantity in question	Tool related to it	Relationship
[grams of Al]	atomic mass	27 g Al = 1 mol Al
[900 molecules] [O_2]	Avogadro's number	6.02×10^{23} molecules O_2 = 1 mol O_2
[Al_2O_3] [O_2]	chemical formula	2 mol Al = 3 mol O 1 mol O_2 = 2 mol O

The information in the column at the right is what we use to obtain the answer. As you will learn, we can use a method called the factor label method to make sure the units of the answer work out correctly. The proper setup of the solution is

$$900 \text{ molecules } O_2 \times \frac{1 \text{ mole } O_2}{6.02 \times 10^{23} \text{molecules } O_2} \times \frac{2 \text{ mol O}}{1 \text{ molecule } O_2} \times \frac{2 \text{ mol Al}}{3 \text{ mole O}} \times \frac{27 \text{ g Al}}{1 \text{ mol Al}} = \text{answer}$$

Notice that we have not actually calculated the answer. Nevertheless, we really have *solved* the problem; we just haven't done the dirty work of doing the calculation. At the end of most chapters in the textbook, and in some of the sections in the Study Guide, you will find questions titled Thinking It Through. These questions ask you to figure out what you need to know to solve various problems, but not what the answers are. The goal is to make you *think* about how to solve the problems without having to worry out the answer. They are worthwhile exercises and you should be sure to work on them. As you will see, some are pretty difficult. But as they say, "No pain, no gain!"

We realize, of course, that many problems have more than one path to the answer. We understand that after correctly analyzing a problem and after recognizing what tools must be used, intermediate calculations and thought processes can validly follow more than one *order*. Therefore, you might choose a path in which the order of the steps is different from ours. This is why we provide answers to the thinking it through exercises, so that you can have the reinforcement (and the reward) of comparing answers when your method differs from ours.

Time to Begin

As you begin your study of chemistry, we wish you well. Move on to the course now, and good luck on getting that A!

Chapter 1
INTRODUCTION

This chapter introduces you to some basic concepts which you will need to understand future discussions in class and the textbook, and to function effectively in the laboratory part of your course. We begin by explaining what chemistry is about—namely, chemicals and chemical reactions. You will also learn about the scientific method, which describes how scientists learn about nature. And you will learn about the subjects that will be the principal focus of our study—matter and energy.

In the second half of the chapter we examine the importance of measurements and the units used to express them. You will study the modern version of the metric system and learn the concept of significant figures. Finally, we discuss density and specific gravity to illustrate how measurement and calculation combine to give us useful, quantitative properties of matter.

If you've had a prior course in chemistry, much of what is discussed in this chapter will seem familiar. Nevertheless, be sure you really understand it fully and can do the assigned homework. In particular, be sure you've learned the meanings of the bold-faced terms in the text as well as equations and other relationships that are placed between thick light-blue lines, such as Equation 1.1 on page 6. Important equations are summarized at the end of this Study Guide chapter.

Learning Objectives

As you study of this chapter, keep in mind the following objectives:

1 To learn the meaning of a chemical reaction.

2 To learn how science develops through the application of the scientific method. In particular, you should learn the distinction between a law and a theory.

3. To learn the definitions of matter and energy, the difference between kinetic and potential energy, the law of conservation of energy, and the difference between heat and temperature.

4 To learn how matter is identified by its characteristics, or properties, and how properties are classified.

5 To learn the units used for expressing measurements in the sciences and how to convert among differently sized units.

6 To learn the kinds of measurements normally made in the laboratory, the apparatus used to obtain them, and the units used to express them.

7 To learn how the number of digits (significant figures) reported in a measurement relates to the reliability of the measurement. Be sure you know the difference between accuracy and precision.

8 To learn how to use the units associated with quantities as a tool for setting up the arithmetic in a problem.

9 To learn about density and specific gravity and to use them in calculations.

1.1 What is Chemistry?

Review

This section starts with a discussion of the way chemistry has affected our lives and the way chemists respond to studying this subject. To begin the course with the proper attitude, you might take a few moments to imagine what life would really be like if we had to do without the materials created through chemical research.

An important point made in this section is that when chemical changes (chemical reactions) occur, the characteristics (properties) of the substances involved change, often dramatically. This is because a chemical reaction transforms substances into new chemicals, which have properties that differ from the chemicals present initially. Observing such changes is what makes chemistry so fascinating, especially in the laboratory.

Self-Test

1. A simple experiment you can perform in your kitchen at home or in an apartment is to add a small amount of milk of magnesia to some vinegar in a glass. Stir the mixture and observe what happens. Then add some milk of magnesia to the same amount of water and stir. What evidence did *you* observe that suggests that there is a chemical reaction between the milk of magnesia and the vinegar?

____

____

2. Drop an Alka Seltzer tablet into a glass of water. Observe what happens. What evidence is there that a chemical reaction is taking place? ______________________________

____

New Terms

Write the definitions of the following terms, which were introduced in this section. If necessary, refer to the Glossary at the end of the text.

chemistry

chemical reaction

1.2 Chemistry and the Scientific Method

Review

The sequence of steps described by the scientific method is little more than a formal description of how people logically analyze any problem, scientific or otherwise. Observations are made in order to collect data (empirical facts), which are then analyzed in a search for generalizations. Generalizations often lead to laws, which are concise statements about the behavior of chemical or physical systems. Laws, however, offer no explanations about *why* nature behaves the way it does. Tentative explanations are called hypotheses; tested explanations are called theories. The scientific method consists of collecting data in experiments, formulating theories, and testing the theories by more experimentation. Based on the results of new experiments, the theories are refined, tested further, refined again, and so on.

Self-Test

3. Identify each of the following statements as either a law or a theory.
   (a) In general, what goes up must come down. ______________
   (b) The ice ages resulted from the tilting of the earth's rotation axis which was caused by the earth being hit by very large meteors.
   ____

4. What does *empirical* mean? ______________________________

____

New Terms

Write the definitions of the following terms, which were introduced in this section. If necessary, refer to the Glossary at the end of the text.

natural science	empirical fact	hypothesis
law	data	theory generalization
scientific method		

1.3 Matter and Energy

Review

Matter has mass and occupies space. It includes all the tangible things we encounter. Mass and weight are not the same, although we often use the terms interchangeably. The mass of an object is constant, and refers to the amount of matter in an object. The weight of an object can vary depending on the force of gravity.

Energy is something an object has if it has the capability of performing work. There are two kinds of energy an object can have. Kinetic energy is energy of motion and can be calculated from the object's mass (m) and velocity (v) by the equation $KE = 1/2\ mv^2$. Potential energy is stored energy, and the potential energy stored in chemicals, which can be released in chemical reactions, is sometimes called chemical energy. The law of conservation of energy states that energy cannot be destroyed, but only changed from one form to another.

Energy can be transferred between objects in a variety of ways, but we most commonly observe energy being transferred as heat. (You will learn later that heat is actually a kind of kinetic energy that the individual atoms of a substance possess.) Temperature is a measure of the intensity of heat, and heat always flow spontaneously from hot objects to cool ones.

Self-Test

5. How does the kinetic energy of a 1000-lb car traveling at 60 mph compare with the kinetic energy of a 4000-lb car traveling at the same speed? ____________________

    ____

6. How does the kinetic energy of a 2000-lb car moving at 20 mph compare with the kinetic energy of the same car traveling at 60 mph? ________

7. What is the difference between potential energy and chemical energy?

New Terms

Write the definitions of the following terms, which were introduced in this section. If necessary, refer to the Glossary at the end of the text.

matter	energy	chemical energy
mass	kinetic energy	temperature
weight	potential energy	

1.4 Properties of Matter

Review

In the text, we see that the properties of substances can be classified in two ways. One way is to divide them into either physical properties or chemical properties. The other is to divide them into intensive or extensive properties.

Chemical and Physical Properties

Physical properties are ones that can be observed without changing the chemical makeup of a substance. In general, physical properties can be specified without reference to another chemical substance. Examples are an object's mass, or the temperature at which it melts, or its volume. When we describe a chemical property of a substance, we describe how the substance reacts chemically with something else. Such a chemical reaction produces new chemical substances, so after observing a chemical reaction, the substance has changed into a different substance. A chemical property of iron, for example, is that it rusts when in contact with air and moisture. When we observe this property, the iron changes to rust.

Extensive and Intensive Properties

Extensive properties, such as mass or volume, depend on the size of the sample of matter being examined. Although extensive properties are important for a given sample, they are not especially useful for identifying substances. More useful are intensive properties, because all samples of a given substance have identical values for its intensive properties. For example, if we were asked whether a sample of a liquid was water, we would examine its properties and compare them to those of a known sample of water. If we were to find that the mass of the liquid is 12.0 g, we still would not be any closer to knowing whether or not the sample is water. By itself, the mass is of no value, because different samples of water, or any other liquid, have different masses. However, if we further note that the liquid is clear, has no color, has no odor, and freezes at 0 °C, we would strongly suspect the sample to be water. This is because *all* samples of pure water are clear, colorless, odorless, and freeze at 0 °C.

Self-Test

8. Identify the following as chemical properties or physical properties.

   (a) Nitroglycerine explodes if it is heated. ____________________

   (b) Gold is a yellow metal. ____________________

9. Sodium is a soft, silvery metal that melts at 97.8 °C. It burns with a yellow light in the presence of chlorine gas to give the compound sodium chloride (table salt).

   (a) What are some physical properties of sodium? ____________________

   ____

   (b) Give a chemical property of sodium. ____________________

   ____

10. (a) Give two examples of intensive properties. ____________________

    ____

    (b) Give two examples of extensive properties. ____________________

    ____

New Terms

Write the definitions of the following terms, which were introduced in this section. If necessary, refer to the Glossary at the end of the text.

property
physical property
extensive property
intensive property
chemical property

1.5 Units of Measurement

Review

Quantitative measurements (observations involving numbers) are necessary in the sciences in order to make meaningful progress. Numbers that come from measurement must be expressed in some sort of units. A metric-based system has the advantage that converting among units is accomplished by just moving the decimal point.

The International System of Units (the SI) is founded on a set of carefully defined base units, which are given in Table 1.1. Quantities other than mass, length, time, etc. are obtained from these base quantities by mathematical operations, and their units (called derived units) are obtained from the base units by the same operations. For example, volume is a product of three length units

$$\text{length} \times \text{width} \times \text{height} = \text{volume}$$

The unit for volume is the product of the units for length, width, and height.

$$\text{meter} \times \text{meter} \times \text{meter} = \text{meter}^3$$
$$\text{m} \times \text{m} \times \text{m} = \text{m}^3$$

Similarly, speed is expressed as a ratio of distance divided by time. The SI base unit for distance is the meter and the base unit for time is the second. Therefore

$$\text{speed} = \frac{\text{distance}}{\text{time}} = \frac{\text{meter}}{\text{second}} = \frac{\text{m}}{\text{s}}$$

Often, the base units (or the derived units that come from them) are too large or too small to be used conveniently. For example, if we were to use cubic meters to express the volumes of liquids that we measure in the laboratory, we would find ourselves using very small numbers such as 0.000025 m^3 or 0.000050 m^3. Because they have so many zeros, these values are difficult to comprehend. To make life easier for us, the SI has a simple way of making larger or smaller units out of the basic ones. This is done with the decimal multipliers and SI prefixes given in Table 1.2 on page 10. Be sure you learn the ones in the table at the top of page 14 of the Study Guide.

The SI prefixes are tools we can use to scale units to convenient sizes and for translating between differently sized units. You will see how this is done in Section 1.6. Notice that each prefix stands for a particular decimal multiplier. Thus *kilo* means "$\times$ 1000" or "$\times$ 10^3." This lets us translate a quantity into the value that it has when expressed in terms of the base units. For example, suppose we wanted to know how many meters are in 25 kilometers (25 km). Since kilo (k) means "$\times$ 1000," then

SI Prefixes and Decimal Multipliers

Prefix	Symbol	Multiplication Factor
mega	M	10^6
kilo	k	10^3
deci	d	10^{-1}
centi	c	10^{-2}
milli	m	10^{-3}
micro	m	10^{-6}
nano	n	10^{-9}
pico	p	10^{-12}

$$25 \text{ km} = 25 \times 1000 \text{ m}$$
$$= 25{,}000 \text{ m}$$

Similarly, a length of 25 millimeters (25 mm) would be

$$25 \text{ mm} = 25 \times 0.001 \text{ m}$$
$$= 0.025 \text{ m}$$

Units for Laboratory Measurements

Length, volume, mass and temperature are four commonly measured quantities in the lab. Units usually used for length are millimeters and centimeters. Remember that 10 mm = 1 cm. It is also useful to remember one crossover relationship between the metric and English units for length—for example,

1 in. = 2.54 cm

The liter, which is the traditional metric unit of volume, is a bit too large to be convenient for most of the laboratory measurements that you will encounter. Most laboratory glassware is marked in units of milliliters (mL). Remember:

$$1000 \text{ mL} = 1 \text{ L}$$
$$1 \text{ cm}^3 = 1 \text{ mL}$$

Be sure to practice converting between milliliters and liters; it is an operation you will perform frequently throughout the course.

Mass is normally measured in grams. The SI base unit is 1000 g (1 kilogram). The apparatus used to measure mass is called a balance. Temperature is measured with a thermometer, and in the sciences it is measured in units of degrees Celsius (°C). The Celsius and Fahrenheit degree units are of different sizes; five degree units on the Celsius scale correspond to nine degree units on the Fahrenheit scale. Equation 1.2 enables you to make conversions between °C and °F.

The SI unit of temperature is the kelvin (K). Zero on the Kelvin scale corresponds to –273 °C (rounded to the nearest degree) and is called absolute zero, because it is the coldest temperature. (Notice that the name of the *temperature scale* is capitalized; the name of the *unit* measured on that scale, the kelvin, is not capitalized.) The kelvin and Celsius degree are the same size, so a temperature change of 10 K, for example, is the same as a temperature change of 10 °C. Be sure you can convert from °C to K (Equation 1.3), because when temperature is needed in a calculation, it nearly always must be expressed in kelvins. In mathematical equations, the capital letter T is used to stand for the Kelvin temperature.

Self-Test

11. Give the SI base unit and its abbreviation for

    (a) mass ____________________

    (b) length ____________________

    (c) time ____________________

    (d) electric current ____________________

    (e) temperature ____________________

12. Torque (pronounced tork) is a quantity that expresses a twisting force, such as that applied to a nut or a bolt by a wrench. It is a product of distance × force and in English units is normally given in foot pounds. The SI derived unit for force is the Newton (symbol, N). What is the SI derived unit for torque?

    Answer ____________________

13. Fill in the blanks.

    (a) 1 __________ gram = 0.01 gram

    (b) 1 __________ meter = 10^{-9} meter

    (c) 1 __________ g = 0.001 g

    (d) 1 __________ m = 1000 m

    (e) 1 pm = __________ m

    (f) 1 μg = __________ g

(g) 1 dm = __________ m

14. Fill in the blanks.

    (a) 63 dm = __________ m

    (b) 0.023 Mg = __________ g

    (c) 2450 nm = __________ m

    (d) 2487 cm = __________ m

15. Fill in the blanks.

(a)	______ cm = 1.35 m	(g)	______ mL = 0.022 L
(b)	______ mm = 22.4 cm	(h)	______ L = 346 mL
(c)	______ cm = 32.6 mm	(i)	______ L = 2.41 mL
(d)	______ mL = 1.250 L	(j)	______ K = 25 °C
(e)	______ cm^3 = 246 mL	(k)	______ °C = 265 K
(f)	______ L = 525 cm^3	(l)	______ °C = 300 K

16. (a) What Celsius temperature corresponds to 23 °F? __________

    (b) What Fahrenheit temperature equals 10 °C? __________

17. An object has a mass of 12 g, measured in a laboratory in California. What would its mass be if measured on the moon, where the gravity is only one sixth of Earth's gravity? __________

New Terms

Write the definitions of the following terms, which were introduced in this section. If necessary, refer to the Glossary at the end of the text.

qualitative observation	milliliter (mL)
quantitative observation	weighing
International System of Units	balance
base unit	Celsius scale
derived unit	Kelvin temperature scale
decimal multiplier	kelvin (K)
meter (m)	absolute zero
centimeter (cm)	joule (J)
millimeter (mm)	calorie (cal)

1.6 Scientific Calculations and Significant Figures

Review

When a number is obtained by measurement, the digits in the number are called significant figures. They include all the digits known for sure plus the first digit that contains some uncertainty. Expressing a measurement to the correct number of significant figures allows us to express to someone else who sees that number how precise the measurement is. For example, a measured length of 32.47 cm has four significant figures. It tells us that the 3, 2, and 4 are known with certainty and the measuring instrument allowed the hundredths place to be *estimated* to be a 7. Since no digit is reported in the thousandths place, it is assumed that no estimate of that place could be obtained. In general, the larger the number of significant figures in a measurement, the greater is its precision.

The rules given on page 18 are tools we use to correctly express the number of significant figures in a computed quantity.

Multiplication and Division

The answer cannot contain more significant figures than the factor that has the fewest number of significant figures. Study the example on page 18.

Addition and Subtraction

The answer is rounded to the same number of decimal places as the quantity with the fewest decimal places. See the example on page 19.

When using exact numbers in calculations, they can be considered to have as many significant figures as desired. They contain no uncertainty.

Example 1.1 Calculations Using Significant Figures

Problem

Assume that all of the numbers in the following expression come from measurement. Compute the answer to the correct number of significant figures.

$$(3.25 \times 10.46) + 2.44 = ?$$

Solution

When we perform the multiplication (which must, of course, be done first) we obtain 33.995. This should be rounded to three significant figures because that's how many there are in 3.25. The result is 34.0, which is then added to 2.44

34.0	one decimal place
+ 2.44	two decimal places
36.4	rounded to one decimal place

The correct answer, therefore, is 36.4

The Factor-Label Method

The factor label method is a tool we use to properly set up the arithmetic in numerical problems. It relies on the cancellation of units from the numerator and denominator of fractions. Relationships between units are used to form conversion factors by which the given quantity is multiplied. Successive conversion factors are used until the given units are converted to the units desired for the answer.

Sometimes, the factor label method helps us find the relationships we need to solve a problem. For example, to convert 65.0 gallons to liters we need a relationship between gallons and liters. In Table 1.3 (page 11) we find the following:

$$1 \text{ gal} = 3.786 \text{ L}$$

This can be used to form two different conversion factors.

$$\frac{1 \text{ gal}}{3.786 \text{ L}} \quad \text{and} \quad \frac{3.786 \text{ L}}{1 \text{ gal}}$$

To convert 65.0 gal to liters, we choose to multiply by 3.786 L/1 gal because the unit gal will cancel.

$$65.0 \cancel{\text{gal}} \times \frac{3.786 \text{ L}}{1 \cancel{\text{gal}}} = 246 \text{ L (rounded)}$$

Notice that in choosing which conversion factor to use, we watch the units that must be cancelled.

If you set up problems so that units cancel to give the correct units for the answer, your answer will also be correct, provided you have used valid relationships between the units. For example, using 1 gal = 2 L to make a conversion factor would give the wrong numerical answer because 1 gallon does *not* equal 2 liters. But if you use correct relationships, the units will guide you to the correct answer.

Thinking It Through

For the following, identify the information needed to solve the problem and show what must be done with it.

1. The distance between the earth and the moon is approximately 239,000 miles. Radio waves travel through space at the speed of light, 3.00×10^8 m/s. How many seconds does it take for a radio wave to go to the moon?

Self-Test

18. Which term, accuracy or precision, is related to the number of significant figures in a measured quantity?

 ____

19. Perform the following arithmetic and express the answers to the proper number of significant figures (assume all numbers come from measurements).

 (a) 4.87×3.1 ______________

 (b) $8.4 \div 21.02$ ______________

 (c) $14.35 + 0.022$ ______________

 (d) $145.3 - 4.68$ ______________

20. The distance between two houses was measured to be 0.27 miles. What is the distance expressed in yards (1 mile = 5280 ft)?

 ____

21. What conversion factor could you use to convert

 (a) 62 lb to kg ______________

 (b) 45 in.^{2} to cm^2 ______________

New Terms

Write the definitions of the following terms, which were introduced in this section. If necessary, refer to the Glossary at the end of the text.

significant figures	precision	factor-label method
accuracy	exact number	

1.7 Density and Specific Gravity

Review

Density is the ratio of an object's mass to its volume, as expressed in Equation 1.8 on page 23. Determining the density therefore involves two measurements and a brief calculation. In performing the calculation, be careful about significant figures and the significant-figure rules that we follow in multiplication and division.

Density is an intensive property, meaning that its value for a given substance is the same regardless of the size of the sample. Increasing the sample size increases both the mass *and* the volume, but their ratio remains the same. Density also serves as a tool for converting between the mass and volume for a sample of a substance. For example, methyl alcohol (a substance used in "dry gas" and as the fuel in "canned heat") has a density of 0.791 g/mL, which tells us that each milliliter of the liquid has a mass of 0.791 g. This information can be expressed as a conversion factor in two ways.

$$\frac{0.791\text{ g}}{1.00\text{ mL}} \quad \text{and} \quad \frac{1.00\text{ mL}}{0.791\text{ g}}$$

Suppose we wanted the mass of 12.0 mL of methyl alcohol. We would use the first factor so the units mL will cancel.

$$12.0\text{ mL} \times \frac{0.791\text{ g}}{1.00\text{ mL}} = 9.49\text{ g}$$

If we wanted the volume occupied by, say 1.50 g of methyl alcohol, we would use the second factor.

$$15.0\text{ g} \times \frac{1.00\text{ mL}}{0.791\text{ g}} = 19.0\text{ mL}$$

The factor that we choose simply depends on the units that we want to cancel.

Specific Gravity

A quantity related to density is specific gravity—the ratio of the density of a substance to the density of water. If we have the specific gravity of a substance, we can get its density in any units we wish simply by multiplying the specific gravity of the substance by the density of water in the desired units.

Thinking It Through

For the following, identify the information needed to solve the problem and show what must be done with it.

2. A rectangular piece of steel 12.0 in. long and 18.2 in wide has a mass of 12.5 lb. Steel has a specific gravity of 7.88. Water has a density of 62.4 lb/ft^3. What is the thickness of the piece of steel expressed in inches?

Self-Test

22. Why is density more useful for identifying a substance than either the mass or the volume alone? ________________________________

---

23. A sample of copper was found to have a mass of 34.2 g and a volume of 3.82 cm^3.

    (a) What is the density of copper in g/cm^3? ____________

    (b) What is the specific gravity of copper? ____________

    (c) What is the density of copper in units of lb/ft^3? (Refer to Example 1.7 on page 25 of the text for the necessary data.)

    ____

24. Aluminum has a density of 2.70 g/cm^3.

    (a) What is the mass of 8.30 cm^3 of aluminum? ____________

    (b) What is the volume in mL of 25.0 g of aluminum? ____________

New Terms

Write the definitions of the following terms, which were introduced in this section. If necessary, refer to the Glossary at the end of the text.

density | specific gravity

Answers to Thinking It Through

1. We must relate the distance, 239,000 miles, to time in seconds. The speed of light can be used to convert between meters and seconds, so we need a relationship that will take us from miles to meters. There are several ways to obtain this, but the simplest is to use the relationship between miles and kilometers given in the table of conversions on the inside rear cover of the textbook.

$$1 \text{ mi} = 1.609 \text{ km} = 1.609 \times 10^3 \text{ m}$$

The setup of the problem according to the factor label method is

$$293{,}000 \text{ mi} \times \frac{1.609 \times 10^3 \text{ m}}{1 \text{ mi}} \times \frac{1 \text{ s}}{3.00 \times 10^8 \text{ m}} \Rightarrow \text{answer}$$

Notice that all the units will cancel correctly to give the desired units for the answer. (If you work through to the answer, you should get 1.57 s.)

2. For a rectangular object, volume = length × width × thickness. We are given the length and width, so to calculate thickness we need the volume. We've

been given the specific gravity of the steel and the density of water. To see how we might use these quantities, we go back to the definitions.

$$d = \frac{m}{V} \quad \text{and} \quad \text{sp. gr.} = \frac{d_{\text{substance}}}{d_{\text{water}}}$$

From the density of the water and the specific gravity of water, we can find the density of the steel in lb/ft^3. The density of the steel equals its specific gravity multiplied by the density of water. The density so calculated is then used along with the mass of the steel to calculate the volume of the steel object, and then the thickness is obtained by dividing the volume by the product of the length × width, both expressed in ft. The thickness in units of ft is then changed to in. (The answer to the problem is 0.20 in.)

Answers to Self-Test Questions

1. You should have observed that the milk of magnesia dissolves in the vinegar, but not in plain water.
2. The fizzing is evidence that a reaction is taking place.
3. (a) law (b) theory
4. Based on observation (experiment) or experience.
5. The KE of the 1000-lb car is one-fourth of the KE of the 4000-lb car.
6. At 60 mph the KE is 9 times larger than at 20 mph.
7. They are the same kind of energy.
8. (a) chemical property (b) physical property
9. (a) soft, silvery, melts at 97.8 °C (b) Burns in presence of chlorine to give sodium chloride.
10. (a) color, melting point, boiling point, odor (b) mass, volume
11. (a) kilogram, kg, (b) meter, m, (c) second, s, (d) ampere, A, (e) kelvin, K
12. newton × meter or N m
13. (a) centi (b) nano, (c) milli, (d) kilo, (e) 10^{-12}, (f) 10^{-6}, (g) 10^{-1},
14. (a) 6.3, (b) 23,000, (c) 0.00000245, (d) 24.87
15. (a) 135, (b) 224, (c) 3.26, (d) 1250, (e) 246, (f) 0.525, (g) 22, (h) 0.346, (i) 0.00241, (j) 298, (k) –8, (l) 27
16. (a) –5 °C, (b) 50 °F
17. Its mass would be 12 g, because mass is not affected by gravity.
18. Precision
19. (a) 15, (b) 0.40, (c) 14.37, (d) 140.6
20. 475.2 yd rounds to 480 yd (2 sig. fig.)
21. (a) 1 kg/2.205 lb, (b) (2.54 cm/1 in.)^2 = 6.45 cm^2/1 in.^{2}
22. Density doesn't depend on sample size; mass and volume do depend on the size of the sample.
23. (a) 8.95 g/cm^3, (b) 8.95, (c) 558 lb/ft^3
24. (a) 22.4 g, (b) 9.26 mL

Tools you have learned

Remove this chart from the Study Guide and keep it handy when tackling homework problems.

Tool	*Function*
SI Prefixes mega (10^6), kilo (10^3), deci (10^{-1}), centi (10^{-2}), milli (10^{-3}), micro (10^{-6}), nano (10^{-9}), pico (10^{-12})	Create larger and smaller units. Convert between differently sized units.
Units in laboratory measurements 1 cm = 10 mm, 1 L = 1000 mL	Convert between these commonly used units of measurement.
Rules for counting significant figures.	Determining the number of significant figures in a number.
Rules for arithmetic and significant figures	Rounding answers to the correct number of significant figures.
Factor label method	Setting up the arithmetic in numerical problems.
Density $d = \frac{m}{V}$	Convert between mass and volume for a substance.
Specific gravity $\text{sp.gr.} = \frac{d_{\text{substance}}}{d_{\text{water}}}$	Convert density from one set of units to a different set of units

Summary of Important Equations

Kinetic energy

$$\text{K.E.} = \frac{1}{2}mv^2$$

Converting between °C and °F

$$t_F = \frac{9}{5}t_C + 32\ °\text{F}$$

Converting between °C and K

$$T_K = t_C + 273.15\ \text{K}$$

$$T_K = t_C + 273\ \text{K}$$

Density

$$d = \frac{m}{V}$$

Specific gravity

$$\text{sp. gr.} = \frac{d_{\text{substance}}}{d_{\text{water}}}$$

Chapter 2

THE STRUCTURE OF MATTER: ATOMS, MOLECULES, AND IONS

The central theme of this chapter is *organization*. Here you learn how we classify different kinds of matter according to their properties. You are also introduced to the periodic table, which is one of the most important and useful tools we have for correlating and organizing chemical information. You are not expected to memorize the periodic table, but you should begin to become familiar with it.

Along the way you also will learn a little about atoms—the tiny particles all chemicals are made of—and about the particles that make up atoms. This is just the beginning of your study of atomic structure, but you will see that understanding how atoms are put together helps you remember chemical facts and explain the properties of chemical compounds.

Learning Objectives

Throughout your study of this chapter, keep in mind the following objectives:

1 To be able to differentiate among the three basic classifications of matter: elements, compounds, and mixtures.

2 To learn the law of definite proportions and the law of conservation of mass.

3 To understand the basis for the atomic theory of matter and how it explains the laws of chemical combination.

4 To understand the basis for the assignment of relative atomic masses of the elements. You should know how the average atomic mass is obtained from the masses and relative abundances of isotopes.

5 To learn about the origin of the periodic table.

6 To learn the basics of atomic structure and the properties of the principal particles that make up an atom.

7 To learn how to write the symbol for an isotope.

8 To learn the makeup of the modern periodic table and to learn the names of various sets of elements within the table.

9 To become familiar with the three general classes of elements and their locations in the periodic table, and to learn the physical properties that differentiate elements of one class from those of another.

10 To learn the kinds of compounds formed by metals and nonmetals. You should learn the formulas of the simple hydrogen and oxygen compounds of the nonmetals and you should learn the ions formed by the elements known as the *representative elements*.

11 To learn to write the formulas of ionic compounds.

12 To learn about the properties that are characteristic of ionic and molecular compounds.

13 To learn how to write the names of compounds from their formulas and the formulas from their names.

2.1 Elements, Compounds and Mixtures

Review

In chemistry we deal with three classes of matter. Elements are the basic building blocks of all substances and cannot be decomposed into simpler substances by chemical reactions. Compounds are formed from elements and contain elements in fixed (constant) proportions by mass. Stated differently, the proportions of the elements in a compound cannot be changed. Mixtures are formed from elements and/or compounds and can be of variable composition. A mixture composed of two or more phases is said to be heterogeneous; one composed of just a single phase is a solution and is said to be homogeneous.

Separating a compound into its components (elements) involves a chemical change, just as in the combination of the elements to form the compound. Separation of a mixture usually can be accomplished by a physical change—a change in which the chemical identities of the components aren't altered. Be sure to study Figure 2.4 of page 38.

Chemical Symbols and Chemical Formulas

A chemical symbol is a simple way of identifying an element. Remember that when the symbol consists of two letters, only the first is capitalized. Some symbols are derived from the old Latin names for the elements, but most symbols come from the English name. You will find it useful to know the names and symbols for at least the first twenty elements that appear in the Periodic Table on the inside front cover of the textbook. Your teacher will probably ask you to learn a more extensive list.

A chemical formula is a shorthand way of representing the name of a compound. The subscripts in a formula also specify the numbers of atoms of each kind of element in the simplest unit of the substance. In this capacity, you will often use a chemical formula as a problem solving tool to obtain information about the composition of a substance. To do this you must know how to count the atoms represented by the formula. Note in particular the special way that the formulas of hydrates are written. When counting atoms, be sure to include all the atoms in the water.

Chemical Equations

The reactants in a chemical equation appear on the left side of the arrow and are the substances present before the reaction begins. The products appear on the right side of the arrow and are the substances formed by the reaction. An equation is balanced by placing numbers called coefficients (often called *stoichiometric coefficients*) in front of the chemical formulas so that for each element there is the same number of atoms on both sides of the arrow. You will learn more about balancing equations later; for now you just need to recognize when an equation is balanced.

Sometimes in a chemical equation we specify the physical states of the chemicals involved or whether they are dissolved in a solution. The following symbols are used following the chemical formula:

solid	(s)
liquid	(l)
gas	(g)
aqueous solution	(aq)

Self-Test

1. A student placed some sand and salt in a beaker and stirred them with a glass rod.

   (a) After stirring, does the beaker contain a compound or a mixture? ____________________

   (b) How many phases are present? ____________________

   (c) Can you suggest a method of separating the components of the beaker? Does this involve a chemical or a physical change?

   ____

   ____

   ____

   ____

2. What are the symbols for the following elements?

   (a) aluminum __________ (d) magnesium __________

   (b) sodium __________ (e) chlorine __________

   (c) carbon __________ (f) beryllium __________

3. What are the names of the following elements?

   (a) Ca __________ (d) K __________

   (b) S __________ (e) P __________

   (c) O __________ (f) N __________

4. How many atoms of each element are represented in the formula of...

   (a) Al_2Cl_6 __________

   (b) $CaSiO_3$ __________

   (c) $K_2Cr_2O_7$ __________

   (d) $C_{12}H_{22}O_{11}$ __________

   (e) $Na_2CO_3 \cdot 10H_2O$ __________

   (f) $(NH_4)_2SO_4$ __________

5. How many atoms of each element are found on each side of the following equations? Are the equations balanced?

   (a) $Al_2O_3 + 6HCl \rightarrow 2AlCl_3 + 3H_2O$

   ____

   (b) $2(NH_4)_3PO_4 + 3CaO \rightarrow 3NH_3 + Ca_3PO_4 + 3H_2O$

   ____

6. Rewrite the equation in part (a) of the preceding question to show that the Al_2O_3 is a solid, the HCl and $AlCl_3$ are dissolved in water, and that H_2O is a liquid.

   ____

New Terms

Write the definitions of the following terms, which were introduced in this section. If necessary, refer to the Glossary at the end of the text.

decomposition	solution	hydrate
element	phase	chemical equation
compound	physical change	reactants
pure substance	chemical symbol	products
mixture	atom	coefficients
homogeneous	molecule	balanced equation
heterogeneous	chemical formula	physical state

2.2 Dalton's Atomic Theory

Review

Although the concept of atoms was not new when Dalton presented his atomic theory, the theory nevertheless revolutionized chemistry. This is because it was based on two *experimentally observed* laws relating to the composition of substances and chemical reactions. We call them laws of chemical combination.

Be sure you study the definitions of the law of conservation of mass and the law of definite proportions given on page 41. Notice that the laws refer to experimentally measurable quantities—the masses of substances in compounds and in chemical reactions.

Study the postulates of Dalton's theory. You should be able to describe how they explain the laws of chemical combination mentioned above. The law of multiple proportions (page 43), which was discovered as a result of Dalton's theory, is easily explained on the basis of the atomic theory. Study Figure 2.8 to obtain a clear understanding the this law.

Isotopes

Contrary to Dalton's theory, not all the atoms of a given element have exactly the same mass. Atoms of an element with slightly different masses are said to be isotopes of the element. The existence of isotopes did not affect the validity of Dalton's theory because of two reasons.

1 Any sample of an element large enough to see has an enormous number of atoms.

2 The proportions of the different isotopes in samples of an element are uniform from sample to sample.

As a result, an element behaves as if its atoms have masses corresponding to the *average mass* of the isotopes.

Atomic Masses

Knowing the actual masses of individual atoms is not nearly as important as knowing their *relative* masses. If we know the relative value of the mass of the atoms in one element compared to those of another element, we can always use the weighing operation in the lab to get these two elements in any ratio *by atoms* that we want. The entire purpose of learning about atomic masses is the practical one of being able to use the weighing operation to take samples of elements in known ratios *by atoms*. Atoms are simply too tiny to use a counting operation to get such ratios.

The atomic mass of an element is its relative mass compared to the isotope of carbon called "carbon-12," which is assigned a mass of exactly 12 mass units (u)

$$1 \text{ atom } ^{12}\text{C} = 12 \text{ u (exactly)}$$

For example, from a variety of experiments scientists have found that atoms of hydrogen have, on the average, a mass that is 1/12th the mass of the atoms in carbon-12. Therefore, *relative* to carbon-12 atoms, atoms of hydrogen have a relative atomic mass of 1 (more precisely, 1.0079). This means that if we have a sample of carbon with a mass of 12 g, and we want a sample of hydrogen *with as many atoms*, we have to take a sample of 1.0 g of hydrogen.

Example 2.2 on page 46 illustrates how the average atomic mass is obtained from accurate isotopic masses and relative abundances.

Self-Test

7. State the law of conservation of mass. ____________________

____

8. State the law of definite proportions. ____________________

____

9. In a 4.00 g sample of table salt, NaCl, there is 1.57 g of Na. Use the law of definite proportions to calculate the mass of Cl in a 6.00 g of NaCl.

____

10. If the ratio by mass of the elements in the carbon dioxide in a sample obtained in New York City is 2.66 g O to 1.00 g C, what ratio would be present in a sample of this compound taken in Los Angeles?

____

11. A vinegar sample purchased in one store was found to have a mass ratio of oxygen to carbon of 173 g O to 1.00 g C. A sample bought in a different store had a ratio of 185 g O to 1.00 g C. Do these data suggest that vinegar is a compound or a mixture? Why?

____

____

12. In the compound sodium chloride, there is 1.45 g of chlorine for every 1.00 g of sodium. Suppose that 2.00 g of chlorine were mixed with 1.00 g of sodium in some appropriate reaction vessel. After the reaction between them is as complete as possible:

    (a) What is the total mass of chemicals in the reaction vessel?

    ____

    (b) How many grams of chlorine have reacted?_______________

    (c) How many grams of chlorine remain unreacted?_______________

13. According to Dalton, all atoms of the same element

    (a) are indestructible
    (b) have identical masses
    (c) are very small
    (d) *a* and *b* only
    (e) *a*, *b* and *c*

    ____

14. Covellite is a soft, indigo-blue mineral made of copper combined with sulfur in a ratio of 1.000 g S to 0.506 g Cu. Chalococite is a soft, blackish mineral made of copper and sulfur in a ratio of 1.000 g S to 0.253 g Cu. Do these two minerals illustrate the law of multiple proportions? Explain.

____

____

15. How do isotopes of the same element differ chemically?

____

16. How do isotopes of the same element differ physically?

____

17. The mineral covellite is CuS; copper atoms are combined in it with sulfur atoms in a 1-to-1 ratio by atoms. If you wanted to make covellite from 32.06 g of sulfur, how many grams of copper should be used?

    ____

18. The atomic mass of gold (rounded) is 197. How much heavier are gold atoms than carbon-12 atoms?

    ____

19. The element iridium is composed of 37.3% of ^{191}Ir, which has a mass of 190.960 u, and 62.7% ^{193}Ir, which has a mass of 192.963 u. What is the average atomic mass of iridium?

    ____

New Terms

Write the definitions of the following terms, which were introduced in this section. If necessary, refer to the Glossary at the end of the text.

law of conservation of mass	law of definite proportions
Dalton's atomic theory	law of multiple proportions
atom	isotope

2.4 A Modern View of Atomic Structure

Review

At the center of an atom is a tiny particle called a nucleus, which is composed of subatomic particles called protons and neutrons. Protons are positively charged and neutrons are electrically neutral. Surrounding the nucleus are electrons, which are negatively charged. The amount of charge carried by a proton and an electron is the same; however, the kind of charge (positive and negative) is different. Each proton has one unit of positive charge, and each electron has one unit of negative charge. Because of this, in a neutral atom the number of electrons surrounding the nucleus equals the number of protons in the nucleus.

Protons and neutrons (collectively referred to as nucleons) have nearly the same mass (approximately 1 atomic mass unit) and are much heavier than electrons. As a result, nearly all the mass of an atom is contained within its nucleus.

All of the atoms in a given element, no matter if the element consists of several isotopes, have the same number of protons, and atoms of different elements have different numbers of protons. This makes the number of protons—called the atomic number (Z)—something unique for each element, and it corresponds to the size of the positive charge on the nuclei of the element.

The isotopes of an element differ only in their numbers of neutrons, so they differ only in mass. The sum of the protons and neutrons—called the mass number (*A*)—is one way to specify which isotope of an element is being discussed. Thus, the name carbon-12 identifies the one isotope of carbon that has a mass number of 12. (Since carbon's atomic number is 6, meaning six protons, the nucleus of a carbon-12 atom must have 6 neutrons so that the total number of nucleons is 12.)

Sometimes it is desirable to specify the mass number and the atomic number when writing the symbol of an isotope. This is illustrated on page 50. Some other examples are shown below.

$^{35}_{17}Cl$	$^{190}_{77}Ir$	$^{87}_{37}Rb$
17 protons	77 protons	37 protons
18 neutrons	113 neutrons	50 neutrons
17 electrons	77 electrons	37 protons

Self-Test

20. Name the three subatomic particles. ________________

21. Name the nucleons. ________________

22. Name the particles that are in the nucleus. ________________

23. An atom with a mass number of 25 has an atomic number of 12. How many of the following particles does it have?

    (a) electrons _____ (c) neutrons _____

    (b) protons _____ (d) nuclei _____

24. An atom with 14 electrons has 15 neutrons. What is its approximate atomic mass?

    ____

25. An isotope of oxygen has 10 neutrons. Write its symbol.________________

26. Which is the only atom that has an atomic mass that is exactly a whole number? ________________

New Terms

Write the definitions of the following terms, which were introduced in this section. If necessary, refer to the Glossary at the end of the text.

nucleus nucleon

proton	atomic number
neutron	mass number

2.5 The Periodic Table

Review

The number of facts in chemistry is enormous. Besides the many similarities among various elements, many differences also exist. Progress in understanding and explaining these similarities and differences required a search for order and organization. The product of this search—the modern periodic table—has become our primary tool for this organizing chemical facts.

Mendeleev discovered that similar properties are repeated at regular intervals when the elements are arranged in order of increasing atomic mass. By breaking this sequence at the right places, Mendeleev arranged elements with similar properties in vertical columns (groups) within his periodic table. Mendeleev's genius was his insistence on having elements with similar properties in the same column, even if it sometimes meant leaving empty spaces for (presumably) undiscovered elements.

The Modern Periodic Table

Atomic number forms the basis for the sequence of the elements in the modern periodic table shown on page 55 and on the inside front cover of the textbook. The rows are called periods and the columns are called groups, just as in Mendeleev's original table. Sometimes the term *family* is used when speaking of a group of elements. The periods are numbered with Arabic numerals (1, 2, etc.). In the conventional form of the table, which we shall use throughout the remainder of the text, the groups are given Roman numerals and a letter (e.g., Group IIA). However, the IUPAC has recommended that the groups be numbered sequentially from left to right with Arabic numerals. You should be aware of this because you may encounter this numbering elsewhere.

In the table that we will use, the representative elements are the A-group elements. The B-group elements are the transition elements. The two long rows of elements below the main body of the table are the inner transition elements (the lanthanides and actinides). Examine Figure 2.12 on page 55 to be sure you understand where they properly fit into the periodic table. Also, remember the names of these families:

Group IA	Alkali metals	Group VIA	Halogens
Group IIA	Alkaline earth metals	Group 0	Noble gases

Self-Test

27. What problem did Mendeleev face with the elements iodine and tellurium? How did he solve it?

 ____

28. Argon and potassium do not fit in atomic mass-order in the periodic table. What does this suggest about Mendeleev's basis for constructing the periodic table?

 ____

29. According to the IUPAC system, in which groups are the following elements found?

 (a) lithium ______________
 (b) iron ______________
 (c) silicon ______________
 (d) sulfur ______________

30. Among the elements Mg, Al, Cr, U, Kr, K, Br, and Ce,

 (a) Which are representative elements? ______________
 (b) Which is a transition element? ______________
 (c) Which is a halogen? ______________
 (d) Which is a noble gas? ______________
 (e) Which is an alkaline earth metal? ______________
 (f) Which is an alkali metal? ______________
 (g) Which is an actinide element? ______________
 (h) Which is a lanthanide element? ______________

New Terms

Write the definitions of the following terms, which were introduced in this section. If necessary, refer to the Glossary at the end of the text.

group
period
periodic table
inner transition elements
lanthanide elements
actinide elements

periodic law
family of elements
representative elements
transition elements
alkali metals (alkalis)
alkaline earth metals
noble gases (inert gases)
halogens

2.6 Metals, Nonmetals, and Metalloids

Review

Metals are shiny, are good conductors of heat and electricity, and many are malleable and ductile. They are found in the lower left portion of the periodic table and make up most of the known elements. (Study Figure 2.13 on page 57.)

Nonmetals lack the luster of metals, are nonconductors of electricity, and are poor conductors of heat. Many nonmetals are gases (H_2, O_2, N_2, F_2, Cl_2 and the elements of Group 0) and those that are solids tend to be brittle. Nonmetals are found in the upper right portion of the periodic table.

Metalloids are semiconductors, but resemble nonmetals in many of their properties. They are located on opposite sides of the steplike line running from boron (B) to astatine (At).

We can use the periodic table as a tool to help us compare the metallic and nonmetallic properties of the element. Within the periodic table, elements become less metallic going from left to right in a period and more metallic going from top to bottom in a group.

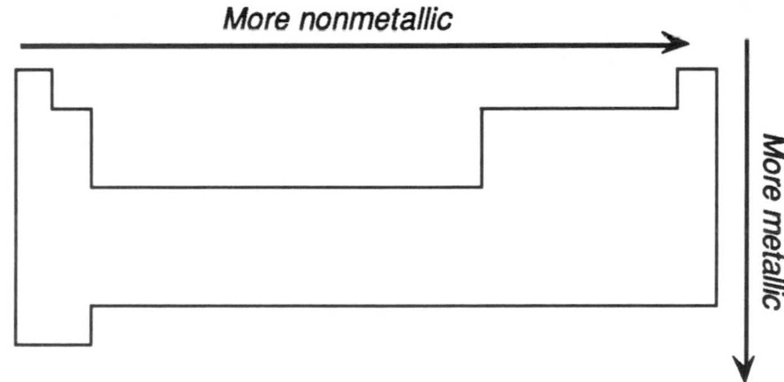

Self-Test

31. Fill in the blanks.

(a) The ability of copper to be drawn into wire depends on its

______________.

(b) Gold can be hammered into very thin sheets because of its ______________.

(c) The property that makes mercury useful in thermometers is ______________.

(d) The reason that tungsten is used as filaments in electric light bulbs is ______________.

(e) A nonmetal having a yellow color is ______________.

(f) Two elements that are liquids at room temperature are ______________ and ______________

(g) The color of bromine is ______________.

(h) The reason that helium is used to inflate the Goodyear blimp rather than hydrogen is ______________

(i) Sodium is rarely seen as a free metal because ______________ ______________.

(j) Two common semiconductors used in electronic devices are ______________ and ______________.

32. Which element is more metallic, Ga or Ge? ______________

33. Which element is more nonmetallic, As or P? ______________

New Terms

Write the definitions of the following terms, which were introduced in this section. If necessary, refer to the Glossary at the end of the text.

metal	malleability
metalloid	ductility
nonmetal	semiconductor

2.7 Reactions of the Elements; Formation of Molecular and Ionic Compounds

Review

Compounds are possible because atoms are able to stick together. In this chapter we don't go into much detail about the ways this can happen (we do that in later chapters); all you need to know now is that the attractions that bind atoms to each other in compounds are called chemical bonds.

Sometimes the atoms link together by sharing electrons to form molecules, which are separate individual particles consisting of two or more atomic nuclei surrounded by enough electrons to make the whole package electrically neutral. Water, H_2O, is an example. Because molecules are never formed from just parts of atoms, they always contain atoms in definite, simple, whole-number ratios. A compound that's composed of molecules is (naturally) called a molecular compound, and we use a molecular formula to show the number of atoms of each of the elements that are present in a molecule of the substance.

Sometimes atoms can transfer electrons when they combine to form a compound and this produces electrically charged particles that we call ions. Ionic compounds always contain positive ions and negative ions in ratios that give electrically neutral substances. In an ionic compound, the ions are arranged around one another in a way that maximizes the attractions between oppositely charged ions and minimizes the repulsions between like-charged ions. Since we can't identify unique molecules in an ionic compound, we always write their formulas with the smallest set of whole-number subscripts. This identifies one formula unit of the substance.

Compounds of Nonmetals with Each Other

It is useful to remember that nonmetals combine with nonmetals to form molecular compounds. Examples presented in this section include the simple hydrogen and oxygen compounds of the nonmetals. Also notice that most of the nonmetals exist in their elemental forms are molecules. Be sure you learn the ones that are diatomic: H_2, N_2, O_2, F_2, Cl_2, Br_2, and I_2. (Only the noble gases occur in nature as single atoms.)

The formulas of the simple hydrogen compounds of the nonmetals are given in Table 2.4 on page 63. Notice that we can use the periodic table to figure out their formulas: For a nonmetal in a particular group in the periodic table, the number of hydrogens in the formula equals the number of steps to the right we have to go to get to Group 0 (the noble gases). Thus, for any element in Group VIIA, we have to move *one* step to the right to get to Group 0; the formula of the hydrogen compound contains *one* hydrogen (e.g., HF). Similarly, for an element in Group IVA, we have to move *four* elements to the right to get to Group 0, and each element in Group IVA forms a hydrogen compound with *four* hydrogens (e.g., CH_4).

For the oxygen compounds of the nonmetals (Table 2.5), there is no simple rule to help remember their formulas. Notice, however, that elements in the same group form oxides with the same simplest formulas. That's one way to simplify the learning of them.

Compounds of Metals

Metals combine with nonmetals to form ionic compounds by transfer of electrons. Cations (positive ions) are formed by metals when their atoms lose electrons. Anions (negative ions) are formed by nonmetals when their atoms acquire electrons.

Table 2.6 on page 65 illustrates how the periodic table can help you remember the ions formed by the representative metals and nonmetals. The number of positive charges on a cation equals the group number; the number of negative charges on an anion equals the number of spaces to the right we have to go to get to a noble gas in the periodic table. Be sure you understand how the formulas of the ions in Table 2.6 correlate with the locations of the elements in the periodic table

Self-Test

34. In terms of their composition, what is the major difference between a molecular compound and an ionic compound?

____

____

35. (a) What name is used for the tiny individual particles that occur in molecular compounds? ____________________

(b) What is the name for the individual particles that make up an ionic compound? ____________________

36. Without referring to anything but the periodic table, write the formula for the simplest compound formed from hydrogen and

(a) sulfur ____________________

(b) bromine ____________________

(c) phosphorus ____________________

(d) carbon ____________________

37. What is the simplest formula for an oxide of
    (a) phosphorus ____________
    (b) sulfur ____________
    (c) silicon ____________
38. Write the formulas for the seven elements that occur in nature as diatomic molecules.

    ____
39. Without looking at Table 2.6, but using the periodic table on the inside front cover of your textbook, write the formula for the ion formed by each of the following elements:

(a) K	________	(e) Ba	________
(b) Al	________	(f) Na	________
(c) N	________	(g) Br	________
(d) Mg	________	(h) Se	________

40. What is the general formula for an ion formed by a metal from
    (a) Group IA ____________
    (b) Group IIA ____________
41. What is the general formula for an ion formed by a nonmetal from
    (a) Group VA ____________
    (b) Group VIIA ____________

New Terms

Write the definitions of the following terms, which were introduced in this section. If necessary, refer to the Glossary at the end of the text.

chemical bond	ion
molecule	ionic compound
molecular compound	cation
molecular formula	anion

2.8 Ionic Compounds

Review

Writing the formulas for ionic compounds correctly is a skill you must master, which is why the Rules for Writing Formulas of Ionic Compounds are identified by the icon in the margin. The rules are simple to apply, so be sure to study Example 2.3 on page 66 and work Practice Exercise 10.

The transition and post-transition metals generally are able to form two or more different ions, and there are no simple rules that enable us to figure them out. Therefore, you must simply memorize the symbols (including charges) for these ions which are given in Table 2.7.

Table 2.8 contains the formulas and names of frequently encountered polyatomic ions—ions composed of more than one atom. These should also be memorized. Practice writing both their names and their formulas. Be sure to learn their charges, too. (It might help to prepare a set of flash cards.) When you feel you have learned the contents of Tables 2.7 and 2.8 and can use the periodic table to write the formulas of the ions in Table 2.6, try the Self-Test below.

Self-Test

42. Write the formula for the ionic compound formed by

    (a) Na and S ____________

    (b) Sr and F ____________

    (c) Al and Se ____________

    (d) Mg and Cl ____________

43. Write the formula for the ionic compound formed from

    (a) sodium ion and perchlorate ion ____________

    (b) barium ion and hydroxide ion ____________

    (c) ammonium ion and dichromate ion ____________

    (d) magnesium ion and sulfite ion ____________

    (e) nickel ion and phosphate ion ____________

    (f) silver ion and sulfate ion ____________

44. Write the formulas for *two* different ionic compounds of

    (a) iron and oxygen ____________ ____________

    (b) copper and chlorine ____________ ____________

    (c) mercury and chlorine ____________ ____________

    (d) tin and sulfur ____________ ____________

    (e) lead and oxygen ____________ ____________

New Terms

Write the definitions of the following terms, which were introduced in this section. If necessary, refer to the Glossary at the end of the text.

cation	post-transition metal
anion	binary compound
formula unit	polyatomic ion

2.9 Properties of Ionic and Molecular Compounds

Review

The physical properties of ionic compounds are the result of the strong attractive forces that exist between ions of opposite charge as well as the strong repulsive forces between ions of the same charge. Ionic solids tend to have high melting points (i.e., they melt at high temperatures), they tend to be brittle and easily shattered when struck. In the solid state, ionic compounds do not conduct electricity, but they do conduct when melted or when dissolved in water. (Substances that dissolve in water to yield electrically conducting solutions are called electrolytes.) This is because melting frees the ions and allows them to move—a requirement for electrical conduction. The ions also become free to move about when an ionic compound dissolves in water.

By contrast, many molecular compounds are relatively soft and easily deformed when struck, and they tend to have relatively low melting points. When melted, they do not conduct electricity because the liquid does not contain charge particles. Molecular compounds that dissolve in water without reacting with the solvent do not give electrically conducting solutions, either.

New Terms

none

2.10 Inorganic Chemical Nomenclature

Review

The following is a summary of the rules for naming simple inorganic compounds. Practice applying them until you feel comfortable with them. Knowing how to name compounds will make chemistry more interesting for you and will make the rest of the course easier to understand.

Binary Compounds of Metals and Nonmetals

These are ionic compounds composed of just two elements, one a cation and one an anion. The cation is named first followed by the anion. Monatomic (one-atom) anions end in the suffix *-ide*. (Hydroxide, OH^-, and cyanide, CN^- are the only two polyatomic ions ending in *-ide*. If it ends in *-ide* and isn't hydroxide or cyanide, then you can be sure it's a monatomic anion.)

Compounds of metals that can form two cations, such as iron, can be named according to two different systems. The old system uses the ending -*ous* for the cation with the lower charge and the ending *-ic* for the one with the higher charge. Except for mercury, if the symbol for the metal is derived from the element's Latin name, then the Latin stem is used in specifying the metal (ferrous ion and ferric ion, for example). The preferred method for naming these kinds of cations is the Stock system, in which the number of positive charges on the metal ion is specified by placing a Roman numeral in parentheses following the English name for the metal [iron(II) ion and iron(III) ion, for example].

Binary Compounds of Two Nonmetals

The number of atoms of each nonmetal is specified by an appropriate Greek prefix (*mono-, di-, tri-,* etc.). Learn the prefixes for 1 through 10 given on page 73 of the text.

Binary Acids and Their Salts

Acids are substances that release H^+ and produce hydronium ions, H_3O^+. Examples include many of the hydrogen compounds of the nonmetals, which react with water to produce H_3O^+. Examples include hydrogen chloride, HCl, and hydrogen bromide.

$$HCl(g) + H_2O \rightarrow H_3O^+(aq) + Cl^-(aq)$$

$$HBr(g) + H_2O \rightarrow H_3O^+(aq) + Br^-(aq)$$

These two acids are called binary acids because they contain only one element besides hydrogen, and so are binary compounds in their pure states. In naming the acids, we add the prefix *hydro-* and the suffix *-ic acid* to the stem of the nonmetal name. Thus hydrogen chloride gives *hydro*chlor*ic acid*, and hydrogen bromide gives *hydro*brom*ic acid*.

Acids react with bases (hydroxide compounds of metals) in a reaction called neutralization. The products are water and an ionic compound. We use the general term salt to mean any ionic compound except one that contains hydroxide ion or oxide ion (which are bases). Salts formed by neutralizing binary acids contain the monatomic anion of the acid and have the *-ide* ending.

Oxoacids and Their Salts

Many nonmetals form more than one oxoacid. If only two of them are possible, the name of the one having the larger number of oxygen atoms ends in *-ic* and the name of the acid with the smaller number of oxygens ends in *-ous*. When there are more than two oxoacids for a given nonmetal (as there are for the halogens), the following prefixes are used (in the order of increasing numbers of oxygen atoms).

hypo ... ous acid	(For example, hypochlorous acid — $HClO$)
... ous acid	(For example, chlorous acid — $HClO_2$)
... ic acid	(For example, chloric acid — $HClO_3$)
per ... ic acid	(For example, perchloric acid — $HClO_4$)

Remember that *...ic* acids give anions that end in *-ate*, and that *...ous* acids give anions that end in *-ite*.

Acid salts

Monoprotic acids such as HCl have just one hydrogen to be neutralized. Polyprotic acids, such as H_2SO_4 and H_3PO_4, have more than one hydrogen to be neutralized, and the neutralization can be incomplete to give anions that still contain "acidic hydrogens." Compounds formed by these anions are called acid salts. In naming them, the hydrogen is specified either as "hydrogen" or, when only one acid salt is possible, by the prefix *bi-* before the name of the anion (for example, HSO_4^- is named as hydrogen sulfate ion or bisulfate ion).

Self-Test

45. Name these compounds.

(a) CaI_2 ____________________

(b) $FeBr_3$ ____________________

(c) $Fe(NO_3)_2$ ____________________

(d) As_4O_{10} ______________________

(e) ClF_5 ______________________

(f) HNO_2 ______________________

(g) $HBrO_3$ ______________________

(h) HIO ______________________

(i) $KHSO_3$ ______________________

(j) $CaHPO_4$ ______________________

46. Write the formulas for these compounds.

(a) barium arsenide ______________

(b) sodium perbromate ______________

(c) manganese(IV) oxide ______________

(d) iodic acid ______________

(e) disulfurdichloride ______________

(f) selenium tetrachloride ______________

(g) cobaltous acetate ______________

(h) dinitrogen pentoxide ______________

47. Acetate ion has the formula, $C_2H_3O_2^-$. What would the name be for the acid $HC_2H_3O_2$. ______________________

48. Formic acid, which we mentioned causes the painful sensation from ant bites, has the formula $HCHO_2$. Write the name of the ion CHO_2^-.

____

49. Tartaric acid occurs in many fruit juices, especially grape juice. It has the formula $H_2C_4H_4O_6$. Give two names for the salt $NaHC_4H_4O_6$.

____

____

50. Why do we call H_2O water instead of dihydrogen oxide?

____

New Terms

Write the definitions of the following terms, which were introduced in this section. If necessary, refer to the Glossary at the end of the text.

nomenclature	salt
inorganic compound	binary acid
organic compound	oxoacid
Stock system	monoprotic acid
acid	polyprotic acid
base	acid salt
neutralization	

Answers to Self-Test Questions

1. (a) a mixture, (b) two, (c) Add water, which dissolves the salt but not the sand. This involves a physical change.
2. (a) Al, (b) Na, (c) C, (d) Mg, (e) Cl, (f) Be
3. (a) calcium, (b) sulfur,
   (c) oxygen, (d) potassium,
   (e) phosphorus, (f) nitrogen
4. (a) 2Al, 6Cl (b) 1Ca, 1Si, 3O (c) 2K, 2Cr, 7O (d) 12C, 22H, 11O
   (e) 2Na, 1C, 13O, 20H (f) 2N, 8H, 1S, 4O
5. (a) On the left: 2Al, 3O, 6H, 6Cl; on the right: 2Al, 3O, 6H, 6Cl (Balanced)
   (b) On the left: 6N, 24H, 2P, 11O, 3Ca; On the right: 3N, 15H, 1P, 7O, 3Ca Not balanced
6. $2Al_2O_3(s) + 6HCl(aq) \rightarrow 2AlCl_3(aq) + 3H_2O(l)$

7. and 8. Consult the definitions at the bottom of page 41 of the textbook.

9. 3.65 g of Cl
10. A ratio of 2.66 g O to 1.00 g C.
11. A mixture. The oxygen-to-carbon ratio is not constant from one sample to another.
12. (a) 3.00 g total,
    (b) 1.45 g Cl react,
    (c) 0.55 g Cl unreacted
13. (e)
14. Yes, the amounts of copper that combine with 1.00 g S in the two compounds are in the ratio of 0.506 to 0.253, which is a ratio of 2 to 1, simple whole numbers.
15. They don't differ chemically.
16. They differ slightly in their masses.
17. 63.54 g of copper

18. 16.4 times as heavy
19. 192
20. electron, proton, neutron
21. proton and neutron
22. proton and neutron
23. (a) 12, (b) 12, (c) 13, (d) 1
24. 29 u
25. $^{18}_{8}O$
26. ^{12}C
27. If placed in the table in order of atomic weight, their properties do not match those of other elements in the same columns. He put them in the table according to their properties rather than according to their atomic weights.
28. Atomic weight is not really the basis for the periodic law.
29. (a) Group 1, (b) Group 8, (c) Group 14, (d) Group 16
30. (a) Mg, Al, Kr, K, Br (b) Cr (c) Br
    (d) Kr (e) Mg (f) K (g) U (h) Ce
31. (a) ductility, (b) malleability, (c) low melting point and fairly high boiling point, (d) it conducts electricity and has a very high melting point.
    (e) sulfur, (f) bromine, mercury, (g) red-brown, (h) helium is very unreactive, (i) it is very reactive toward oxygen, (j) silicon and germanium
32. Ga
33. P
34. A molecular compound is made up of neutral particles (molecules); an ionic compound is composed of charged particles (ions).
35. (a) molecules (b) ions
36. (a) H_2S, (b) HBr, (c) PH_3 (or H_3P), (d) CH_4 (or H_4C)
37. (a) P_2O_3 or P_2O_5 (b) SO_2 or SO_3 (c) SiO_2
38. H_2, N_2, O_2, F_2, Cl_2, Br_2, I_2
39. (a) K^+, (b) Al^{3+}, (c) N^{3-}, (d) Mg^{2+}, (e) Ba^{2+}, (f) Na^+, (g) Br^-, (h) Se^{2-}
40. (a) M^+, (b) M^{2+}
41. (a) X^{3-}, (b) X^-
42. (a) Na_2S, (b) SrF_2, (c) Al_2Se_3
43. (a) $NaClO_4$ (b) $Ba(OH)_2$ (c) $(NH_4)_2Cr_2O_7$,
    (d) $MgSO_3$, (e) $Ni_3(PO_4)_2$, (f) Ag_2SO_4
44. (a) FeO and Fe_2O_3, (b) CuCl and $CuCl_2$,
    (c) Hg_2Cl_2 and $HgCl_2$, (d) SnS and SnS_2, (e) PbO and PbO_2
45. (a) calcium iodide, (b) iron(III) bromide or ferric bromide,
    (c) iron(II) nitrate or ferrous nitrate, (d) tetraarsenic decaoxide,
    (e) chlorine pentafluoride, (f) nitrous acid (g) bromic acid,

(h) hypoiodous acid (i) potassium hydrogen sulfite or potassium bisulfite, (j) calcium monohydrogen phosphate

46. (a) Ba_3As_2, (b) $NaBrO_4$,
(c) MnO_2, (d) HIO_3,
(e) S_2Cl_2, (f) $SeCl_4$,
(g) $Co(C_2H_3O_2)_2$, (h) N_2O_5
47. acetic acid
48. formate ion
49. sodium hydrogen tartrate, sodium bitartrate
50. Because its common name is so well-known.

Tools you have learned

Remove this chart from the Study Guide and keep it handy when tackling homework problems.

Tool	*Function*
Subscripts in a formula	Specifies the number of atoms of each kind in a unit of the compound
Rules for writing formulas of ionic compounds 1 Cation first, then anion. 2 Formulas must represent electrical neutrality 3 Smallest set of subscripts	Permits us to write correct chemical formulas for ionic compounds.
Rules for naming compounds Binary ionic compounds Binary molecular compounds Binary acids Oxoacids Acid salts	Allows us to write a formula from a name, and a name from a formula.

Summary of Rules for Naming Compounds

Binary compound of a metal and nonmetal. Name the metal first, followed by the nonmetal. For the nonmetal, add the suffix *ide* to the stem of the nonmetal name.

For metals that form more than one ion, use ous ending for lower charge, ic ending for higher charge. In Stock system, use Roman numeral in parentheses to give the charge of the metal.

Binary compound of two nonmetals. Use Greek prefixes to specify the number of atoms of each element in the formula. Usually, when only one atom is present, the prefix mono is omitted.

Greek prefixes

1 = mono	2 = di	3 = tri	4 = tetra	5 = penta
6 = hexa	7 = hepta	8 = octa	9 = nona	10 = deca

Binary acids. These are aqueous solutions of hydrogen compounds of nonmetals. They are named as *hydro……ic acid* where the stem of the nonmetal name is inserted in place of the dots.

Oxoacids. Acids of the general formula H_nXO_m, where X is a nonmetal. For a given nonmetal, there is usually more than one acid possible. When there are two, the one with the most oxygens has the ending *ic*. The other has the ending *ous*. For the halogens, there are four acids. The one with the fewest oxygens is the *hypo……ous acid*, the one with the most oxygens is the *per……ic acid*.

Acids that end in *ic* form anions ending in *ate*.

Acids that end in *ous* form anions ending in *ite*.

Acid salts. For acids with two hydrogen (e.g., H_2SO_4), the acid anion can be named by adding the prefix *bi* to the name of the fully neutralized anion (e.g., bisulfate ion).

Acid anions are also named by putting the name hydrogen before the name of the fully neutrallized anion. If more than one hydrogen, then use Greek prefixes to specify number (e.g., HSO_4^- is the hydrogen sulfate ion, $H_2PO_4^-$ is the dihydrogen phosphate ion).

Chapter 3

STOICHIOMETRY: QUANTITATIVE CHEMICAL RELATIONSHIPS

Three concepts are at the heart of this chapter, the *mole concept, stoichiometric equivalency,* and the *balanced equation.* All of the tools studied in this chapter depend on these. Moreover, all three derive from one central fact: chemical reactions occur on a particle to particle basis.

The *mole concept* provides a way to obtain samples of substances in whatever ratios *by particles* we choose. The reference standard for "number of particles" or "number of formula units" is Avogadro's number or one mole.

The term *stoichiometric equivalency,* regardless of its context, concerns a ratio *by moles* and so a ratio by particles. The term can apply to a specific *formula* and thus refer to the ratios by moles of the atoms in a compound as disclosed by the formula's *subscripts.* Or stoichiometric equivalency can apply to a specific *balanced equation* and refer to the ratios by moles of reactants or products that can be written using the equation's *coefficients.*

The *balanced equation,* the third major concept, is possible because mass is conserved and atoms are not broken up in *chemical* reactions.

Learning Objectives

In this chapter, you should keep in mind the following goals.

1. To learn what "mole of a substance" means in terms of a specific number of formula units (Avogadro's number) and in terms of a specific mass of the substance (the formula or molecular mass in grams).
2. To learn to construct conversion factors, using a formula's subscripts, that express any stoichiometric equivalencies possible—the ratios *by moles* between the atoms of any two elements in the formula.
3. To learn how to calculate formula or molecular masses using atomic masses and see these values as giving the grams per mole of the substance.

4 To learn how to convert a given chemical formula into the percentage composition, element by element.

5 To learn how to convert a percentage composition into an empirical formula.

6 To learn how to use a molecular mass and an empirical formula to write a molecular formula.

7 To learn how to convert information about reactants and products into a balanced chemical equation.

8 To learn how to use the coefficients of a balanced equation to construct the stoichiometric equivalencies between any two substances in the reaction.

9 To learn how to use formula masses and stoichiometric equivalencies to calculate the mass relationships between any two substances whose formulas are given in a balanced equation.

10 To learn how recognize a limiting reactant problem and to tell from given masses of *two* reactants if either is present in stoichiometric excess and by how much.

11 To learn how to convert given data about masses of reactants and the masses of products actually obtained into the percentage yield of a given product.

12 To learn how to do the calculations either for preparing a predetermined volume of a solution of known molar concentration or for finding out how much of a solute is obtained from a given volume of such a solution.

13 To learn how to do stoichiometry problems when a reaction is carried out in solution and the measurements are of volumes and molar concentrations.

14 To learn how to do the calculations needed to prepare a dilute solution from a more concentrated solution of known molarity.

3.1 The Mole Concept

Review

The *mole* is a counting number equal to 6.02×10^{23} and called Avogadro's number. It equals the number of atoms in 0.012 kg (exactly) of carbon-12. Being a counting number, it follows that one mole of one substance must have the identical number of particles (formula units) as one mole of any other substance.

Chemical formulas give the identities of the atoms present as well as their ratios *by atoms*. If you want to know what ratio by atoms must be taken to have molecules of ammonia, NH_3, for example, just look at the subscripts, 1 and 3. They tell us that *in ammonia* we have 1 atom of N to 3 atoms H. In *ammonia* 1 atom of N has 3 atoms of H as its *chemical equivalent,* so we can write

$$1 \text{ atom N} \Leftrightarrow 3 \text{ atom H}$$

We don't say that 1 atom of N *equals* 3 atoms of H; cups do not *equal* saucers. We do say, however, that if you want to have one molecule of NH_3, then by taking 1 atom of N you are *required* to take 3 atoms of H. It's the stoichiometric "law" about NH_3, and all of the specific data you need is right in front of you, in the formula NH_3. From this "law," we can write the following important stoichiometric ratios that hold for NH_3.

$$\frac{1 \text{ atom N}}{3 \text{ atom H}} \quad \text{and} \quad \frac{3 \text{ atom H}}{1\text{atom N}}$$

However, the major point of this section is that ratios by atoms are identical to ratios by moles. When you grasp this point, you have made a great stride toward being able to function rationally (sensibly) in experimental work. We can write for NH3,

$$\frac{1 \text{ mol N}}{3 \text{ mol H}} \quad \text{and} \quad \frac{3 \text{ mol H}}{1 \text{ mol N}}$$

Another binary compound of H and N is hydrazoic acid, HN_3. A question that could come up is "How many moles of H are combined with, say, 4.5 moles of N in hydrazoic acid?" Any time you see a question that asks "How many moles of ....are combined with.....(so many).....moles of ....?" you can be certain that this is a stoichiometric equivalency problem. The problem could be stated as follows.

For HN_3,

$$4.5 \text{ mol N} \Leftrightarrow \text{? mol H}$$

Once you recognize the *kind* of problem, you see that the tool you need is one that connects the number of moles of N to the number of moles of H for the given substance, HN_3. So you "reach for the tool," here a stoichiometric ratio prepared from the subscripts in the formula, HN_3.

$$\frac{3 \text{ mol N}}{1 \text{ mol H}} \quad \text{and} \quad \frac{1 \text{ mol H}}{3 \text{ mol N}}$$

The formula thus provides what is needed to write the appropriate conversion factor. Then comes the arithmetic. The "given," 4.5 mol N, is multiplied by the second factor. The ability to cancel units correctly is a good guide to the choice of which factor, but *to reach for any kind of factor or tool you must first recognize what the essential problem is.* Thus,

$$4.5 \text{ mol N} \times \frac{1 \text{ mol H}}{3 \text{ mol N}} = 1.5 \text{ mol N}$$

Thinking It Through

For the following, identify the information needed to solve the problem and show (or explain) what must be done with it.

1 Compound *X* was known to be either ammonia (NH_3), hydrazine, (N_2H_4), or hydrazoic acid (HN_3). In a sample of 0.012 mol of compound *X* there was found to be 0.024 mol of N atoms and 0.048 mol of H atoms. What was compound *X*?

Self-Test

1. How many atoms of iron, Fe, are in 1.00 mol of iron?

   ___

2. How many mol of O and how many mol of H are present in 0.800 mol of H_2O?

   ___

3. How many mol of Ca, of O, and of H are present in 0.600 mol of $Ca(OH)_2$?

   ___

New Terms

Write the definitions of the following terms, which were introduced in this section. If necessary, refer to the Glossary at the end of the text.

Avogadro's number
mole
stoichiometric equivalency
stoichiometry

3.2 Measuring Moles of Elements and Compounds

Review

The major point of this section concerns *the relationship between moles and mass for a given substance.* We need this relationship because laboratory balances let us measure substances in units of mass, not in units of moles. In *planning* experiments, numbers of *moles* are the critical data. To *carry out* an experiment, however, we must have mole data expressed in equivalent numbers of grams. Otherwise, we can't do actual measurements.

For any specified substance, the combined mass of Avogadro's number of its formula units—one mole—equals the formula (or molecular) mass expressed in grams. This relationship is what connects the size of a sample in *moles* to its size in *grams*. It's a connection without which no rational experimental work is possible. Being able to calculate moles from grams and grams from moles are thus skills *that must be mastered.* Do not fail to accomplish this before going on.

The mass of one mole of a substance is found simply by calculating its formula mass and writing "grams" after it. Remember the policy cited in the text of rounding atomic masses to their second decimal place (H to its third decimal place) before using them in a calculation. Be sure to count all of the atoms given in a formula. Calcium nitrate has the formula $Ca(NO_3)_2$, so one formula unit has 1 Ca, 2 N, and 6 O atoms. The formula mass of $Ca(NO_3)_2$ is found by the following.

$$1 \times 40.08 + 2 \times 14.01 + 6 \times 16.00 = 164.10$$

we can now write (where "1" in "1 mol" is an exact number),

$$1 \text{ mol } Ca(NO_3)_2 = 164.10 \text{ g } Ca(NO_3)_2$$

We do not use the equivalency symbol, $\Leftrightarrow$, for this relationship any more than we use it in "1 in. = 2.54 cm." Just as the later defines the identical *length* in different units, so the equation, 1.00 mol $Ca(NO_3)_2$ = 164.10 g $Ca(NO_3)_2$, defines the identical *amount of substance* in different units. Both refer to Avogadro's number of the formula units defined by the formula $Ca(NO_3)_2$.

From the mass-mole relationship for $Ca(NO_3)_2$, we can prepare two conversion factors to use as needed:

$$\frac{164.10 \text{ g } Ca(NO_3)_2}{1 \text{ mol } Ca(NO_3)_2} \quad \text{or} \quad \frac{1 \text{ mol } Ca(NO_3)_2}{164.10 \text{ g } Ca(NO_3)_2}$$

We also know from the meaning of Avogadro's number that

$$1 \text{ mol} = 6.02 \times 10^{23} \text{ formula units}$$

So this relationship gives us the following conversion factors

$$\frac{6.02 \times 10^{23} \text{ formula units}}{1 \text{ mol}} \quad \text{or} \quad \frac{1 \text{ mol}}{6.02 \times 10^{23} \text{ formula units}}$$

Once we know a molar mass, such as 164.10 g/mol for $Ca(NO_3)_2$, we can substitute such mass per mole in these conversion factors. Thus, because

$$164.10 \text{ g } Ca(NO_3) = 6.02 \times 10^{23} \text{ formula units of } Ca(NO_3$$

we can write the following conversion factors.

$$\frac{6.02 \times 10^{23} \text{ formula units of } Ca(NO_3)}{164.10 \text{ g } Ca(NO_3)} \quad \text{or} \quad \frac{164.10 \text{ g } Ca(NO_3)}{6.02 \times 10^{23} \text{ formula units of } Ca(NO_3)}$$

Thinking It Through

For the following, identify the information needed to solve the problem and show (or explain) what must be done with it.

2 How many atoms and how many grams are in 0.500 mol of gold?

Self-Test

4. Calculate the formula mass of $(NH_4)_3PO_4{\cdot}3H_2O$, and write the two conversion factors relating grams to moles for this compound.

___

5. For ammonia, 17.0 g $NH_3 = 6.02 \times 10^{23}$ molecules NH_3. Write the two conversion factors that the equation makes possible.

___

6. How many atoms of iron, Fe, are in 0.400 mol of iron?

___

7. How many g of O are present in 0.800 mol of H_2O?

___

8. How many g of H are present in 0.600 mol of $Ca(OH)_2$?

___

9. If an experiment calls for 0.500 mol of H_2SO_4 (sulfuric acid), how many grams of H_2SO_4 do we have to weigh out?

___

10. If we expect 110 g of CO_2 to be made in an experiment, how many mol of CO_2 will be made?

___

11. How many grams of phosphorus are needed to make 24.00 g of tetraphosphorous hexaoxide, P_4O_6? How many grams of O?

New Terms

Write the definitions of the following terms, which were introduced in this section. If necessary, refer to the Glossary at the end of the text.

formula mass　　　　molecular mass

3.3 Percentage Composition

Review

Qualitative analysis is used in the lab to find out what elements are present in a substance. Then *quantitative analysis* is used to obtain their relative amounts. If a compound is known to consist, say, of 60.26% carbon, then to make 100.00 g of the compound from its elements requires 60.26 *grams* of carbon. Another way of looking at a percentage like 60.26% carbon is is that it is the number of grams of carbon that could be obtained by changing 100.00 g of the compound back into its elements.

We can find the percentage of an element in a compound in two ways that involve two different kinds of data. One way is experimental; it uses the actual masses of the individual elements found in a weighed sample of the compound. The other is theoretical; it uses the formula to calculate the percentages. If the formula of a compound is E_nZ_m. Then the percentage of element E is

$$\frac{n \times (\text{atomic mass of } E)}{\text{formula mass of } E_nZ_m} \times 100\%$$

Thinking It Through

For the following, identify the information needed to solve the problem and show (or explain) what must be done with it.

3 A compound consisting only of nitrogen and oxygen was found to have 36.85% N. What is the percent O? Do the percentages correspond to N_2O_4 or N_2O_3?

Self-Test

12. A 0.9278-g sample of a bright orange compound was found to consist of 0.3683 g of chromium, 0.1628 g of sodium and 0.3967 g of oxygen. Calculate the percentage composition.

____

13. An analysis of a sample believed to be photographers hypo, $Na_2S_2O_3$, gave the following percentage composition: Na, 29.10%; S, 40.50%; and O, 30.38%. Do these data correspond to the formula given? Use a separate sheet of paper to solve the problem.

____

New Terms

Write the definitions of the following terms, which were introduced in this section. If necessary, refer to the Glossary at the end of the text.

percentage by mass
percentage composition
qualitative analysis
quantitative analysis

3.4 Empirical and Molecular Formulas

Review

When the smallest whole-number subscripts are used to describe the ratio of the atoms of different elements in a compound the resulting formula is an *empirical formula.* The empirical formula of caffeine, for example, is $C_4H_5N_2O$, but its molecular formula, the actual composition of one molecule, is $C_8H_{10}N_4O_2$.

The percentages by mass of the elements in a compound numerically equal the *grams* of each element in 100 grams of the compound. The subscripts in an empirical formula, however, numerically stand in ratios by *moles* (or, to say the same, ratios by atoms). We can therefore take the grams of each element in a 100-g sample (as given by the percentages), multiply each mass by a conversion factor obtained from the atomic masses, and find the corresponding moles of each element in the sample. The numbers of moles we calculate this way are generally not whole numbers, so we proceed to find what the whole numbers are, as in Example 3.11 in the text. Then we can think of the subscripts not only as giving ratios by moles but also ratios by whole numbers of intact atoms.

For compounds that can be made to burn to CO_2 and H_2O, the experimental data are masses not of elements but of compounds, so the masses of the elements present in the sample that is burned have to be calculated from the masses of CO_2 and H_2O. But this is easy because we know that 1 mol of CO_2 corresponds to 1 mol of C; and 1 mol of H_2O corresponds to 2 mol of H. In other words, the calculations are nothing more than grams-to-moles-to-grams calculations, as Example 3.13 showed.

Often the experimentally measured formula mass is not the same as the one we calculate from an empirical formula. It is some simple multiple of it, such as 2× or 3× or 4× and so forth. To find out which multiple, we divide the experimental formula mass by the calculated formula mass (from the empirical formula). That will give the *whole* number multiple we need (or one so close to a whole number we can round to it). Finally, multiply each subscript in the empirical formula by the whole number and the result is the molecular formula. For example, the true molecular mass of benzene is 78 and its empirical formula is CH (with a formula mass of 13). We simple divide 78 by 13 to get 6. This tells us that there are really 6 units of CH in each benzene molecule, and so the molecular formula is $C_{1\times6}H_{1\times6} = C_6H_6$.

Thinking It Through

For the following, identify the information needed to solve the problem and show (or explain) what must be done with it.

4 A compound was known to contain only potassium and oxygen. When a sample with a mass of 0.2564 g was analyzed, there was found to be 0.2128 g of potassium. How many grams of oxygen were also present? What is the empirical formula of the compound?

Self-Test

14. Carotene, the pigment responsible for the color of carrots, has a percentage composition of 89.49% C and 10.51% H. Its molecular mass was found to be 546.9 Calculate its empirical formula and its molecular formula.

---

15. When 0.8788 g of a liquid isolated from oil of balsam was burned completely, 2.839 g of CO_2 and 0.9272 g of H_2O were obtained. The molecular mass of the compound was found to be 136.2. Calculate the empirical and molecular formulas of this compound.

New Terms

Write the definitions of the following terms, which were introduced in this section. If necessary, refer to the Glossary at the end of the text.

empirical formula
molecular formula

3.5 Writing and Balancing Chemical Equations

Review

In balancing a chemical equation:

1. Be sure all the chemical formulas are correct—both the atomic symbols and the subscripts.
2. Remember that all atoms among the reactants must end up somewhere among the products and that a coefficient is a multiplier for all of the atoms in a formula.
3. If groups of atoms, which sometimes are clustered within parentheses in formulas, are intact on both sides of the arrow, balance them as units in themselves, as if they were atoms.
4. Let subscripts in one formula suggest coefficients for another.

Self-Test

16. Balance each of these equations.

(a) $Li + H_2 \rightarrow LiH$ ____________________

(b) $Na + O_2 \rightarrow Na_2O$ ____________________

(c) $Sr + O_2 \rightarrow SrO$ ____________________

(d) $HCl + SrO \rightarrow SrCl_2 + H_2O$ ____________________

(e) $HBr + Na_2O \rightarrow NaBr + H_2O$ ____________________

(f) $Al + S \rightarrow Al_2S_3$ ____________________

(g) $CH_4 + F_2 \rightarrow CF_4 + HF$ ____________________

(h) $CO_2 + H_2 \rightarrow CH_4 + H_2O$ ____________________

New Terms

Write the definitions of the following terms. Although they were not first introduced in this section, be sure now that you can write definitions and give examples. If necessary, refer to the Glossary at the end of the text.

balanced equation — subscript
coefficient

3.6 Using Chemical Equations in Calculations

Review

The key concept in this section, which has also been stated earlier, is that the coefficients in a balanced equation provide stoichiometric equivalencies by *moles* and so lead us to *ratios by moles* between any two substances. If the equation, for example, is

$$2\text{Al} + 3\text{S} \rightarrow \text{Al}_2\text{S}_3$$

then we can write, *for this particular reaction:*

$$2\ \text{Al} \Leftrightarrow 3\ \text{S}$$
$$2\ \text{Al} \Leftrightarrow 1\ \text{Al}_2\text{S}_3$$
$$3\ \text{S} \Leftrightarrow 1\ \text{Al}_2\text{S}_3$$

From the first equivalency, for example, we can write the following two ratios by moles.

$$\frac{2\ \text{mol Al}}{3\ \text{mol S}} \quad \text{or} \quad \frac{3\ \text{mol S}}{2\ \text{mol Al}}$$

It is *essential* to include the units "mol Al" and "mol S"; just writing $\frac{2}{3}$ or $\frac{3}{2}$ is meaningless because without the units of "mol Al" and "mol S" we'd have no idea what these ratios are all about. It is just as meaningless to write

$$\frac{2\ \text{mol}}{3\ \text{mol}} \quad \text{or} \quad \frac{3\ \text{mol}}{2\ \text{mol}}$$

because these also beg the questions: "2 mol of what?" or "3 mol of what?"

If we write the conversion factors correctly, then it's easy to handle a question such as "How many moles of sulfur are going to be needed if we start with 1.8 mol of Al?" Remember, whenever a problem asks something like "how many moles of.....combine with (or are needed for) so many moles of ....?" we have a stoichiometry kind of problem. Once we have recognized the kind of problem, we reach for the tools we have learned. Here, we first need the *mole-ratio* tool, which connects the number of moles of one species to the number of moles of another using the coefficients.

What we are *converting* through the use of a conversion factor is the number of given moles of Al into the number of needed moles of sulfur, the number chemically equivalent to the Al in the reaction specified.

$$1.8 \text{ mol Al} \times \frac{3 \text{ mol S}}{2 \text{ mol Al}} = 2.7 \text{ mol S}$$

Don't go beyond Examples 3.17 and 3.18 as well as Practice Exercises 18-19 in the text until you have this phase of stoichiometry mastered. Once that's done, the shift from mole quantities to grams quantities, or the other way around, is easy.

Once you figure out numbers of moles and are ready (at least in principle) to move from the experiment planning to the executing stage, you reach for the next tool, the *moles-to-grams tool* already learned (Section 3.2). This tool, a conversion factor, is a ratio prepared from the formula mass either of grams to moles or of moles to grams for a specific substance. For example, if we need 2.7 mol S (atomic mass 32.06), we can find the number of grams by:

$$2.7 \text{ mol S} \times \frac{32.06 \text{ g S}}{1 \text{ mol S}} = 87 \text{ g S (properly rounded)}$$

Thinking It Through

For the following, identify the information needed to solve the problem and show (or explain) what must be done with it.

5 Sulfur dioxide will react with oxygen (O_2) under special conditions to form sulfur trioxide. How many moles of SO_2 are needed to react with 4.5 mol of O_2, and how many moles of SO_3 will form?

Self-Test

17. Phosphorus burns according to the following equation.

$$4P + 5O_2 \rightarrow P_4O_{10}$$

(a) How many moles of O_2 react with 10.0 mol of P?

____

(b) How many moles of P_4O_{10} can be made from 2.00 mol of P?

____

(c) To make 4.00 mol of P_4O_{10}, how many moles of O_2 are needed?

____

(d) To make 3.60 mol of P_4O_{10}, how many moles of P are needed?

____

How many moles of O_2?

____

18. Arsenic combines with oxygen to form As_2O_3 according to the following equation.

$$4As + 3O_2 \rightarrow 2As_2O_3$$

(a) To make 9.68 g As_2O_3, how many grams of As are needed?

____

(b) How many grams of O_2 are needed to make 8.92 g As_2O_3?

____

(c) If 5.85 g As are used, how many grams of As_2O_3 form?

____

New Terms

none

3.7 Limiting Reactant Calculations

Review

When the quantities of *two* reactants are given in a stoichiometry problem, you can be sure that at least one part of the problem will be to determine which reactant limits the amount of a product that can be made.

To discover which reactant limits, arbitrarily pick one reactant and convert its given mass into the corresponding number of moles (mass-to-moles tool using the formula mass). Then do a moles-to-moles conversion, the number of moles of the reactant you picked to the number of moles required of the other reactant according to the coefficients in the balanced equation. Now change the number of moles you just calculated into the number of grams of the second reactant (the moles-to-grams tool given by the formula mass). Is there enough of the other reactant provided to completely consume the reactant you picked? If the answer is "yes," you have discovered the limiting reactant—the one you picked. The limiting reactant is always the one that is completely consumed.

If your answer is "no," the first reactant picked is not completely consumed—some is left over—then the second reactant is limiting. Until you have gained confidence from experience, repeat the same kinds of calculations for the second reactant and verify that it is the one completely consumed.

Thinking It Through

For the following, identify the information needed to solve the problem and show (or explain) what must be done with it.

6 Aluminum and sulfur, when strongly heated, react to give aluminum sulfide, Al_2S_3. How many grams of Al_2S_3 can form from a mixture of 5.65 g of Al and 12.4 g of S?

Self-Test

19. To prepare some $AlBr_3$, a chemist mixed 3.13 g of Al with 28.8 g of Br_2. The equation is

$$2Al + 3Br_2 \rightarrow 2AlBr_3$$

(a) Identify the limiting reactant. __________

(b) How many grams of $AlBr_3$ can be made? __________

(c) Of the reactant that was taken in excess, how many grams of it are left over?

____

New Term

Write a definition of the following term, which was introduced in this section. If necessary, refer to the Glossary at the end of the text.

limiting reactant

3.8 Theoretical Yield and Percentage Yield

Review

We can calculate the percentage yield of a product by the following equation:

$$\frac{\text{(quantity of product actually obtained)}}{\left(\begin{array}{c}\text{quantity of product predicted from the} \\ \text{stoichiometry calculation based on} \\ \text{the limiting reactant}\end{array}\right)} \times 100\% = \text{percentage yield}$$

"Quantity" may be in moles or in grams, but it must be in the same unit in both the numerator and denominator. The calculation must be based on the limiting reactant (Section 3.7).

Thinking It Through

For the following, identify the information needed to solve the problem and show (or explain) what must be done with it.

7 Aluminum and bromine combine to give aluminum bromide. When 10.00 g of Al was mixed with 80.00 g of Br_2 there was obtained 80.00 g of $AlBr_3$. Calculate the percentage yield of $AlBr_3$.

Self-Test

20. Ethyl alcohol can be converted to diethyl ether (an anesthetic) by the following reaction.

$$\underset{\text{ethyl alcohol}}{2C_2H_5OH} \rightarrow \underset{\text{diethyl ether}}{C_4H_{10}O} + H_2O$$

Some of the diethyl ether is lost during its synthesis because it evaporates very easily. If 40.0 g of ethyl alcohol gives 28.2 g of diethyl ether, what is the percentage yield? ______________

21. Solid sodium carbonate, Na_2CO_3, was sprinkled onto a spill of hydrochloric acid, HCl, to destroy the acid by the following reaction.

$$Na_2CO_3 + 2HCl \rightarrow 2NaCl + H_2O + CO_2$$

If the spill contained 50.0 g of HCl (dissolved in water), was the HCl entirely destroyed by 50.0 g of Na_2CO_3? If not, how many grams of HCl were destroyed? If more Na_2CO_3 was used than necessary, then how many grams of it were in excess?

____

New Terms

Write the definitions of the following terms, which were introduced in this section. If necessary, refer to the Glossary at the end of the text.

by-product
competing reaction
main reaction
side reaction
yield, percentage
yield, theoretical

3.9 Reactions in Solution

Review

Some of the terms in this section answer the "what's this?" kind of question—terms such as "solution," "solute," and "solvent." Other terms address the question, "How concentrated is it?" "Concentration" always refers to a ratio expressed in one set of units or another—a ratio of solute to solvent or a ratio of solute to solution. When the precise value of that ratio isn't known (or doesn't matter too much), we have such qualitative expressions as *dilute* and *concentrated*, or *saturated* and *unsaturated*.

At any given temperature, one particular ratio of solute to solvent—the (maximum) *solubility* of the solute, often given in grams of solute per 100 grams of solvent—is an important physical constant for the solute. A solution with this concentration is a *saturated solution* (at the given temperature).

When the solubility of some solute is very small, then a saturated solution is still described as very dilute. But if the solubility is very large, then even a fairly unsaturated solution may still be quite concentrated. Remember that these expressions are qualitative. Often they are used only to imply a warning. "Be careful with this sulfuric acid solution; it's concentrated." Or they may be used to reassure. "If you spill some of this on your hand, don't panic; it's only a dilute solution. Calmly wash it off with water (but don't dwaddle too long)."

Self-Test

22. At 30 °C the solubility of boric acid in water is 6.35 g/100 g H_2O. A solution containing 0.00600 g boric acid in 100 g of H_2O is

    (a) unsaturated (b) concentrated (c) dilute (d) *a* and *c* ______

23. The solubility of potassium chloride in water is 34.7 g/100 g H_2O at 20 °C and 56.7 g/100 g H_2O at 100 °C. If a saturated solution is cooled from 100 °C to 80 °C and excess KCl precipitates during the cooling process, the solution at 80 °C can be described as

    (a) saturated (b) concentrated (c) *a* and *b* ______

24. An unsaturated solution of potassium chloride in water might become a saturated solution by
    (a) cooling it
    (b) boiling off some of the water
    (c) precipitating some of the potassium chloride
    (d) *a* and *b*
    (e) *a* and *c* ______

25. If you dissolve a small single-portion serving of sugar in a glass of iced tea, you can describe the resulting sugar solution as
    (a) dilute
    (b) unsaturated
    (c) concentrated
    (d) *b* and *c*
    (e) *a* and *b*
    ____

26. When a hot, saturated solution is cooled and some of the solute separates as a solid, this solid is called the

    (a) precipitation (b) precipitate (c) solvent (d) *a* and *c* ______

New Terms

Write the definitions of the following terms, which were introduced in this section. If necessary, refer to the Glossary at the end of the text.

concentrated solution
concentration
dilute solution
precipitate
precipitation reaction
saturated solution
solubility
solution
solute
solvent
standard solution

3.10 Molar Concentration

Review

In nearly all reactions done in the lab the reactants are in solution, and instead of taking a particular mass of a reactant directly, a calculated volume of a solution of known molar concentration is used. We have to know how to prepare solutions of known molarity, and we have to be able to measure out whatever volume we need to obtain some specified number of moles of solute. In many situations, determining the molar concentration of an "unknown" is the goal of an analysis. The calculations learned in this section have to be made for this work, too.

The molarity of a solution is the number of moles of solute per liter of solution. We can use "1000 mL" for 1 L in this context without necessarily implying four significant figures for the value of the *concentration*. If we have a solution with a concentration of 0.482 M NaOH, we can use either of the following two conversion factors for mole calculations:

$$\frac{0.482 \text{ mol NaOH}}{1000 \text{ mL NaOH solution}} \quad \text{or} \quad \frac{1000 \text{ mL NaOH solution}}{0.482 \text{ mol NaOH}}$$

The "0.482 mol NaOH" refers to pure NaOH used as the original solute. The "1000 mL NaOH solution" refers, of course, to the aqueous *solution* not to the volume of the *solvent* used.

If we want to prepare a solution of known molar concentration, we have in mind a particular final volume, so the calculation we have to do is to find first the moles of solute we need and then the grams.

In an analysis designed to measure the molarity of an unknown, the moles of solute in a carefully measured volume of the solution are determined. Then the molarity is simply the ratio of the moles to the volume (taken in liters).

When a dilute solution is to be made from adding solvent to a more concentrated solution, the molarities of both being known (or predetermined), the equation to use is

$$V_{dil}M_{dil} = V_{concd}M_{concd}$$

Thinking It Through

For the following, identify the information needed to solve the problem and show (or explain) what must be done with it.

8 How grams of NaOH are in 125.0 mL of 0.3540 *M* NaOH?

Self-Test

27. How would 500 mL of a 0.125 *M* $CaCl_2$ solution be prepared?

    ____

28. How many moles and how many grams of NaCl are in 250 mL of 1.38 *M* NaCl?

    ____

29. How many milliliters of 0.124 *M* HCl are needed to obtain 0.00244 mol HCl?

    ____

30. How would you prepare 100 mL of 0.500 *M* H_2SO_4 from 0.800 *M* H_2SO_4? ______________________________

    ____

31. How would you prepare 250 mL of 0.100 *M* HCl from 2.00 *M* HCl?

    ____

    ____

New Terms

Write the definitions of the following terms, which were introduced in this section. If necessary, refer to the Glossary at the end of the text.

molar concentration

molarity

3.11 Stoichiometry of Reactions in Solution

Review

The ability to convert volume and molar concentration data into moles of solute, learned in Section 3.10, is applied as the only new tool needed to work problems about stoichiometry when one or more reactants are in solution. Like all stoichiometry problems, the use of the balanced equation's coefficients is the key step.

Thinking It Through

For the following, identify the information needed to solve the problem and show (or explain) what must be done with it.

9 How many milliliters of 0.112 M HBr are needed to react with the KOH in 34.2 mL of 0.142 M KOH by the following reaction

$$KOH + HBr \rightarrow KBr + H_2O$$

Self-Test

32. How many milliliters of 0.135 M HCl are needed to react completely with the K_2CO_3 in 22.8 mL of 0.108 M K_2CO_3 by the following reaction?

$$2HCl + K_2CO_3 \rightarrow 2KCl + H_2O + CO_2$$

---

33. What is the molar concentration of H_2SO_4 if 18.2 mL react exactly with the $KHCO_3$ in 24.8 mL of 0.148 M $KHCO_3$ according to the following equation?

$$H_2SO_4 + 2KHCO_3 \rightarrow K_2SO_4 + 2CO_2 + 2H_2O$$

New Terms

none

Answers to Thinking It Through

1 The formulas give the ratios by atoms (or by moles) of N and H in compound X. The experimental data tell us that the observed ratio is 0.024 mol N to 0.048 mol N, which is the same as 1:2. This ratio fits only N2H4, because 2:4 is the same as 1:2. Moreover, N2H4 is the only candidate for which the ratio of formula units (N2H4) to the atoms of N is 1:2 (0.012 : 0.024).

2 The connection between moles and atoms is by means of Avogadro's number

$$1 \text{ mol gold} = 6.02 \times 10^{23} \text{ atom gold}$$

So 0.500 mol of gold would have half as many atoms, 3.01×10^{23} atom gold.

The connection between grams and moles of gold is by means of the atomic mass of gold, which is (from the Table of Atomic Masses and Numbers) 196.97.

$$1 \text{ mol gold} = 196.97 \text{ g gold}$$

So half a mole of gold would have half of this mass. (The answer is 98.48 g gold.)

3 The percentages must add up to 100%, so when only N and O are present and we're given that the compound is 36.85% N, then the percentage of O is the difference: 100.00 – 36.85 = 63.15%. To find out which formula has 36.85% N, you have to take as a percentage the number of grams of N in 1 mol of the compound (expressed in grams). So formula masses have to be computed for the candidates; they are 92.02 for N_2O_4 and 76.02 for N_2O_3. We test the data against N_2O_4.

$$1 \text{ mol } N_2O_4 = 92.02 \text{ g } N_2O_4$$

and this much N_2O_4 has 2 mol N or 2×14.01 g N = 28.02 g N. The percentage of N, then, is

$$\frac{28.02 \text{ g}}{92.02 \text{ g}} \times 100.00\% = 30.45\% \text{ N}$$

This percentage of N does not match what was given, so the compound must be the other one, N_2O_3 (which you should verify by calculating its percentage N).

4 To find the mass of O in the sample, simply do a subtraction of the mass of K from the total mass.

$$0.2564 \text{ g} - 0.2128 \text{ g} = 0.0436 \text{ g O}$$

We now know the *mass* ratio of K to O, but we need the *mole* ratio to make an empirical formula. We thus need the mass-to-moles tools provided by the atomic masses of K and O, 39.10 for K and 16.00 for O. When you carry out the operation on each mass, you find that there is 0.005442 mol K and

0.00272 mol O. To work our way toward a ratio in *whole* numbers, we divide both by the smaller, 0.00272. (The formula works out to be K_2O.)

5 No problem in reaction stoichiometry can be worked without a balanced equation, so this is the first step.

$$2SO_2 + 3O_2 \rightarrow 2SO_3$$

Now we know (from the coefficients) that the ratio of moles of SO_2 to moles of O_2 is 2:3. So,

$$4.5 \text{ mol } O_2 \times \frac{2 \text{ mol } SO_2}{3 \text{ mol } O_2} = 3.0 \text{ mol } SO_2$$

The coefficients tell us that for each mole of SO_2 used, one mole of SO_3 will form; the mole ratio is 1:1. So 3.0 mol of SO_3 can form.

6 We recognize this as a limiting reactant problem because the masses of *two* reactants are given. But first we need the balanced equation, which we can write from the information given as

$$2Al + 3S \rightarrow Al_2S_3$$

We must convert the masses given into moles so that we can compare the *mole* ratios of the reactants as they were taken.

5.65 g Al = ? mol Al. We need the atomic mass of Al, 26.98

$$5.65 \text{ g Al} \times \frac{1 \text{ mol Al}}{26.98 \text{ g Al}} = 0.209 \text{ mol Al}$$

Doing the same kind of calculation on 12.4 g S (atomic mass 32.07) tells us that 12.4 g S = 0.387 mol S. Now we have to do the comparison. The mole ratio of Al to S is supposed to be 2 Al : 3 S. Arbitrarily beginning with Al, we can see how many moles of S are needed for 0.209 mol of Al.

$$0.209 \text{ mol Al} \times \frac{3 \text{ mol S}}{2 \text{ mol Al}} = 0.314 \text{ mol S}$$

More than 0.314 mol of S was taken, so all of the Al will be used up. Al limits the yield, so we calculate the mass of Al_2S_3 based on Al.

$$0.209 \text{ mol Al} \times \frac{1 \text{ mol } Al_2S_3}{2Al} = 0.104 \text{ mol } Al_2S_3$$

Finally, using a mole-to-mass tool—from the formula mass of Al_2S_3 (150.17) on 0.104 mol of Al_2S_3 would show that a mass of 15.6 g of Al_2S_3 can be obtained.

7 We first need the equation.

$$2Al + 3Br_2 \rightarrow 2AlBr_3$$

Because we're given the masses of *two* reactants, we have to do a limiting reactant calculation first. We find the following moles of reactants using their formula masses.

$$10.00 \text{ g Al} = 0.3706 \text{ mol Al}$$

$$80.00 \text{ g } Br_2 = 0.5006 \text{ mol } Br_2$$

We can tell using the equation's coefficients that 0.3706 mol Al needs 3/2 times as many moles of Br_2 or 0.5559 mol of Br_2. But this much was *not* used, so there will be aluminum left over. It's the Br_2 that limits. Now we see how much $AlBr_3$ we could obtain in theory from 0.5006 mol of Br_2.

$$0.5006 \text{ mol } Br_2 \times \frac{2 \text{ mol } AlBr_3}{3 \text{ mol } Br_2} = 0.3337 \text{ mol } AlBr_3$$

Because the formula mass of $AlBr_3$ calculates to be 166.68, we find the mass of $AlBr_3$ as follows.

$$0.3337 \text{ mol } AlBr_3 \times \frac{166.68 \text{ g } AlBr_3}{1 \text{ mol } AlBr_3} = 88.99 \text{ g } AlBr_3$$

There was obtained only 80.00 g $AlBr_3$, so the percentage yield is

$$\frac{80.00 \text{ g}}{88.99 \text{ g}} \times 100\% = 89.90\%$$

8 We need the volume-moles tool to find the moles of NaOH in the given volume. The molarity is 0.3540 mol NaOH/L, or 0.3540 mol NaOH/1000 mL soln. So the number of moles of NaOH is found by

$$125.0 \text{ mL NaOH soln} \times \frac{0.3540 \text{ mol NaOH}}{1000 \text{ mL NaOH soln}} = 0.04425 \text{ mol NaOH}$$

To find the number of grams of NaOH. we now apply the moles-to-grams tool provided by the formula mass of NaOH, 40.00.

$$0.04425 \text{ mol NaOH} \times \frac{40.00 \text{ g NaOH}}{1 \text{ mol NaOH}} = 1.770 \text{ g NaOH}$$

9 We find the moles of KOH taken in 34.2 mL of 0.142 *M* KOH and this will be identical to the moles of Br needed (because the coefficients of KOH and HBr are equal in the equation). So

$$\text{mol HBr} = \text{mol KOH} = 34.2 \text{ mL KOH soln} \times \frac{0.142 \text{ mol KOH}}{1000 \text{ mL KOH soln}}$$

$$= 0.00486 \text{ mol}$$

Using a mole-volume conversion factor made from the molarity of the HBr, we find the volume of the HBr solution as follows.

$$0.00486 \text{ mol HBr} \times \frac{1000 \text{ mL HBr soln}}{0.112 \text{ mol HBr}} = 43.4 \text{ mL HBr soln}$$

Answers to Self-Test Questions

1. 6.02 x 1023 atoms of F
2. 0.800 mol O; 1.60 mol H
3. 0.600 mol Ca; 1.20 mol O; 1.20 mol H
4. Formula mass = 203.14.

$$\frac{1 \text{ mol } (NH_4)_3PO_4{\cdot}3H_2O}{203.14 \text{ g } (NH_4)_3PO_4{\cdot}3H_2O} \text{ and } \frac{203.14 \text{ g } (NH_4)_3PO_4{\cdot}3H_2O}{1 \text{ mol } (NH_4)_3PO_4{\cdot}3H_2O}$$

5. $$\frac{6.02 \times 10^{23} \text{ molecules } NH_3}{17.0 \text{ g } NH_3} \text{ and } \frac{17.0 \text{ g } NH_3}{6.02 \times 10^{23} \text{ molecules } NH_3}$$

6. 2.41×10^{23} atoms Fe
7. 12.8 g O
8. 1.21 g H
9. 49.04 g H_2SO_4
10. 2.50 mol CO_2
11. 13.52 g P; 10.48 g O. (Note the sum of these; it's 24.00 g.)
12. 39.70% Cr; 17.55% Na; 47.76% O
13. The formula mass of $Na_2S_2O_3$ is 158.12. The calculated percentages are Na, 29.08; S, 40.56; and O, 30.36, which are very close to the experimental percentages, so the formula is correct.
14. C_5H_7
15. C_5H_8
16. (a) $2Li + H_2 \rightarrow 2LiH$
    (b) $4Na + O_2 \rightarrow 2Na_2O$
    (c) $2Sr + O_2 \rightarrow 2SrO$
    (d) $2HCl + SrO \rightarrow SrCl_2 + H_2O$
    (e) $2HBr + Na_2O \rightarrow 2NaBr + H_2O$
    (f) $2Al + 3S \rightarrow Al_2S_3$
    (g) $CH_4 + 4F_2 \rightarrow CF_4 + 4HF$
    (h) $CO_2 + 4H_2 \rightarrow CH_4 + 2H_2O$
17. (a) 12.5 mol O_2, (b) 0.500 mol P_4O_{10}, (c) 20.0 mol O_2
    (d) 14,4 mol P, 18.0 mol O_2
18. (a) 7.33 g As, (b) 2.16 g O_2, (c) 7.72 g As_2O_3
19. (a) Al limits, (b) 30.9 g $AlBr_3$, (c) 1 g Br_2 left over (rounded from 0.959 g)
20. 87.9%
21. 15.5 g HCl not consumed
22. d
23. c
24. d

25. e
26. b
27. Dissolve 6.94 g $CaCl_2$ in water and make the final volume 500 mL.
28. 0.345 mol NaCl; 20.2 g NaCl
29. 19.7 mL HCl solution
30. Add water to 62.5 mL of 0.800 M H_2SO_4 to a final volume of 100 mL.
31. Add water to 12.5 mL of 2.00 M HCl to a final volume of 250 mL.
32. 36.4 mL HCl solution
33. 0.101 M H_2SO_4

Tools you have learned

Remove this chart from the Study Guide and keep it handy when tackling homework problems.

Tool	Function
Atomic mass	g of element ⇔ mol of element
Formula mass (molecular mass)	g of compound ⇔ mol of compound
Chemical formula (formula subscripts)	mole ratios among elements in a compound
Percentage composition	Find the mass of an element in a sample of a compound
Percentage yield $\% \text{ yield} = \frac{\text{actual yield}}{\text{theoretical yield}} \times 100\%$	Relate actual yield of product to theoretical yield
Chemical equation (coefficients of reactants and products)	Relate moles of substances involved in a chemical reaction
Avogadro's number 6.02×10^{23} particles/mol	number of particles ⇔ mol of substance
Molar concentration $\text{molarity} = \frac{\text{mol solute}}{\text{L soln}}$ $M \times V(\text{L}) = \text{mol solute}$ $M \times V(\text{mL}) = \text{mmol solute}$	mol solute ⇔ liters of solution mmol solute ⇔ mL of solution

Summary of Important Equations

Percentage by mass

$$\text{percentage by mass} = \frac{\text{mass of part of sample}}{\text{mass of sample}} \times 100\%$$

Percentage yield

$$\text{Percentage yield} = \frac{\text{actual yield}}{\text{theoretical yield}} \times 100\%$$

Molarity

$$\text{Molarity } (M) = \frac{\text{moles solute}}{\text{liters of solution}}$$

$$\text{Molarity} \times \text{volume (L)} = \text{moles of solute}$$

$$\text{Molarity} \times \text{volume (mL)} = \text{mmol of solute}$$

Dilutions

$$V_{\text{dil}}M_{\text{dil}} = V_{\text{concd}}M_{\text{concd}}$$

Chapter **4**

ENERGY AND THERMOCHEMISTRY

We turn from the materials budget of a chemical reaction (Chapter 3) to the energy budget. Some chemical reactions will not occur until the reactants have received energy—usually heat energy but sometimes electrical energy, sometimes other forms. Other reactions occur and spontaneously give off energy as heat, light, sound, electricity, or as mechanical energy (as from an explosion). We here introduce the important concepts needed to describe such energy changes.

Learning Objectives

In this chapter, keep in mind the following goals.

1. To learn the distinctions between *kinetic* and *potential energy* and to understand the roles played by attractions and repulsions in potential energy.
2. To learn the relationship between molecular kinetic energy and heat and temperature.
3. To learn the concepts of chemical systems, surroundings, boundaries, and exothermic and endothermic changes.
4. To learn the thermal properties of matter—heat capacity, specific heat, and molar heat capacity—how they are defined and the equations used for calculations.
5. To learn how data obtain employing calorimeters can be used to calculate thermal properties.
6. To learn what is meant by the state of a system and a state function.
7. To learn what is meant by an energy change and by an enthalpy change, an how the sign of ΔH determines whether a change is exothermic or endothermic.
8. To learn the relationship between ΔH and the heat of reaction at constant pressure, and to learn the reference temperature and pressure that are used as standards.

9 To prepare, manipulate, and interpret thermochemical equations and enthalpy diagrams.

10 To use Hess's law to calculate enthalpy changes.

11 To use standard enthalpies of formation and Hess's law to calculate enthalpy changes.

4.1 Kinetic and Potential Energy Revisited

Review

Energy is the ability to do work, which is often seen as the *mechanical work* of moving an object from one place to another. The energy of a moving object is called its *kinetic energy*, where K.E. = $(1/2)mv^2$. A battery has energy when it can cause the *electrical work* of moving electrons through a circuit. The energy of a fire appears in forms to which we give the names *light, heat*, and (usually) *sound*.

A match book and a firecracker have *potential energy* because each has the potential to release any of these forms of energy. Potential energy is stored energy. Potential energy exists in a stretched spring, a compressed spring, or in a book raised above the desk, examples that illustrate some of the ways by which potential energy can be stored. There are two important situations where we can tell that changes in potential energy have occurred.

1. Potential energy increases by pulling apart objects that attract each other. (Example, lifting a book above a desk against the force of gravitational attraction.)
2. Potential energy increases by pushing together objects that repel each other. (Example, compressing a string.)

The existence of positively and negatively charged particles in atoms, ions, and molecules means that there are repelling and attracting forces in substances. The net force of attraction that more or less permanently holds one atom near another in a substance is a *chemical bond*. The potential energy in substances that resides in the attracting and repelling forces within a substance is called its *chemical energy*. It can be released or stored by reactions that use or that make the substances.

The total energy of a system is made up of two terms, the kinetic energy and the potential energy. For the universe as a whole, the total energy is conserved (law of conservation of energy).

Self-Test

1. Describe how the kinetic energy and the potential energy of a stone thrown upward change during the upward motion, at the top of the trajectory, and during the downward flight.

   ____

2. Which actions decrease the potential energies of the systems?

   (a) Two electrons move away from each other.

   (b) A stretched rubber band is released.

   (c) Coal burns.

   (d) All of the above. ______________

3. The potential energy in substances by virtue of attractions and repulsions of the particles making up atoms, ions, and molecules is called

   ____

4. The force of attraction that holds the pieces of a molecule together in a given substance is called

   ____

5. To what in nature does the law of conservation of energy apply?

   ____

New Terms

Write the definitions of the following terms, which were introduced in this section. If necessary, refer to the Glossary at the end of the text.

chemical bonds

chemical energy

4.2 Kinetic Theory

Review

The formula units making up substances are in motion and have *molecular kinetic energy*. The combined molecular kinetic energies of all units makes up the *heat* contained within a system. The average molecular kinetic energy is proportional to the *Kelvin temperature* of the system.

Self-Test

6. How does the average molecular kinetic energy of molecules change during an exothermic reaction?

   ____

7. Considering the equation for kinetic energy, why does kinetic energy never have a negative value?

   ____

8. Why is it impossible for a system to have a temperature lower than 0 K?

   ____

9. When a system cools, what becomes of the average molecular kinetic energy of its chemical formula units?

   ____

10. How do the values of each of the two physical quantities that make up the equation for kinetic energy change when the temperature increases?

    ____

New Terms

Write the definitions of the following terms, which were introduced in this section. If necessary, refer to the Glossary at the end of the text.

heat

kinetic theory

4.3 Energy Changes in Chemical Reactions

Review

Bond formation decrease potential energy; bond breaking increases potential energy. When a chemical reaction involves both the breaking and formation of bonds, there is a net overall change in potential energy. If it decreases, then the molecular kinetic energies of the particles increase (because potential energy changes to kinetic energy). If the potential energy increases, there is a net decrease in molecular kinetic energy.

A *system* is any part of the universe we choose to isolate for study by either real or imaginary *boundaries,* which separate the system from everything else in the universe, the system's *surroundings.* When energy cannot move across the boundary, the system is *insulated.*

In an *exothermic change,* the system becomes warmer and there is a net decrease in potential energy within it. In an *endothermic change,* the system becomes cooler, so there is a net increase in potential energy.

Self-Test

11. When a reacting system in an insulated container becomes warmer, does its total energy change? Explain.

    ____

12. When a system changes in such a way that heat flows across its boundaries into the surroundings, the change is described as

    (a) exothermic (b) endothermic (c) noninsulated ____________

13. When photosynthesis occurs within a plant cell, the system, is the change exothermic or endothermic?

    ____

14. If the wall of a plant cell is the boundary for the system inside that undergoes photosynthesis, in which direction must energy flow for photosynthesis to happen?

    ____

4.4 Heats of Reaction; Calorimetry

Review

Because all forms of energy can be converted quantitatively into heat, methods to determine heat become ways to measure quantities of energy available in other forms. These methods take advantage of the *thermal properties* of matter.

We study *specific heat* and *heat capacity* not only to learn about these thermal properties but also to get ready for the study of ΔH in the next section.

Think of "specific heat" as a shortened term for "specific heat capacity." Therefore think of it as the capacity of a specific amount of some substance—one gram or one mole—to store heat, the form of energy that can cause temperatures to change.

The specific heat of a substance is one of its physical constants, like its melting or boiling points or its density. The units for specific heat are J g^{-1} $°C^{-1}$ (J/g °C). Water has a specific heat of 4.184 J g^{-1} $°C^{-1}$, so it takes 4.184 joules of energy to change the temperature of 1 g of water by 1 °C. If you heat 1 g of water to make its temperature rise by 2 °C, you have put 2 × 4.184 J—twice as much energy—into storage in the 1 g water sample. Or if you heat 2 g of water to make its temperature increase by 1 °C, you have similarly put twice as much energy into storage in water, now a 2 g sample.

When the specific amount of substance is the mole, then the specific heat is called the *molar heat capacity* with units J mol^{-1} $°C^{-1}$ (J/mol °C).

The concept of "heat capacity" differs from that of "specific heat" because it refers to a whole sample, or a whole object, not just to one gram or one mole of it. Heat capacity is an extensive property; specific heat is an intensive property. The units of heat capacity, therefore, do not carry the "per gram" or "per mole" identifications involved with specific heats. The units of heat capacity are simply those of energy per degree Celsius, J $°C^{-1}$ (J/°C).

When we know the heat capacity of an object we know its capacity to "soak up" heat from a warmer object. The larger the heat capacity of the object, the more heat it will absorb for each degree Celsius increase in temperature. If the object undergoes a temperature decrease after being in contact with a cooler system, then we can tell from its heat capacity how much energy it releases for each degree of temperature decrease.

When the value of the heat capacity of a calorimeter is given, remember that this refers to the entire apparatus—the water and everything sticking into or under the water. Different calorimeters have different heat capacities even though they might all use water as the operating fluid.

Thinking It Through

For the following, identify the information needed to solve the problem and show (or explain) what must be done with it.

1 When 0.1000 mol of diethyl ether, an anesthetic, was burned in a bomb calorimeter, the temperature of the system rose from 24.000 °C to 26.332 °C. The heat capacity of the calorimeter was 118.0 kJ $°C^{-1}$. The equation for the reaction is

$$C_4H_{10}O(l) + 6O_2(g) \rightarrow 4CO_2(g) + 5H_2O(l)$$

(a) How many kilojoules were liberated by the combustion of this sample?

(b) How many kilojoules were liberated per mole of diethyl ether?

Self-Test

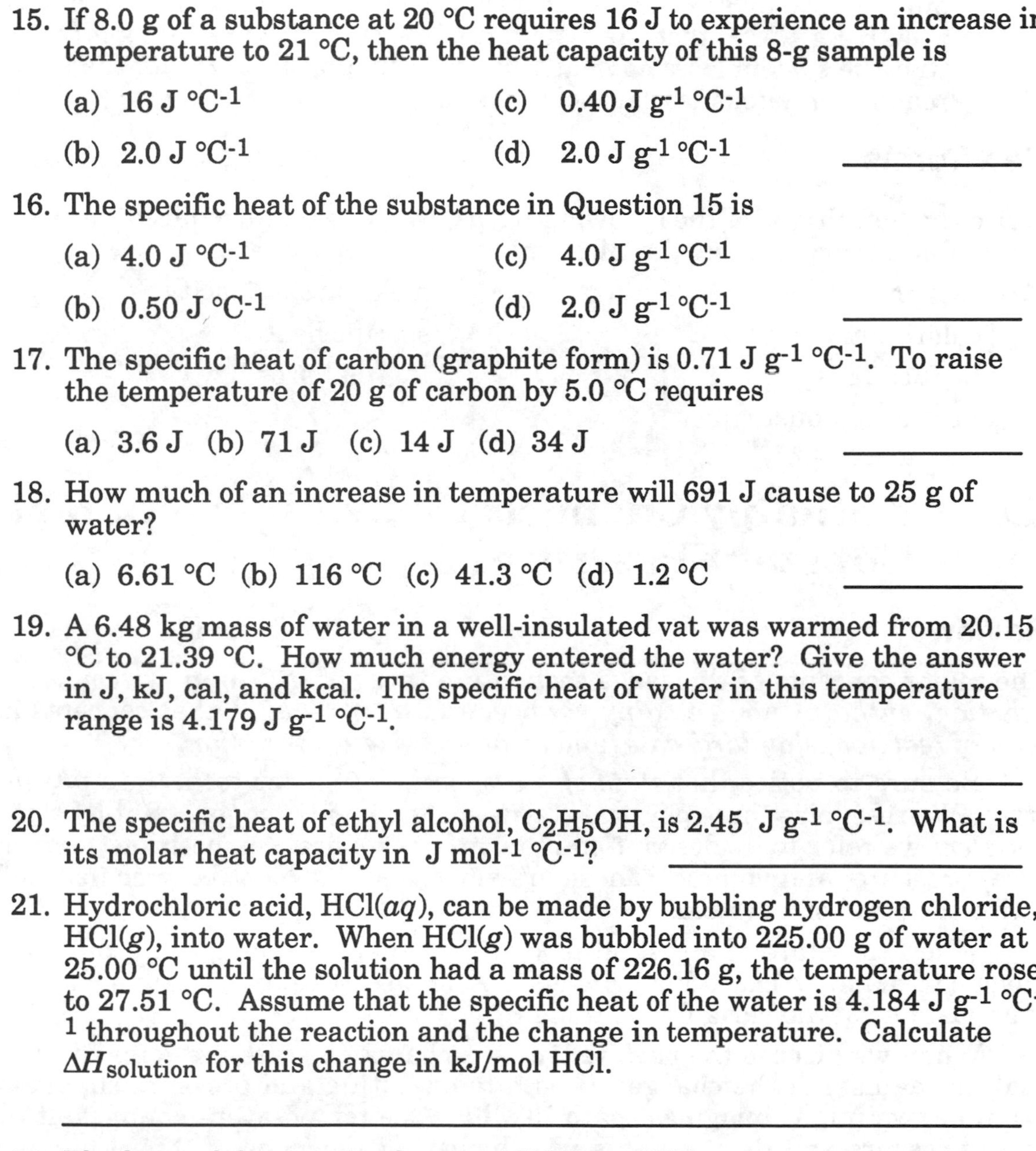

15. If 8.0 g of a substance at 20 °C requires 16 J to experience an increase in temperature to 21 °C, then the heat capacity of this 8-g sample is

    (a) 16 J $°C^{-1}$ (c) 0.40 J g^{-1} $°C^{-1}$

    (b) 2.0 J $°C^{-1}$ (d) 2.0 J g^{-1} $°C^{-1}$ __________

16. The specific heat of the substance in Question 15 is

    (a) 4.0 J $°C^{-1}$ (c) 4.0 J g^{-1} $°C^{-1}$

    (b) 0.50 J $°C^{-1}$ (d) 2.0 J g^{-1} $°C^{-1}$ __________

17. The specific heat of carbon (graphite form) is 0.71 J g^{-1} $°C^{-1}$. To raise the temperature of 20 g of carbon by 5.0 °C requires

    (a) 3.6 J (b) 71 J (c) 14 J (d) 34 J __________

18. How much of an increase in temperature will 691 J cause to 25 g of water?

    (a) 6.61 °C (b) 116 °C (c) 41.3 °C (d) 1.2 °C __________

19. A 6.48 kg mass of water in a well-insulated vat was warmed from 20.15 °C to 21.39 °C. How much energy entered the water? Give the answer in J, kJ, cal, and kcal. The specific heat of water in this temperature range is 4.179 J g^{-1} $°C^{-1}$.

    ____

20. The specific heat of ethyl alcohol, C_2H_5OH, is 2.45 J g^{-1} $°C^{-1}$. What is its molar heat capacity in J mol^{-1} $°C^{-1}$? __________

21. Hydrochloric acid, HCl(*aq*), can be made by bubbling hydrogen chloride, HCl(*g*), into water. When HCl(*g*) was bubbled into 225.00 g of water at 25.00 °C until the solution had a mass of 226.16 g, the temperature rose to 27.51 °C. Assume that the specific heat of the water is 4.184 J g^{-1} $°C^{-1}$ throughout the reaction and the change in temperature. Calculate $\Delta H_{solution}$ for this change in kJ/mol HCl.

    ____

22. The heat of the reaction between KOH and HBr solutions was determined using a coffee cup calorimeter. The reaction was

$$KOH(aq) + HBr(aq) \rightarrow KBr(aq) + H_2O$$

A solution of 45.0 mL of 1.00 *M* HBr at 23.5 °C was mixed with a solution of 45.0 mL of 1.00 *M* KOH, also at 23.5 °C. The temperature quickly rose to 30.2 °C. Assume that the specific heats of each solution is 4.18 J g^{-1} $°C^{-1}$, that the densities of the solutions are 1.00 g mL^{-1}, and that the system loses no heat to its surroundings. Calculate the heat of reaction in kilojoules per mole of HCl.

New Terms

Write the definitions of the following terms, which were introduced in this section. If necessary, refer to the Glossary at the end of the text.

calorimeter
calorimetry
heat capacity
heat of combustion
molar heat capacity
specific heat
thermal property

4.5 Enthalpy Changes: Heats of Reaction at Constant Pressure

Review

The major concepts of this large section are those of *state of a system*, a *state function, enthalpy* and *enthalpy change*, and *Hess's law*. Other concepts are *path of reaction, standard state,* and *standard heat of formation.*

Be sure to notice that *state of a system* is a broader term than *physical state*. We recognize three physical states—solid, liquid, and gas. But by *state of a system* we refer to its description in terms of chemical composition, temperature, pressure, and volume. These are variables that must be specified to describe the state of a system.

When the state of a system is specified, there is enough information for anyone to make an exact duplicate of it. A *change in state* refers to a change in even just one of the variables defining it.

When we discuss the *path* taken by a change in state, we refer to the actual mechanism of the change. If both temperature and pressure change, for example, we might imagine a path in which the temperature changes at constant pressure and then the pressure changes at the (constant) new temperature. Or we would envision first a pressure change at constant temperature followed by the temperature change at the (constant) new pressure. Either way we get from the initial state to the same final state.

Some energy features of a change in state depend on the path taken and some do not. It is very important that you learn to distinguish these. When a battery discharges, both heat and electrical work are usually evolved. The total of these two kinds of energy is the same regardless if the discharge occurs suddenly or slowly because, either way, the system changes to the identical final state. But how much of this total energy appears as heat and how much as electrical work does depend on the speed of discharge—the path of the overall reaction. If the chemical reaction in the battery is allowed to go as rapidly as possible, more heat than electrical work is generated. But if a slow discharge is used, then more electrical work and less heat is obtained.

The energy of a system, E, is a state function— one that depends only on the initial and final states of the system and is independent of the path from one to the other. The value of E cannot be known; only the difference in energy, ΔE, between the initial and final states is knowable. The difference is always understand to be taken in the following way (always "final minus initial").

$$\Delta E = E_{\text{final}} - E_{\text{initial}}$$

The value of ΔE does depend on the openness of the system. The heat determined with a bomb calorimeter (a closed system in which the pressure can increase with the reaction) is not necessarily the same as the heat determined with a calorimeter open to the atmosphere. The reactions studied in this section are always open to the atmosphere and so are under constant pressure..

The total energy of a system at constant pressure is called the system's *enthalpy*, which is also a state function. Only an *enthalpy change*, ΔH, can be known, not an absolute value of H. The path of the change does not determine ΔH, only how the energy of ΔH will be divided between heat and some form of work.

When a path is found where all the energy change at constant pressure appears as heat, then $\Delta H = q$, the heat of the reaction at constant pressure. In an exothermic change, the final enthalpy of the system is less than the initial enthalpy. Since

$$\Delta H = H_{\text{final}} - H_{\text{initial}}$$

the value of ΔH for an exothermic change is negative, so q is also negative. In an endothermic change, one that must import energy across the boundary from the surroundings to keep the initial and final temperatures constant, ΔH is positive, so q is also positive.

When $H_{\text{final}} > H_{\text{initial}}$, ΔH is positive, and the change is endothermic. Thus at its most fundamental level, an endothermic change is any that increases the enthalpy of the system. On the other hand, when $H_{\text{final}} < H_{\text{initial}}$, ΔH is negative, the change is exothermic, and the system suffers a loss in enthalpy.

Remember that ΔH refers just to the system, not the system plus the surroundings. If the system loses energy, the surroundings gains energy, so no energy is lost. This is the essence of the law of conservation of energy — the energy of the universe is constant.

Because the value of ΔH for a given reaction depends on both the temperature and the pressure at which the system is kept, we need reference conditions of T and P. For enthalpy experiments, the standard conditions are 25 °C and 1 atm of pressure. Values of ΔH measured under these conditions are called *standard heats of reaction* (or *standard enthalpies of reaction*), and the associated symbol is ΔH°.

The actual value of ΔH° depends on the *scale* of the reaction, the actual mole quantities of substances as disclosed by the coefficients in the balanced equation. When the equation includes ΔH°, the equation is called a *thermochemical equation*. The following are all valid thermochemical equations for the reaction of hydrogen and oxygen that gives water.

(1) $2H_2(g) + O_2(g) \rightarrow 2H_2O(l)$ $\Delta H^\circ = -571.5$ kJ

(2) $3H_2(g) + \frac{3}{2}O_2(g) \rightarrow 3H_2O(l)$ $\Delta H^\circ = -857.3$ kJ

(3) $4H_2(g) + 2O_2(g) \rightarrow 4H_2O(l)$ $\Delta H^\circ = -1143$ kJ

The value of ΔH° differs for each because, as the coefficients indicate, the scale changes. Always be careful to notice the physical states of the substances involved (*s, l,* or *g*), because enthalpy data depend on these, too.

Hess's law goes back to the law of conservation of energy and to the fact that ΔH is a *state* function. We can concoct any sort of tortuous path we wish (or find essential) to get the reactants converted into the products *as long as the overall summation of these steps results in the overall reaction for which we want to find the value of* ΔH°.

Because of the law of conservation of energy, the value of ΔH° for a reaction in one direction can simply be given the opposite sign for the same reaction run in the opposite direction.

Finally, remember that if we multiply or divide all of the coefficients in a thermochemical equation by some number, we must do the identical operation to the value of ΔH.

Be sure to catch the seemingly slight change in units when shifting from *standard heat of reaction*, ΔH°, which has units of energy (just energy and not energy per mole of something) to *standard heat of formation*, ΔH_f°, which has units of energy *per mole.* The "per mole" part refers to the compound whose formation *from its elements in their standard states* is described by the associated equation. If you are very careful about the units and the physical states, the calculations are no more difficult than any other Hess's law calculations. Standard heats of formation let us calculate standard enthalpy changes for many other kinds of reactions, including many that cannot really be measured in any direct way.

Example 4.1 Manipulating Thermochemical Equations

Problem

Is it energetically feasible to obtain copper metal from copper sulfide, $CuS(s)$, by reaction with hydrogen? Calculate $\Delta H°$ for the following equation from the thermochemical equations provided below.

$$CuS(s) + H_2(g) \rightarrow Cu(s) + H_2S(g)$$

(1)	$2CuS(s) + 3O_2(g) \rightarrow 2CuO(s) + 2SO_2(g)$	$\Delta H° = -807.18$ kJ
(2)	$2CuO(s) + C(s) \rightarrow 2Cu(s) + CO_2(g)$	$\Delta H° = -83.05$ kJ
(3)	$3H_2(g) + SO_2(g) \rightarrow H_2S(g) + 2H_2O(l)$	$\Delta H° = -92.32$ kJ
(4)	$C(s) + O_2(g) \rightarrow CO_2(g)$	$\Delta H° = -393.51$ kJ
(5)	$2H_2(g) + O_2(g) \rightarrow 2H_2O(l)$	$\Delta H° = -571.70$ kJ

Solution

Before you get experience with these kinds of problems the biggest difficulty is knowing where to start. There aren't any pat rules, but probably the best advice is to pick one of the compounds in the target equation, the one we want to work towards, *that appears in only one of the listed equations.* If we have more than one such choice, then just select one of them and begin. It will all work out. Thus, we might pick $CuS(s)$. It appears only in equation (1), and it's on the left side of the arrow just as it is in the target equation. So we rewrite (1), as is, on scratch paper, but we won't do anything yet about the values of $\Delta H°$ until we've worked out the chemical equations. Put a circle or box around $CuS(s)$ just to mark it. (We wouldn't want to cancel it later!)

(1) $2\boxed{CuS(s)} + 3O_2(g) \rightarrow 2Cu(s) + 2SO_2(g)$

Notice that none of the unboxed formulas appears in the target equation, so now, if possible, we pick another chemical out of our target, another one that appears only once anywhere in the listed equations. We won't pick $H_2(g)$ yet; it's in both (3) and (5). But $H_2S(g)$ is only in (3) — and on the correct side to boot — and $Cu(s)$ is only in (2) — also on the correct side. Let's pick $Cu(s)$. On the scratch paper we'll write (2) below the one already set down, equation (1). Then we'll put a box around $Cu(s)$ and cancel what we can.

(1) $2\boxed{CuS(s)} + 3O_2(g) \rightarrow 2CuO(s) + 2SO_2(g)$

(2) $2CuO(s) + C(s) \rightarrow 2\boxed{Cu(s)} + CO_2(g)$

Notice that if we added these, we could cancel $2CuO(s)$ from the left side of (2) and the right side of (1). It's not necessary to write intermediate summations, but we'll do it to see where we are. Equation (6) is the sum of (1) and (2), simplified.

(6) $2\boxed{CuS(s)} + 3O_2(g) + C(s) \rightarrow 2\boxed{Cu(s)} + 2SO_2(g) + CO_2(g)$

At one or another of these intermediate stages it's a good idea to keep an eye out for chances to do some significant cancelling of unboxed chemicals. The only "rule" is to "keep an eye out." When this will work is a matter of judgement. But notice in (6) how it includes the chemicals needed to make one unboxed chemical, $CO_2(g)$, from others on the opposite side of the arrow, $C(s)$ and $O_2(g)$. All three of these have to be cancelled eventually to trim things down to the target equation. Now look at equation (4). If we reverse it and add it to (6), we'll be able to cancel several things. (We'll mark this as "(4)-reverse" so that we'll have a signal later when we figure out the net ΔH°.)

(6) $2\boxed{CuS(s)} + 3O_2(g) + C(s) \rightarrow 2\boxed{CuS(s)} + 2SO_2(g) + CO_2(g)$

(4)-reverse $CO_2(g) \rightarrow C(s) + O_2(g)$

Although we can't cancel $3O_2(g)$ by $O_2(g)$, we can at least cancel one of the three oxygen molecules in $3O_2(g)$. Now if we add (6) and (4)-reverse, we get (7).

(7) $2\boxed{CuS(s)} + 2O_2(g) \rightarrow 2\boxed{Cu(s)} + 2SO_2(g)$

The unboxed chemicals have to go; and equation (3) has one of them, $SO_2(g)$, besides having two, $H_2S(g)$ and $H_2(g)$, that we have to get into the equation. Equation (7) shows two $SO_2(g)$ but equation (3) has only one. So we multiply (3) through by 2, and box certain formulas.

(7) $2\boxed{CuS(s)} + 2O_2(g) \rightarrow 2\boxed{Cu(s)} + 2SO_2(g)$

$2 \times (3)$ $6\boxed{H_2(g)} + 2SO_2(g) \rightarrow 2\boxed{H_2S(g)} + 4H_2O(l)$

The sum of these two is (8).

(8) $2\boxed{CuS(s)} + 2O_2(g) + 6\boxed{H_2(g)} \rightarrow 2\boxed{CuS(s)} + 2\boxed{H_2S(g)} + 4H_2O(l)$

The boxes indicate that we've finally gotten all of the target chemicals in place; and we have only one unused equation, (5). We have to retain some $H_2(g)$ for the target equation but we have to cancel all of the $2O_2(g)$ and $4H_2O(l)$. If we reverse (5) and multiply it by 2, everything will all work out.

(8) $2\boxed{CuS(s)} + 2O_2(g) + 6\boxed{H_2(g)} \rightarrow 2\boxed{CuS(s)} + 2\boxed{H_2S(g)} + 4H_2O(l)$

2 × (5)-reverse $4H_2O(l) \rightarrow 2O_2(g) + 4H_2(g)$

The sum of these two is (9).

(9) $2CuS(s) + 2H_2(g) \rightarrow 2Cu(s) + 2H_2S(s)$

We could at this stage divide all of the coefficients by 2, but that would complicate the calculation of the net $\Delta H°$. Let's do that calculation next.

Equation Used		$\Delta H°$
(1) as is		−807.18 kJ
(2) as is		−83.05 kJ
(4)-reverse	Change sign of $\Delta H°$	+393.51 kJ
2 × (3)	Double (3)'s $\Delta H°$	−584.64 kJ
2 × (5)-reverse	Double (5)'s $\Delta H°$ and change sign	+1143.40 kJ
Net = [target equation] × 2	Net $\Delta H°$ =	+62.04 kJ

Therefore, for the target equation we divide everything by 2:

$CuS(s) + H_2(g) \rightarrow Cu(s) + H_2S(g)$ $\Delta H° = +31.02$ kJ

Thinking It Through

For the following, identify the information needed to solve the problem and show (or explain) what must be done with it.

2 Calculate $\Delta H°$ for the target reaction in the worked example:

$$CuS(s) + H_2(g) \rightarrow Cu(s) + H_2S(g)$$

Use the following $\Delta H_f°$ data.

$CuS(s)$ $\Delta H_f° = -48.53 \text{ kJ mol}^{-1}$

$H_2S(g)$ $\Delta H_f° = -17.506 \text{ kJ mol}^{-1}$

Does your result agree with the answer obtained in the worked example (making allowances for errors caused by rounding or in experimental data)?

Self-Test

23. ΔH_{system} is defined as

    (a) Hreactants – Hproducts

    (b) Hproducts – Hreactants

    (c) ΔHproducts – ΔHreactants

    (d) the change in the heat capacity of the system __________

24. If we had some way of knowing that $H_{products}$ = 400 kcal and that $H_{reactants}$ = 600 kcal, then ΔH =

    (a) 1000 kcal (c) –200 kcal

    (b) 200 kcal (d) –1000 kcal __________

25. If the value of ΔH_{system} = –200 J, and the system's temperature before and after the change is the same, then the value of $\Delta H_{surroundings}$ is

    (a) –200 J (c) 0

    (b) 200 J (d) not measurable __________

26. What are the experimental conditions to which all standard enthalpy data refer? ____________________

27. The thermochemical equation for the combustion of 2 mol of CO(*g*) is

    $$2CO(g) + O_2(g) \rightarrow 2CO_2(g) \qquad \Delta H^\circ = -566 \text{ kJ}$$

    Write the thermochemical equation for the combustion of 4 mol of CO(*g*).

    ____

28. The direct reaction of methane (CH_4) with oxygen to give carbon monoxide is hard to accomplish without also producing carbon dioxide. However, we can use a Hess's law calculation to calculate ΔH°, anyway. Calculate ΔH° in kcal for the following reaction from the thermochemical equations given.

    $$2CH_4(g) + 3O_2(g) \rightarrow 2CO(g) + 4H_2O(l)$$

    Use the following thermochemical equations:

    (1) $2C(s) + O_2(g) \rightarrow 2CO(g) \qquad \Delta H^\circ = -221.08 \text{ kJ}$

    (2) $CH_4(g) + 2O_2(g) \rightarrow CO_2 g) + 2H_2O(l) \qquad \Delta H^\circ = -890.4 \text{ kJ}$

(3) $C(s) + O_2(g) \rightarrow CO_2(g)$ $\Delta H^\circ = -393.51$ kJ

____

29. Acetic acid can be made from acetylene by the following overall reaction:

$$\underset{\text{acetylene}}{2C_2H_2(g)} + 2H_2O(l) + O_2(g) \rightarrow \underset{\text{acetic acid}}{2HC_2H_3O_2}$$

Calculate ΔH° for the formation of *one* mole of acetic acid by this reaction using the following thermochemical equations:

(1) $C_2H_4(g) \rightarrow H_2(g) + C_2H_2(g)$ $\Delta H^\circ = 174.464$ kJ
ethylene

(2) $C_2H_5OH(l) \rightarrow C_2H_4(g) + H_2O(l)$ $\Delta H^\circ = 44.066$ kJ
ethyl alcohol

(3) $C_2H_5OH(l) + O_2(g) \rightarrow HC_2H_3O_2(l) + H_2O(l)$ $\Delta H^\circ = -495.22$ kJ

(4) $2H_2(g) + O_2(g) \rightarrow 2H_2O(l)$ $\Delta H^\circ = -571.70$ kJ

____

30. Any substance in its most stable chemical form at 25 °C and 1 atm is said to be in its

(a) thermochemical state (b) standard state

(c) STP state (d) zero H state ____________

31. Write the thermochemical equations together with ΔH_f° values (Table 4.2 in the textbook) for the formation of each of the following compounds from its elements. Use units of kJ/mol.
(a) $H_2SO_4(l)$ (b) $Al_2O_3(s)$ (c) $CO(NH_2)_2(s)$

____

____

____

32. Write the thermochemical equations, including correct values of ΔH° in kJ, for the formation of each of these compounds from its elements. Use Table 4.2 in the textbook for needed data.

(a) 2 mol of Fe_2O_3

____

(b) 3 mol of $C_2H_6(g)$

____

33. On a separate sheet of paper, draw an enthalpy diagram for the formation of one mole of H_2O_2 from its elements. Use data in Table 4.2 of the textbook in units of kJ as needed.

34. The value of ΔH_f° for acetylene, $C_2H_2(g)$, is fairly high *and positive.*

    (a) On a separate sheet of paper, draw an enthalpy diagram for the formation of 1 mol of this compound from its elements. Use units of kJ mol-1.

    (b) Is the decomposition of acetylene to its elements an exothermic or an endothermic process? ________________

35. Use the Hess law equation and ΔH_f° data in Table 4.2 of the textbook to calculate ΔH° for the reaction of methane with oxygen given in Problem 28. (Having done 28 the long way you'll see the powerful advantage of the Hess law equation.)

    ____

36. Use the Hess law equation, data in Table 4.2 of the textbook, and the results of the calculations in Problem 29 to calculate ΔH_f° for acetic acid in kJ mol^{-1}.

    ____

37. Calculate $\Delta H^\circ_{combustion}$ for acetylene, $C_2H_2(g)$ in kJ mol^{-1} from ΔH_f° data from Table 4.2 of the textbook. The equation is

$$2C_2H_2(g) + 5O_2(g) \rightarrow 4CO_2(g) + 2H_2O(l)$$

____

New Terms

Write the definitions of the following terms, which were introduced in this section. If necessary, refer to the Glossary at the end of the text.

enthalpy
enthalpy change
heat of reaction at constant pressure
Hess's law of heat summation
standard atmosphere
standard enthalpy of formation
standard heat of formation
standard heat of reaction
standard state
state function
state of a system
thermochemical equation

Answers to Thinking It Through

1 (a) What we first want is the relationship between the temperature change and the amount of heat released:

$$\Delta t \Leftrightarrow \text{? kilojoules}$$

The connection between Δt and kilojoules is the heat capacity of the calorimeter, 118.0 kJ °C^{-1}.

$$\Delta t = 26.332\ °C - 24.000\ °C = 2.332\ °C$$

$$118\ \frac{\text{kJ}}{°C} \times 2.332\ °C = 275.2\ \text{kJ}$$

(b) To calculate the heat liberated per mole, we take the ratio of 275.2 kJ to 0.1000 mol, which gives units of kJ/mol. The answer is obtained as shown below.

$$\frac{275.2\ \text{kJ}}{0.1000\ \text{mol}} = 2751\ \text{kJ mol}^{-1}$$

2 Hess's law equation is applicable:

$$\Delta H° = (\text{sum of } \Delta H_f° \text{ of products}) - (\text{sum of } \Delta H_f° \text{ of reactants})$$

$$\Delta H° = [-17.506\ \text{kJ mol}^{-1} \times 1\ \text{mol (H}_2\text{S)}] - [-(-48.53\ \text{kJ mol}^{-1} \times 1\ \text{mol (CuS))}]$$

If you work through to the answer you should obtain $\Delta H° = +31.02$ kJ

Answers to Self-Test Questions

1. On the upward flight, kinetic energy changes to potential energy. At the top, where the stone is motionless, all of the initial kinetic energy has become potential energy. On the way down, potential energy changes to kinetic energy as the stone picks up speed.
2. d
3. chemical energy
4. a chemical bond
5. the universe
6. It increases.
7. KE = $1/mv^2$; mass (m) is never negative and even if velocity (v) were negative, the square of the velocity must be a positive number. Hence, KE is always positive (or zero). (Note. As you may have already studied in a physics course, velocity can be negative because it is a vector quantity.)
8. Because there is no such thing as negative kinetic energy and because 0 K corresponds to the cessation of all motions associated with molecular kinetic energy.
9. The average molecular kinetic energy decreases.

10. Only the v term (velocity) changes—it increases.
11. The total energy is a constant; but potential energy changes into molecular kinetic energy.
12. a
13. endothermic
14. From the surroundings into the system.
15. a
16. d
17. b
18. a
19. 33.6×10^3 J, 33.6 kJ, 8.03×10^3 cal, 8.03 kcal
20. 113 J mol^{-1} °C.
21. 74.9 kJ/mol HCl
22. 55 kJ/mol HCl
23. b
24. c
25. b
26. 25 °C and 1 atm
27. $4CO(g) + 2O_2(g) \rightarrow 4CO_2(g)$ $\Delta H° = 1132$ kJ 28. −1214.9 kJ
29. −427.90 kJ 30. b
31. (a) $H_2(g) + S(s) + 2O_2(g) \rightarrow H_2SO_4(l)$ $\Delta H_f° = -811.32$ kJ mol^{-1}

    (b) $2Al(s) + \frac{3}{2}O_2(g) \rightarrow Al_2O_3(s)$ $\Delta H_f° = -1669.8$ kJ mol^{-1}

    (c) $C(s) + \frac{1}{2}O_2(g) + N_2(g) + 2H_2(g) \rightarrow CO(NH_2)_2(s)$ $\Delta H° = -333.19$ kJ mol^{-1}
32. (a) $4Fe(s) + 3O_2(g) \rightarrow 2Fe_2O_3(s)$ $\Delta H° = -1644.4$ kJ

    (b) $6C(s) + 9H_2(g) \rightarrow 3C_2H_6(g)$ $\Delta H° = -254.001$ kJ
33. $H_2(g) + O_2(g)$

    $\Delta H° = -187.6$ kJ

    $H_2O_2(l)$

34. (a)

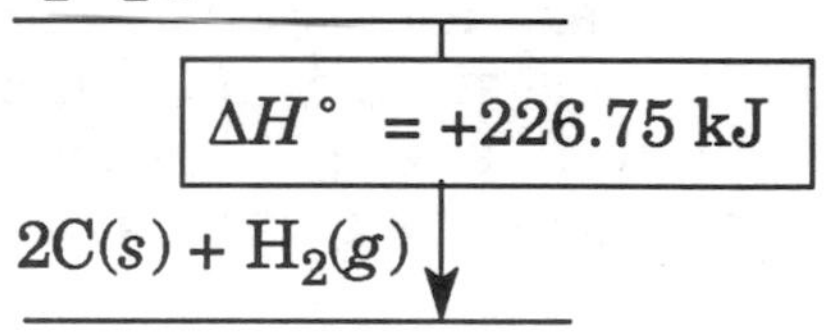

(b) Its decomposition is exothermic.

35. −1214.9 kJ
36. −487.05 kJ
37. 1299.7 kJ mol^{-1}

Tools you have learned

Remove this chart from the Study Guide and keep it handy when tackling homework problems.

Tool	*Function*
Heat Capacity heat capacity $= \dfrac{\text{heat}}{\Delta t}$	Calculate amount of heat involved in a reaction from calorimetry data (heat capacity of calorimeter and temperature change)
Specific Heat The ratio of heat capacity to mass specific heat $= \dfrac{\text{heat capacity}}{\text{mass}}$ $= \dfrac{\text{heat}}{\text{mass } \Delta t}$	Calculate heat energy from mass and temperature change: specific heat $\times$ mass $\times$ Δt = heat
Hess's Law For reactions that can be written in steps, ΔH° is the same as the sum of the values of ΔH° for the individual steps.	Calculate the enthalpy change for a given reaction from values of ΔH° or ΔH_f° of other reactions that, taken together, can be manipulated to give the stated reaction as a result.

Summary of Important Equations

Heat capacity

$$\text{Heat capacity} = \frac{\text{heat absorbed}}{\Delta t}$$

Specific heat

$$\text{Specific heat} = \frac{\text{heat capacity}}{\text{mass}(g)}$$

Heat energy

$$\text{Heat energy} = \text{specific heat} \times \text{mass} \times \Delta t$$

Hess's law

$$\Delta H^\circ = \left(\text{sum of } \Delta H_f^\circ \text{ of products}\right) - \left(\text{sum of } \Delta H_f^\circ \text{ of reactants}\right)$$

Chapter **5**

ATOMIC AND ELECTRONIC STRUCTURE

In Chapter 2 you learned that atoms are not the simplest particles of matter. They are composed of still simpler parts called protons, neutrons, and electrons. The protons and neutrons of an atom are found in the nucleus and the electrons surround the nucleus in the atom's remaining volume. Because the nucleus is so small and so far from the outer parts of an atom, it has very little direct influence on ordinary chemical and physical properties. Its indirect role is in determining the number of electrons a neutral atom will have. As you will learn in this chapter, it is the number of electrons and their energy distribution (which we call the atom's electronic structure) that is the primary factor in determining an atom's chemical and physical properties.

We begin the chapter with a discussion of the nature of light and the energy carried by light waves. As mentioned in the caption of the opening photograph on page 170, the study of the light emitted by atoms provided the clues to the internal arrangements of the electrons in atoms. The remainder of the chapter will be devoted to descriptions of the electronic structures of atoms and how they correlate with observable properties of atoms.

You will find that much of this chapter deals with theory, and you may find that a lot of it seems remote from everyday experience. If you find it difficult to understand at first, don't be discouraged. Reread the discussions, study the worked examples, and do the exercises. It may take some time, but it should all fit together eventually.

Learning Objectives

Throughout your study of this chapter, keep in mind the following objectives:

1. To learn about the properties of light, and to learn how the light emitted by an atom when it is excited, or energized, gives clues to how the electrons are distributed in the space outside the atom's nucleus.
2. To see how the theoretical model of the electronic structure of the atom developed, historically.
3. To see that matter behaves not only as though it were composed of discrete particles, but also as waves.

4 To learn the conditions under which the wave properties of electrons are most noticeable.

5 To learn how waves interact with each other to produce a diffraction pattern.

6 To learn how the theoretical study of matter waves leads to a more sophisticated explanation of atomic structure.

7 To learn the names, symbols, and allowed values for the three quantum numbers associated with electron waves in atoms.

8 To examine the relationship between the quantum numbers and the energies of the electron waves in atoms.

9 To see how the magnetic behavior of the electron leads to the idea that the electron is spinning, and to see how this electron spin affects the electronic structure of an atom.

10 To see how electrons are arranged among the orbitals of an atom, and to predict such arrangements using Figure 5.15.

11 To see how the structure of the periodic table correlates with predicted electron configurations and to learn how to use the periodic table as tool for predicting electron configurations.

12 To learn the electron configuration of chromium and copper, which are not predicted by the rules for writing electron configuration of atoms.

13 To learn the shapes and directional properties (relative orientations) of *s*, *p*, and *d* orbitals.

14 To learn how electrons close to the nucleus shield outer electrons from the full effects of the nuclear charge.

15 To see how atomic and ionic size are related to electronic structure and to learn how these properties vary within the periodic table.

16 To learn how the size of an atom changes when it gains or loses electrons to become an ion.

17 To learn how the energy needed to remove an electron from an atom can be related to the atom's electronic structure, and how this quantity varies according to the atom's position in the periodic table.

18 To see how the energy change associated with the addition of an electron to an atom varies according to the atom's position in the periodic table.

5.1 Electromagnetic Radiation and Atomic Spectra

Review

Electromagnetic energy is energy that is carried by electromagnetic radiation (light waves). These waves are characterized by their frequency, ν (the number of oscillations per second), which is given in units called hertz (Hz). Remember,

$$1 \text{ Hz} = 1 \text{ s}^{-1}$$

(second raised to the minus one power, which is the same as 1/second).

A wave is also characterized by its wavelength, λ, which is the distance between any two successive peaks. In the SI, wavelength is expressed in meters (or submultiples of meters, such as nanometers).

The product of wavelength and frequency is the speed of the wave, and for electromagnetic radiation traveling through a vacuum, this speed is a constant called the *speed of light* (symbol, c). Be sure you know the equation

$$\lambda \times \nu = c$$

Check with your teacher to find out whether you are expected to know the value of c, 3.00×10^8 m/s (or 3.00×10^8 m s^{-1}). Study Examples 5.1 and 5.2 to be sure you can change wavelength to frequency and frequency to wavelength.

Electromagnetic Spectrum

Electromagnetic radiation comes in a range of frequencies or wavelengths that we call the electromagnetic spectrum. Visible light constitutes only a narrow band of wavelengths ranging from approximately 400 to 700 nm. Infrared, microwaves, and radio and TV broadcasts are waves of longer wavelength (lower frequency) than visible light. Ultraviolet, X rays, and gamma radiation are waves of shorter wavelength (higher frequency) than visible light.

The Energy of Light

Light often behaves as if it comes in tiny packets (also called *quanta*) of energy that we call photons. The energy of a photon is proportional to the frequency of the light. The equation is

$$E = h\nu$$

where E is the energy, ν is the frequency, and h is a proportionality constant called Planck's constant.

Atomic spectra

A continuous spectrum (Figure 5.5) contains all wavelengths of light and is produced by the glow of a hot object, such as the filament in a light bulb, or the sun. The spectrum emitted by energized (*excited*) atoms is called an atomic spectrum or emission spectrum and is not continuous. It contains only a relatively few wavelengths, as illustrated in Figure 5.6.

The atomic spectrum is different for each element, and can be used to identify the element; it's like the element's "fingerprint." Atomic spectra suggest that the energy if an electron in an atom is quantized, which means it can have only certain specific amounts of energy. These energies correspond to a characteristic set of energy levels possessed by atoms of the element. When an electron goes from one energy level to a lower one, the difference in energy ΔE appears as light having a frequency determined by the equation $\Delta E = h\nu$.

Thinking It Through

For the following, identify the information needed to solve the problem and show (or explain) what must be done with it.

1. How many microwave photons having a wavelength of 1.05 cm are needed to raise the temperature of a cup of coffee (250 mL) from room temperature (25 °C) to 80 °C? Assume the coffee has a density of 1.0 g/mL and a specific heat of 4.18 J g^{-1} $°C^{-1}$.

Self-Test

1. Radio station WGBB in New York broadcasts at a frequency of 1240 kHz on the AM radio band. What is the wavelength in meters of these radio waves?

   ____

2. A certain compound is found to absorb infrared radiation strongly at a wavelength of 3.0×10^{-6} m. What is the frequency of this radiation?

   ____

3. What is the wavelength in nanometers of electromagnetic radiation that has a frequency of 6.0×10^{15} Hz?

   ____

4. Calculate the energy in Joules of a photon of green light that has a wavelength of 546 nm.

   ____

5. What is the energy in kilojoules of one mole of photons having a wavelength of 546 nm?

   ____

New Terms

Write the definitions of the following terms, which were introduced in this section. If necessary, refer to the Glossary at the end of the text.

electronic structure	visible spectrum
electromagnetic energy	photon
electromagnetic radiation	quantum
frequency	continuous spectrum
hertz	atomic spectrum
wavelength	energy level
electromagnetic spectrum	quantized energy

5.2 The Bohr Model of the Hydrogen Atom

Review

The Rydberg equation is an *empirical equation*, which means it was derived from data obtained by experimental measurements. It can be used to calculate the wavelengths of the lines in the spectrum of hydrogen as illustrated in Example 5.3. (You probably don't need to memorize the Rydberg equation, but check with your instructor to be sure.)

Bohr developed his model of the atom to explain the spectrum of hydrogen. You should know the basic features of the model:

1 The electron was believed to travel in circular orbits.

2 The energy and size of an orbit can have only certain values, which are related to the value of the quantum number n.

3 When the atom absorbs energy, the electron is raised to a higher, more energetic orbit.

4 When the electron drops from a higher to a lower orbit, a photon is emitted whose energy equals the energy difference between the two orbits. The frequency of the photon is determined by the relationship $E = h\nu$.

Bohr was able to use his model to derive an equation that matched the Rydberg equation almost exactly. This was the theory's greatest success.

Although Bohr's model of the atom was later shown to be incorrect, he was the first to recognize the existence of quantized energy levels in atoms. He was also the first to introduce the idea of a quantum number—an integer related to the energy of an electron in an atom.

Self-Test

6. According to Bohr's model of the hydrogen atom, what was the value of the quantum number for the lowest-energy orbit?

   ____

7. In terms of b in Equation 5.3, what is the energy of the electron in the first Bohr orbit ($n = 1$)?

   ____

   What is the energy of the electron in the second Bohr orbit ($n = 2$)?

   ____

   In terms of b, how much energy is emitted when the electron drops from the orbit with $n = 2$ to the orbit with $n = 1$?

   ____

8. Calculate the wavelength in nanometers of the spectral line produced when an electron drops from the 6th Bohr orbit to the second. What color is this spectral line?

   ____

New Terms

Write the definitions of the following terms, which were introduced in this section. If necessary, refer to the Glossary at the end of the text.

quantum number
ground state

5.3 Wave Properties of Matter and Wave Mechanics

Review

De Broglie proposed that particles have wave properties, with a wavelength that is inversely proportional to the product of the particle's mass and velocity. Only very light particles, such as the electron, proton, and neutron, have wavelengths large enough to be observed experimentally. Diffraction—the constructive and destructive interference of waves—can be used to demonstrate the wave nature of these particles.

Wave mechanics (quantum mechanics) is a theory about electronic structure that is based on de Broglie's hypothesis. In this theory, the electron in an atom is considered to be a standing wave (one in which the positions of the peaks and troughs don't move.) Electron waves are called orbitals and each one can be identified by a set of values for three quantum numbers. You should learn the names of the quantum numbers and their allowed values (including the restrictions on their values).

The principal quantum number, n, can only have integer values that range from 1 to ∞.

$$n = 1 \text{ or } 2 \text{ or } 3 \text{ or } \ldots\infty$$

In categorizing the energies of the orbitals, n identifies the shell. An orbital with $n = 1$ is in the first shell, and so forth.

The secondary quantum number, l, divides the shells into subshells and determines the shapes of the orbitals. For a given n, allowed values of l range from 0 to $(n - 1)$. Remember the letter designations for the subshells.

value of l	0	1	2	3
letter	*s*	*p*	*d*	*f*

The magnetic quantum number, m_l, divides the subshells into individual orbitals and determines the orientations of the orbitals relative to each other. In a given subshell, the values of m_l range from $+l$ to $-l$. These values determine the number of orbitals in a given kind of subshell.

type of subshell	*s*	*p*	*d*	*f*
number of orbitals	1	3	5	7

These relationships are illustrated in Figure 1 below, which should be studied in conjunction with Table 5.1 on page 189 of the text. The relative energies of the various subshells and orbitals are described by Figure 5.15 of the text. You need not memorize this figure, although we will use it in Section 5.5 to determine how the electrons in an atom are distributed among the various orbitals.

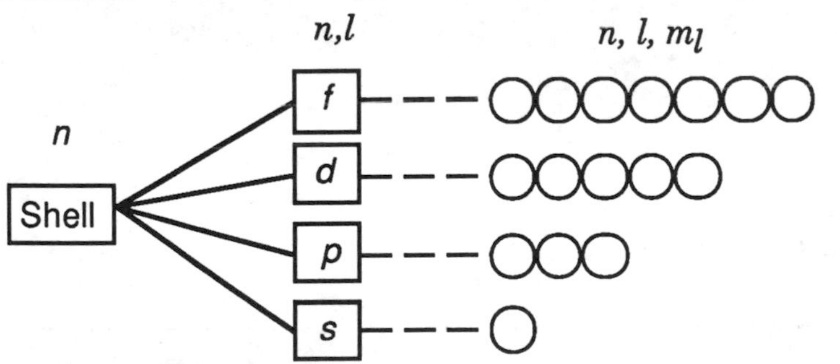

Figure 1
The way the quantum numbers for electrons correspond to shells, subshells and individual orbitals in an atom.

In Figure 5.15, there are several points to note. First, we see that every shell has an *s* subshell. Shells above the first shell each have a *p* subshell. Any shell above the second shell has a *d* subshell as well, and beyond the third shell, each also has an *f* subshell. Second, notice that the energy of the shells increase with increasing value of *n*. Third, notice that within a shell the energies of the subshells are in the order $s < p < d < f$, and that all orbitals of a given subshell are of equal energy. Finally, notice how the subshells of one shell begin to overlap with subshells of other shells as *n* becomes larger.

Self-Test

9. What do the terms *in phase* and *out of phase* mean?

____

____

10. What would be the subshell designation (e.g., 1*s*) corresponding to the following sets of values of *n* and *l*?

    (a) $n = 2, l = 1$ ____________

    (b) $n = 4, l = 0$ ____________

    (c) $n = 3, l = 2$ ____________

    (d) $n = 5, l = 3$ ____________

11. Which of the following sets of quantum numbers are unacceptable?

    (a) $n = 2, l = 1, m_l = 0$

    (b) $n = 2, l = 2, m_l = 1$

    (c) $n = 2, l = 1, m_l = -2$

    (d) $n = 3, l = 2, m_l = -2$

    (e) $n = 0, l = 0, m_l = 0$ ____________

12. What subshells are found in the fourth shell?

____

13. Which subshell is higher in energy?

    (a) 3*s* or 3*p* ________

    (b) 4*p* or 4*d* ________

    (c) 3*p* or 4*p* ________

New Terms

Write the definitions of the following terms, which were introduced in this section. If necessary, refer to the Glossary at the end of the text.

- diffraction
- diffraction pattern
- wave mechanics
- quantum mechanics
- traveling wave
- standing wave
- node
- orbital
- principal quantum number (n)
- shell
- secondary quantum number (l)
- subshell
- magnetic quantum number (m_l)

5.4 Electron Spin and the Pauli Exclusion Principle

Review

The electron behaves like a tiny magnet. A way of explaining this is to imagine that it spins like a top. Two directions of spin are possible, which are identified by the values of the spin quantum number, m_s. The Pauli exclusion principle states that no two electrons in the same atom can have the same values for all four of their quantum numbers. The result is that no more than two electrons can occupy any one orbital. (No more than two electrons can have the same wave form.) When all of the electrons in an atom are not *paired*, a residual magnetism occurs and the atom is paramagnetic. On the other hand, if all the electrons are paired, the atom is diamagnetic.

Self-Test

14. How many electrons can occupy

    (a) a 2*s* subshell? __________

    (b) a 3*d* subshell? __________

    (c) the shell with $n = 3$? __________

    (d) the shell with $n = 4$? __________

    (e) the subshell with $n = 3$ and $l = 1$ __________

New Terms

Write the definitions of the following terms, which were introduced in this section. If necessary, refer to the Glossary at the end of the text.

- electron spin
- spin quantum number (m_s)
- Pauli exclusion principle
- paramagnetism
- diamagnetism

5.5 Electronic Structures of Multielectron Atoms

Review

The distribution of electrons among an atom's orbitals is the electronic structure, or electron configuration of the atom. Electrons always fill orbitals of lowest energy first (Figure 5.15). When filling a set of equal-energy orbitals, Hund's rule requires that the electrons spread out over the orbitals as much as possible and that the number of unpaired electrons be a maximum. This gives the ground state (lowest energy) configuration. In Section 5.6 you will see that the periodic table can also be used in place of Figure 5.15 to obtain an atom's electron configuration.)

Learn to draw orbital diagrams. When indicating the orbitals of a *p*, *d*, or *f* subshell, be sure to show *all* of the orbitals of that subshell, even if some of them are unoccupied. For example, for boron we should write

B (↑↓) (↑↓) (↑)( )( ) (correct)
1*s* 2*s* 2*p*

The following is *incorrect*, because only one of the three 2*p* orbitals is shown.

B (↑↓) (↑↓) (↑) (incorrect)
1*s* 2*s* 2*p*

Self-Test

15. Use Figure 5.15 to write the electron configuration for

(a) Al ____________________

(b) Br ____________________

(c) V ____________________

16. Give orbital diagrams for

    (a) phosphorus

    (b) silicon

New Terms

Write the definitions of the following terms, which were introduced in this section. If necessary, refer to the Glossary at the end of the text.

electronic structure	orbital diagram	Hund's rule
electron configuration	core electrons	
orbital diagram		

5.6 Electronic Configurations and the Periodic Table

Review

Figure 5.16 illustrates how we can use the periodic table as a guide in writing electron configurations. Following the aufbau principle, we assign electrons to subshells beginning with the 1*s* (corresponding to period 1 in the periodic table). Then, as we cross the periodic table row after row, we let the table tell us which subshells become occupied.

When we cross through Groups IA and IIA, we fill an *s* subshell; across Groups IIIA through Group 0 we fill a *p* subshell. For the *s* and *p* subshells, the value of *n* that goes with the *s* and *p* designations equals the period number. Crossing a row of transition elements corresponds to filling a *d* subshell with *n* equal to *one less than the period number*. Crossing a row of inner transition elements (the lanthanides or actinides) corresponds to filling an *f* subshell with *n* equal to *two less than the period number*. Be sure to study Example 5.4 which begins on page 194 of the text.

Basis for the Periodic Table

The main point of this discussion is that atoms of a given group in the periodic table have similar electron configurations for their outer shells. Thus, all elements in Group IA have an outer shell with one electron in an *s* orbital. Similarly, the elements in Group IIA each have two electron in their outer shell *s* orbital.

Abbreviated Configurations

In the shorthand electron configuration of an element, we only show subshells above those that are filled in an atom of the preceding noble gas. For example, for potassium (atomic number $Z = 19$), we only show orbitals beyond those that are filled in argon, the preceding noble gas. Argon ($Z = 18$) has completed $1s$, $2s$, $2p$, $3s$, and $3p$ subshells ($2 + 2 + 6 + 2 + 6 = 18\ e^-$). The next subshell is the $4s$, so for potassium we write

$$\text{K}\ \ [\text{Ar}]\,4s^1$$

and we write the orbital diagram as

K [Ar] (↑)
 4s

Similarly, for the element nickel we write

$$\text{Ni}\ \ [\text{Ar}]\,3d^8 4s^2$$

and give its orbital diagram as

Ni [Ar] (↑↓)(↑↓)(↑↓)(↑)(↑) (↑↓)
 3d 4s

Valence Shell Configurations

In writing the valence shell configuration for an element, we only list subshells that are in the *outer shell* of the atom. (Valence shell configurations are only of interest to us for the representative elements.) The valence shell configuration for the element sulfur, for example, is

$$\text{S}\qquad 3s^2 3p^4$$

Self-Test

17. What is the shorthand electron configuration for iodine?

____

18. What is the shorthand electron configuration of cobalt?

____

19. How many unpaired d electrons are found in an atom of cobalt?

____

20. Use the periodic table to write the valence shell electron configuration of
    (a) oxygen ____________________
    (b) barium ____________________
    (c) indium ____________________
    (d) bismuth ____________________
21. Use the periodic table to predict the shorthand electron configuration for
    (a) Tc ____________________
    (b) Fe ____________________
    (c) Gd ____________________

New Terms

Write the definitions of the following terms, which were introduced in this section. If necessary, refer to the Glossary at the end of the text.

shorthand configuration	valence electrons
valence shell	valence shell configuration

5.7 Some Unexpected Electron Configurations

Review

Half-filled subshells and (especially) filled subshells have extra stability that causes the electron configurations of some elements such as chromium, copper, silver, and gold to deviate from the configurations that we would predict by following the procedures of the last section. You should learn the electron configurations of these elements as exceptions.

Self-Test

22. From what you learned in this section, predict the electron configuration of europium, Eu ($Z = 63$).

    ____

New Terms

none

5.8 Shapes of Atomic Orbitals

Review

The Heisenberg uncertainty principle and the wave nature of the electron make it impossible for us to know where the electron is in an atom. Instead, we refer to relative probabilities of finding the electron at different locations. In effect, we imagine the electron as a blur or cloud with a greater electron density in some places than in others.

In general, the size of a given kind of orbital increases with increasing n. The s orbitals are all spherical in shape, meaning that if we examine a surface on which the probability of finding the electron is everywhere the same, the surface is a sphere. In Figure 5.19, we see that the 2s, 3s, and higher s orbitals have spherical nodes, too, but that is not really important for our discussions.

The p orbitals are sometimes described as being dumbell-shaped. Each p orbital has *two* regions of electron density located on opposite sides of the nucleus (Figure 5.20) and the three p orbitals of a p subshell are oriented perpendicular to each other (Figure 5.21).

The d orbitals are more complex, still, than the p orbitals. Four of them in a given subshell have the same shape, each consisting of four lobes of electron density. The fifth consists of a pair of lobes along the (arbitrary) z axis, plus a donut-shaped ring of electron density in the xy plane.

Self-Test

23. On a separate piece of paper, practice sketching the shapes of s and p orbitals.

24. On the axes below, sketch the shapes of d_{xy}, d_{xz}, d_{yz}, and $d_{x^2-y^2}$ orbitals.

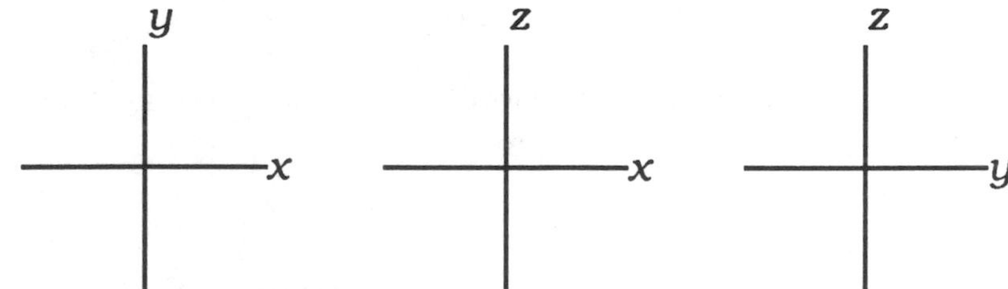

New Terms

Write the definitions of the following terms, which were introduced in this section. If necessary, refer to the Glossary at the end of the text.

uncertainty principle

electron density

electron cloud

5.9 Variation of Properties with Electronic Structure.

Review

The key to understanding the way the properties discussed in this section vary within the periodic table is understanding the concept of effective nuclear charge. The notion here is that the outer electrons do not feel the full effect of the charge on the nucleus because the charge of the electrons in shells below the outer shell partially offset the nuclear charge. The positive charge that the outer electrons do feel is called the *effective nuclear charge*.

Sizes of Atoms

Although in a strict sense atoms and ions have no true outer boundary, they often behave as if they have nearly a constant size. The size of an atom or ion is generally given in terms of its radius and is expressed in units of picometers, nanometers, or angstroms. Even though the angstrom isn't an SI unit, it is still widely used for expressing small distances. Remember: 1 Å = 0.1 nm = 100 pm.

Within the periodic table, size *decreases* from left to right in a period and from bottom to top in a group. Going across a period, the amount of positive charge felt by the outer electrons increases because electrons in the same shell are not very effective at shielding each other from the nuclear charge. Going from top to bottom in a group, the outer-shell orbitals feel about a constant effective nuclear charge and become larger as n becomes larger. As a result, atoms become larger as we descend a group.

Sizes of Ions

When ions are formed from atoms, their sizes increase as electrons are added and decrease as electrons are removed. Negative ions are always larger, and positive ions are always smaller than the atoms from which they are formed.

Ionization Energy

The ionization energy (IE) is the energy needed to pull an electron from an isolated atom or ion. It is a measure of how tightly held the electrons are. Atoms with more than one electron have a series of ionization energies corresponding to the removal of the electrons one by one. For any given atom, successive ionization energies increase.

You should remember that, in general, the ionization energy increases from left to right across a period and decreases from top to bottom within a group. The increase across a period occurs because of the increasing effective nuclear charge felt by the outer electrons. The decrease down a group occurs because the outer electrons become farther from the nucleus (from which they are well shielded) and are held less tightly. You should also note the special stability of the noble gas configuration. It is also helpful to remember that the IE of an atom becomes *larger* as the atom's size becomes *smaller*.

Electron Affinity

Energy is normally released when an electron is added to an isolated gaseous atom to form a negative ion. This energy is the electron affinity (EA). Most values are given with a negative sign because the process is usually exothermic.

The changes in EA within the periodic table parallel the changes in IE, and for the same reasons. Both increase from left to right in a period and from bottom to top in a group.

When more than one electron is added to an atom (i.e., when an ion with a charge of 2– or 3– is formed) the attachment of the second and third electron is always endothermic, which means that they require an input of energy.

Self-Test

25. Which is the larger ion, Co^{2+} or Co^{3+}? __________

26. Explain your answer to Question 25. ____________________

____

27. The following particles each have the same number of electrons: N^{3-}, O^{2-}, F^{-}. Their nuclear charges increase from N to O to F. How would you expect their sizes to vary? Explain.

____

____

____

28. In each pair, choose the species with the larger IE.

(a) K or Ca __________

(b) Ca or Sr __________

(c) Al or C __________

(d) Fe or Fe^{2+} __________

(e) Ne or Na __________

29. Which elements, metals or nonmetals, tend to have the larger ionization energies?

____

30. Choose the species with the more exothermic EA.

(a) S or Se __________

(b) S or Cl __________

(c) Te or Br __________

(d) S or S^{-} __________

31. From which ion, O^{2-} or O^-, would you expect it to be easier to remove an electron?

____

New Terms

Write the definitions of the following terms, which were introduced in this section. If necessary, refer to the Glossary at the end of the text.

effective nuclear charge
ionization energy
angstrom (Å)
electron affinity

Answers to Thinking It Through

1 We can calculate the energy of one photon from the equation $E = h\nu$, but first we must convert the wavelength to frequency with the equation $\lambda\nu = c$. (Alternatively, we can combine these two equations to give $E = hc/\lambda$.) To find the number of such photons required, we have to calculate the total energy needed. We obtain this from the product of the specific heat of the coffee (4.18 J g^{-1} °C^{-1}), the mass of the coffee (250 g), and the temperature change that the coffee undergoes (55 °C).

$$\text{specific heat} \times \text{mass} \times \text{temp. change} = \text{energy}$$

Once we know the total amount of energy needed (in joules), we divide by the energy per photon to calculate the number of photons needed. (The answer is 3.04×10^{31} photons

Answers to Self-Test Questions

1. 242 m
2. 1.0×10^{14} Hz
3. 50 nm
4. 3.64×10^{-19} J
5. 219 kJ
6 $n = 1$
7. $E = -b$ for $n = 1$, $E = -b/4$ for $n = 2$, change in energy is $0.75b$.
8. 410 nm, violet
9. *in phase*—amplitudes add to give a new wave with a larger amplitude.
   out of phase—amplitudes cancel to give a new wave with a smaller, or even zero amplitude.
10. (a) 2*p*, (b) 4*s*, (c) 3*d*, (d) 5*f*
11. (b), (c), (e) are unacceptable.
12. 4*s*, 4*p*, 4*d*, 4*f*
13. (a) 3*p*, (b) 4*d*, (c) 4*p*
14. (a) 2, (b) 10, (c) 18, (d) 32, (e) 6

15. (a) $1s^22s^22p^63s^23p^1$, (b) $1s^22s^22p^63s^23p^63d^{10}4s^24p^5$
(c) $1s^22s^22p^63s^23p^63d^34s^2$

16. (a) P (↑↓) (↑↓) (↑↓)(↑↓)(↑↓) (↑↓) (↑)(↑)(↑)
$1s$ $2s$ $2p$ $3s$ $3p$
(b) Si (↑↓) (↑↓) (↑↓)(↑↓)(↑↓) (↑↓) (↑)(↑)( )

17. [Kr] $4d^{10}5s^25p^5$
18. Co [Ar] $3d^74s^2$
19. three
20. (a) $2s^22p^4$, (b) $6s^2$ (c) $5s^25p^1$, (d) $6s^26p^3$
21. (a) Tc [Kr] $4d^55s^2$, (b) Fe [Ar] $3d^64s^2$, (c) Gd [Xe] $4f^75d^16s^2$
22. Eu [Xe] $6s^24f^7$ (the $4f$ subshell becomes half-filled)
23. Check your sketches against Figures 5.19 and 5.20.

24. 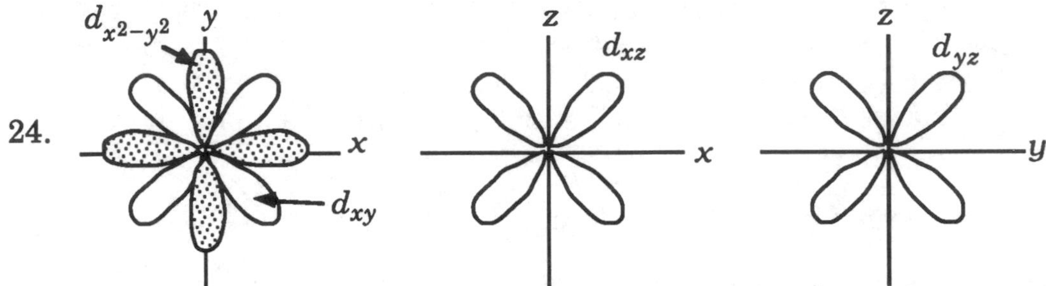

($d_{x^2-y^2}$ and d_{xy}, are concentrated in the xy plane, the d_{xz}, is concentrated in the xz plane, and the d_{yz}, is concentrated in the yz plane.)

25. Co^{2+}
26. In Co^{3+} there is one less e^- than in Co^{2+} and therefore less inter-electron repulsion, allowing the e^- in Co^{3+} to be pulled closer to the nucleus.
27. $N^{3-} > O^{2-} > F^-$. As the nuclear charge increases, the electrons are pulled closer to the nucleus and the size decreases.
28. (a) Ca (b) Ca (c) C (d) Fe^{2+} (e) Ne
29. Nonmetals
30. (a) S (b) Cl (c) Br (d) S
31. O^{2-}

Tools you have learned

Remove this chart from the Study Guide and keep it handy when tackling homework problems.

Tool	*Function*
Wavelength-frequency relationship $\lambda\nu = c$	Convert between wavelength and frequency.
Energy of a photon $E = h\nu$	Calculate the energy carried by a photon of frequency ν.
Periodic Table Remember how the subshells become filled crossing row after row of the periodic table.	Use as a device to write electron configurations of the elements.
Electron configurations These are obtained by using the periodic table.	Enables you to write the orbital diagram for an atom of an element from which other information is obtained.
Periodic Trends in Atomic and Ionic Size Atomic size increases from top to bottom in a group and decreases from left to right in a period.	Compare the sizes of atoms and ions.
Periodic Trends in Ionization Energy IE becomes larger from left to right in a group and from bottom to top in a group. For a given element, successive IE's become larger.	Compare the ease with which atoms of the elements lose electrons.
Periodic Trends in Electron Affinity EA becomes more exothermic from left to right in a period and from bottom to top in a group.	Compare the tendency of atoms or ions to gain electrons.

Summary of Important Equations

The most important equations in this chapter involve the wavelength-frequency relationship and Planck's equation for the energy of a photon.

Wavelength-frequency relationship

$$\lambda \times \nu = c = 3.00 \times 10^8 \text{ m s}^{-1}$$

Energy of a photon of frequency ν.

$$E = h\,\nu$$

where h = Planck's constant (6.626×10^{34} J s)

Your teacher may also ask you to learn the following:

The Rydberg equation,

$$\frac{1}{\lambda} = R_H\left(\frac{1}{n_1^2} - \frac{1}{n_2^2}\right)$$

where $R_H = 109{,}687 \text{ cm}^{-1}$

De Broglie's equation for the wavelength of a matter wave,

$$\lambda = \frac{h}{mv}$$

were h is Planck's constant, m is the mass of the particle, and v is the velocity of the particle.

Chapter 6

CHEMICAL BONDING I

This is the first of two chapters that deal with the attractions that hold atoms to each other in chemical compounds. In this chapter we examine chemical bonds on a rather elementary level. Yet, even these simple explanations provide us with many useful tools for understanding chemical and physical properties.

Learning Objectives

Throughout your study of this chapter, keep in mind the following objectives:

1 To learn how ionic compounds are formed from their elements, what factors cause elements to form ionic compounds, and what determines the charges that atoms acquire when they form ions.

2 To learn to use a simple device called Lewis symbols to represent the valence electrons of an atom or ion of the representative elements.

3 To learn what happens to the energy of two atoms when they share electrons in a covalent bond, and to learn how we represent covalent bonds using Lewis symbols.

4 To learn how the tendency of atoms to acquire a noble gas electron configuration often determines the number of electrons they share with other atoms in covalent bonds, and how this leads to a general rule that many atoms tend to acquire eight electrons in their valence shell by electron sharing.

5 To learn how atoms in some molecules are able to be surrounded by more than or less than an octet of electrons.

6 To see how atoms can use unshared pairs of electrons to form additional covalent bonds.

7 To learn how electrons may be shared unequally in a covalent bond and to see how this affects the distribution of electric charge in a molecule.

8 To learn to draw Lewis structures for molecules and polyatomic ions.

9 To learn a method for selecting the best Lewis structure for a molecule when several are possible.

10 To learn how the structure of a molecule or ion is represented when a single Lewis structure does not adequately describe it.

6.1 Electron Transfer and the Formation of Ionic Compounds

Review

There are two broad classifications of chemical bonds—ionic bonds that occur when atoms transfer electrons between them and covalent bonds that occur when atoms share electrons. The ionic "bond" is really just the attraction that exists between oppositely charged particles (ions).

For a bond to be formed between atoms, there must be a net lowering of the potential energy of the particles. For ionic bonding, the three most important factors that contribute to the overall potential energy change are (1) the ionization energy of the element that forms the cation, (2) the electron affinity of the element that forms the anion, and (3) the lattice energy, which is the potential energy lowering that is produced by the attractions between the ions. For most elements, (1) and (2) taken together produce a net increase in potential energy, so it is the stabilizing (energy-lowering) influence of the lattice energy that enables ionic compounds to exist. But the lattice energy can do this *only* if it is larger than the net PE increase caused by (1) and (2). This restriction leads to certain generalizations about the kinds of elements that form ionic compounds and the charges on the ions that are created in the reaction:

1 Ionic compounds tend to be formed between metals (low IE) and nonmetals (relatively large exothermic EA).

2 The representative metals of Groups IA and IIA plus aluminum lose electrons until they have achieved an electron configuration corresponding to that of a noble gas.

3 Nonmetals gain electrons until they have also achieved an electron configuration that is the same as that of a noble gas.

The tendency of these elements to achieve a noble gas configuration gives rise to the octet rule, which states that *many elements tend to gain or lose electrons until they have achieved a valence shell that contains eight electrons.*

The transition metals and the metals that follow the transition elements in periods 4, 5, and 6 (the post-transition metals) do not follow any particular rule. They lose electrons, but often more than one cation is possible, depending on conditions. The charges on the ions of the common transition metals were given in Table 2.7 on page 68 of the text; if necessary, review them.

Self-Test

1. Show what happens to the valence shells of Ba and I when these elements react to form an ionic compound. What is the formula of this compound?

2. What happens to the electron configuration of a manganese atom when it forms (a) the Mn^{2+} ion and (b) the Mn^{3+} ion?

3. When Rb and Cl_2 react, why doesn't the compound $RbCl_2$ form? What compound does form?

New Terms

Write the definitions of the following terms, which were introduced in this section. If necessary, refer to the Glossary at the end of the text.

chemical bond
ionic bond
lattice energy
octet rule

6.2 Electron Bookkeeping: Lewis Symbols

Review

An element's Lewis symbol is constructed by placing one dot for each valence electron around the chemical symbol for the element. The number of valence electrons, and therefore the number of dots in the Lewis symbol, is equal to the element's group number. (In general, Lewis symbols are only used for the representative elements.)

A simple way to draw Lewis symbols is as follows. First write the chemical symbol. Then imagine a diamond with the symbol at its center. Place a dot at one of the corners of the diamond (it doesn't matter where you start). Move around the diamond from corner to corner as you place additional dots, until the required number are shown. For example, for sulfur (Group VIA), there must be six dots around the symbol S.

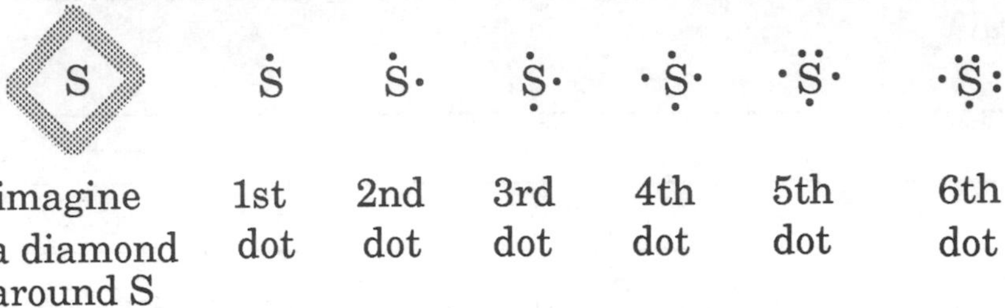

The Lewis symbol for sulfur is ·S: (or any other arrangement such as ·S: which shows two pairs of dots and two single dots).

Example 6.3 on pate 223 illustrates how Lewis symbols can be used to diagram the transfer of electrons that takes place during the formation of an ionic substance. Notice that when we write the Lewis symbol for an anion, we enclose it in brackets to show that all the electrons belong to the ion.

Self-Test

4. Write Lewis symbols for the following:

   (a) P ________________ (c) Cl ________________

   (b) Te ________________ (d) B ________________

5. Diagram the reaction that occurs when Li reacts with Se to form the compound Li_2Se.

   ____

New Terms

Write the definition of the following term, which was introduced in this section. If necessary, refer to the Glossary at the end of the text.

Lewis symbol

6.3 Electron Sharing: The Formation of Covalent Bonds

Review

Covalent bonds are formed when ionic bonds are not energetically favored. In the formation of a covalent bond there is also a potential energy lowering, but it is achieved by a different method than in ionic bonding.

When atoms approach each other to form a covalent bond, there is an attraction of the electrons of each atom toward both nuclei, which leads to an overall lowering of the potential energy. The two nuclei also repel each other because they are of the same charge, so the atoms cannot approach each other too closely. The bond length is determined by the balance of these attractive and repulsive forces, and the net decrease in PE that occurs when the bond is formed is called the bond energy, which is usually expressed in units of kilojoules per mole of bonds formed. The bond energy is also the energy that would be necessary to break the bond.

One of the most important features of covalent bonding is the pairing of electrons that occurs; a covalent bond is the sharing of a *pair* of electrons with their spins in opposite directions. When Lewis symbols are used to represent a molecule, a shared pair of electrons is shown as a pair of dots or a dash between the chemical symbols.

When counting electrons in the valence shells of atoms attached to each other by a covalent bond, we count both electrons of the shared pair as belonging to both atoms joined by the bond. The octet rule applied to covalent bonds states that *atoms tend to share sufficient electrons so as to obtain an outer shell with eight electrons*. An exception to the octet rule is hydrogen, which completes its valence shell when it has a share of two electrons (i.e., when it forms one covalent bond).

Atoms can share one, two, or three pairs of electrons to give single, double, and triple bonds. On page 227, notice how the numbers of electrons needed by atoms of carbon, nitrogen, and oxygen to achieve octets are related to the number of covalent bonds these atoms tend to form.

When one atom donates a pair of electrons to another atom in order to form a single bond, we call the bond a coordinate covalent bond. Once formed it is really no different than any other single bond because electrons can't "remember" where they came from. This is really just a bookkeeping device. When we want to note the origin of the electrons of a coordinate covalent bond, we use an arrow instead of a dash to represent the bond. The arrow points from the donor to the acceptor of the electrons.

Self-Test

6. What happens to the electron density between the two atoms when they form a covalent bond?

 ____

 ____

7. On a separate piece of paper, sketch a diagram that shows how the potential energy varies with the distance between the nuclei of a pair of atoms becoming joined by a covalent bond. On the diagram indicate the bond energy and the bond length. Refer to Figure 6.3 to check your answer.

8. Predict the formulas of the simplest compound that might be formed between hydrogen and each of the following:

 (a) Ge ____________

 (b) Te ____________

 (c) I ____________

9. Draw the structural formula for each of your answers to Question 8.

 (a) (b) (c)

10. Formaldehyde, used in preserving biological specimens, has the formula H_2CO. A molecule of this substance has two hydrogen atoms and an oxygen atom bonded to the carbon. Use Lewis symbols to write a structural formula for H_2CO. (Hint: There's a double bond in the structure.)

11. Count the number of electrons around the atoms in the following molecule (called urea).

```
   H :O: H
   |  ||  |
H—N—C—N—H
  ••     ••
```

---

12. Boron trifluoride, BF_3, can react with a fluoride ion, F^-, to form the tetrafluoroborate ion, BF_4^-. Diagram this reaction using Lewis symbols to show the formation of a coordinate covalent bond.

13. BF_3 reacts with organic chemicals called ethers to form addition compounds. Use Lewis formulas to diagram the reaction of BF_3 with dimethyl ether. The structure of dimethyl ether is

```
   H      H
   |  ••  |
H—C—O—C—H
   |  ••  |
   H      H
```

New Terms

Write the definitions of the following terms, which were introduced in this section. If necessary, refer to the Glossary at the end of the text.

octet rule	triple bond	bond distance
structural formula	covalent bond	bond energy
single bond	bond length	electron pair bond
double bond	coordinate covalent bond	

6.4 Electronegativity and the Polarity of Molecules

Review

Because different atoms have different attractions for electrons, the electrons in a covalent bond can be shared unequally. When this happens, the electron density in the bond is shifted toward that atom with the greater attraction for electrons, so this atom acquires a partial negative charge, $\delta-$. The atom at the other end of the bond acquires a deficiency of electrical charge and carries a partial positive charge, $\delta+$. Bonds in which electrons are shared unequally are polar bonds and the bond is a dipole (two poles of equal but opposite charges separated by the bond distance). The degree of polarity of a bond is expressed quantitatively by the bond's dipole moment.

Electronegativity is the attraction an atom has for the electrons in a bond. The greater the electronegativity *difference* between two atoms that are bonded to each other, the more polar the bond is and the greater is the partial negative charge on the more electronegative of the two atoms.

It is important to remember that there is no sharp dividing line between covalent and ionic bonding. If we arranged all sorts of bonds in order of increasing polarity, we would find a gradual transition from nonpolar covalent bonds to bonds that are essentially 100% ionic. All degrees of ionic character (polarity) are possible.

You should be sure to learn how electronegativity varies in the periodic table. This is summarized in the margin figure at the bottom of page 230. Metals, in the lower left corner of the table, have low electronegativities and nonmetals, in the upper right, have high electronegativities.

Self-Test

14. Use Figure 6.5 to arrange the following bonds in order of increasing percent ionic character: Li—I, Ba—F, N—O, As—Cl.

____

New Terms

Write the definitions of the following terms, which were introduced in this section. If necessary, refer to the Glossary at the end of the text.

partial charge	dipole
polar bond	dipole moment
polar covalent bond	electronegativity
polar molecule	ionic character of a bond

6.5 Drawing Lewis Structures

Review

Although the octet rule is often a useful tool in constructing the Lewis structure for a molecule or polyatomic ion, there are some instances in which it is not obeyed. Whenever an atom is bonded to more than four other atoms, the octet rule cannot be obeyed. Examples are PCl_5 and SF_6 illustrated on page 231. Remember that only atoms below period 2 are able to exceed an octet; their valence shells are able to accommodate more than eight electrons. (Period 2 elements never go beyond eight electrons in the outer shell because the second shell has only *s* and *p* subshells.) Beryllium and boron are unusual elements because they sometimes have *less* than an octet in compounds.

Procedure for Drawing Lewis Structures

To draw a Lewis structure for a molecule or ion, first decide on the skeletal structure. Follow the guidelines on page 232.

Remember:

- In oxoacids, hydrogen is bonded to oxygen, which in turn is bonded to the other nonmetal.
- The central atom in a molecule is usually the least electronegative atom. In the formulas for most simple molecules and polyatomic ions, the central atom is written first.
- When all else fails, the most symmetrical arrangement of atoms is the "best guess."

Next, count all of the valence electrons. Remember that the number of the group in which a representative element (A-group element) is found is the same as the number of valence electrons that it contributes. Sulfur, for example, is in Group VI, so a sulfur atom contributes six valence electrons. Also remember that if the species whose Lewis structure you are working on is a negative ion, add one electron for each negative charge. If it is a positive ion, subtract one electron for each positive charge.

Finally, follow the sequence of steps outlined in Figure 6.7 on page 232. In applying the method, remember that hydrogen never has more than two electrons.

Self-Test

15. If an atom forms more than four bonds, it definitely does not obey the octet rule. Why?

____

16. What is the maximum number of electrons that can be held in the orbitals in the valence shell of

    (a) nitrogen ____________________

    (b) phosphorus ____________________

    (c) What kind of orbitals does the valence shell of phosphorus have that the valence shell of nitrogen does not?

    ____

17. How many valence electrons are in each of the following?

    (a) PO_4^{3-} ____________ (d) IF_5 ____________

    (b) IF_4^- ____________ (e) CO_3^{2-} ____________

    (c) NO_2^+ ____________ (f) C_2^{2-} ____________

18. Write Lewis structures for each substance in Question 17.

    (a) (b) (c)

    (d) (e) (f)

19. Write the Lewis structures for (a) $HClO_2$ and (b) H_2CO_3.

    (a) (b)

New Terms

none

6.6 Formal Charge and the Selection of Lewis Structures

Review

The bond length between a given pair of atoms *decreases* going from a single to a double to a triple bond. At the same time, the bond energy increases. These trends, along with experimentally measured values of these bond properties, help us "fine tune" our descriptions of the bonding in various molecules. One of the observations we can make is that the best Lewis structure for a molecule or ion is one in which the formal charges on the atoms are a minimum.

To determine the formal charge on an atom in a Lewis structure, add up the number of bonds it forms plus the number of unshared electrons. Subtract this number from the number of electrons an isolated atom of the element has. The difference is the formal charge. (This is what Equation 6.1 tells us to do.) To select the preferred ("best") Lewis structure for a molecule or ion, we seek to minimize the number of formal charges.

Example 7.1 Using formal charges to select Lewis structures

Problem

What is the best Lewis structure for chlorous acid, $HClO_2$?

Analysis

We know that the preferred Lewis structure is the one that has the minimum number of formal charges. Therefore, the first step is to construct the Lewis structure following the general procedure given on page 232. Then we calculate the formal charge on each atom. Since the best Lewis structure is the one with the minumum number of formal charges, we attempt to reduce the formal charges by moving electrons.

Solution

At the top of the next page we have the Lewis structure for $HClO_2$ drawn according to the procedure outlined earlier. Notice that we have counted the number of bonds and unshared electrons, which we need to know to calculate the formal charges according to the formula

$$\text{Formal charge} = \begin{pmatrix} \text{number of electrons} \\ \text{an isolated atom} \\ \text{of the element has} \end{pmatrix} - \begin{pmatrix} \text{sum of the number of} \\ \text{bonds the atoms forms} \\ \text{plus the number of} \\ \text{unshared electrons} \end{pmatrix}$$

$$\mathrm{H{-}\ddot{\underset{\cdot\cdot}{O}}{-}\ddot{\underset{\cdot\cdot}{Cl}}{-}\ddot{\underset{\cdot\cdot}{O}}\!:}$$

1 bond + 6 unshared e^- (right oxygen)

2 bonds + 4 unshared e^- (left oxygen and chlorine)

For the oxygen between the H and Cl, the sum of 2 bonds and 4 unshared e^- gives a total of 6. An isolated oxygen atom has 6 valence electrons, so the formal charge on this oxygen is zero. For the chlorine, the sum of bonds and unshared e^- is also 6, and when subtracted from 7 (the number of valence e^- an isolated Cl atom has) we obtain +1. This is the formal charge on the Cl. For the oxygen at the right, the sum of bonds and unshared e^- is 7. When 7 is subtracted from 6 (the number of valence e^- an oxygen atom has), we get –1. This is the formal charge on the oxygen at the right. Placing these formal charges in circles alongside the atomic symbols gives us

$$\mathrm{H{-}\ddot{\underset{\cdot\cdot}{O}}{-}\overset{\oplus}{\ddot{\underset{\cdot\cdot}{Cl}}}{-}\overset{\ominus}{\ddot{\underset{\cdot\cdot}{O}}}\!:}$$

The next step is to attempt to reduce formal charges to obtain a better structure. We do this by shifting an unshared pair of electrons on the more negative atom into the bond it forms to the more positive atom.

$$\mathrm{H{-}\ddot{\underset{\cdot\cdot}{O}}{-}\overset{\oplus}{\ddot{\underset{\cdot\cdot}{Cl}}}{-}\overset{\ominus}{\ddot{\underset{\cdot\cdot}{O}}}\!:} \dashrightarrow \mathrm{H{-}\ddot{\underset{\cdot\cdot}{O}}{-}\ddot{\underset{\cdot\cdot}{Cl}}{=}\ddot{O}\!:}$$

This gives a Lewis structure with no formal charges, so we select it as the preferred Lewis structure for the molecule.

Self-Test

20. Phosphorous acid has the Lewis structure shown at the left below when constructed as described in the preceding section. Assign formal charges to the atoms in this structure and, if possible, draw a better Lewis structure for the molecule.

$$\begin{array}{c} :\ddot{O}: \\ | \\ \mathrm{H{-}\ddot{\underset{\cdot\cdot}{O}}{-}P{-}\ddot{\underset{\cdot\cdot}{O}}{-}H} \\ | \\ \mathrm{H} \end{array}$$

New Terms

Write the definitions of the following terms, which were introduced in this section. If necessary, refer to the Glossary at the end of the text.

bond length   bond energy   formal charge

6.7 Resonance: When Lewis Structures Fail

Review

For some molecules and ions, a single Lewis structure cannot be drawn that adequately explains experimental bond lengths and bond energies. In these instances, we view the true structure of the particle as a sort of average of two or more Lewis structures, and we call this true structure a resonance hybrid of the contributing structures.

When you have a choice as to where to place a double bond in a Lewis structure, as when writing the Lewis structure for the HCO_2^- ion, the number of resonance structures you should draw is equal to the number of choices that you have. This rule works in most cases.

A resonance hybrid is more stable that any of its individual structures would be, if they were to actually exist. The extra stability is called the resonance energy.

Self-Test

21. The oxalate ion, $C_2O_4^{2-}$, has the skeletal structure shown below

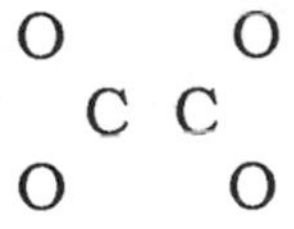

```
O     O
  C C
O     O
```

On a separate sheet of paper, draw the resonance structures for this ion.

22. How would the C—O bond lengths and bond energies in the $C_2O_4^{2-}$ ion (Question 21) compare to those in the following:

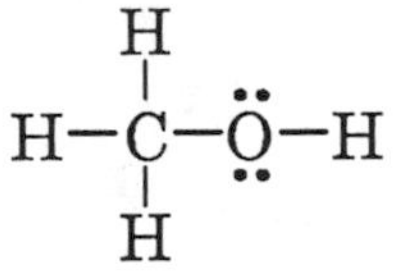

```
   H
   |  ..
H—C—O—H     and    :O=C=O:
   |  ..            ..   ..
   H
```

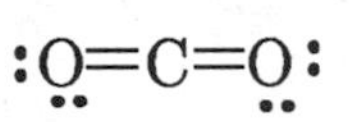

New Terms

Write the definitions of the following terms, which were introduced in this section. If necessary, refer to the Glossary at the end of the text.

resonance

resonance hybrid

Answers to Self-Test Questions

1. Ba ([Xe] $6s^2$) → Ba^{2+} ([Xe]) + $2e^-$; I ($5s^25p^5$) + e^- → I^- ($5s^25p^6$)
   The compound is BaI_2
2. (a) Mn ([Ar] $3d^54s^2$) → Mn^{2+} ([Ar] $3d^5$) + $2e^-$
   (b) Mn^{2+} ([Ar] $3d^5$) → Mn^{3+} ([Ar] $3d^4$)+ e^-
3. Removal of an electron from Rb^+ requires too much of a PE increase. The compound that does form contains Rb^+ and has the formula RbCl.
4. (a) ·Ṗ̤: (b) ·Ṫe̤: (c) ·C̈l̤: (d) ·Ḃ·
5. Li· ↷ ·S̤̈e· ↶ ·Li ⟶ $2Li^+$ [:S̤̈e:]$^{2-}$
6. It shifts toward the region between the two atoms.
7. See Figure 6.3, page 225.
8. (a) GeH_4 (b) H_2Te (c) HI
9. (a) H–Ge(–H)(–H)–H (b) H—T̤̈e—H (c) H—Ï̤:
10. H₂C=Ö̤ (H–C(–H)=O)

    Hydrogen can complete its valence shell by forming one bond, carbon by four bonds,and oxygen by forming two bonds. This is the only arrangement of atoms that satisfies this condition.
11. Two around each H, eight around each N, C, and O.

12. 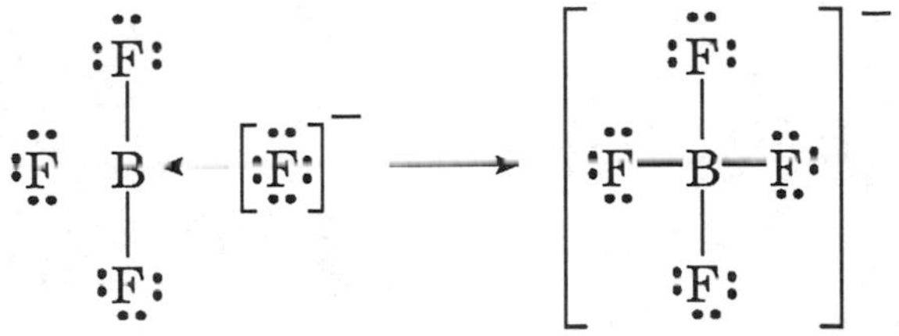 $BF_3 + [:\ddot{F}:]^- \longrightarrow [BF_4]^-$

13. 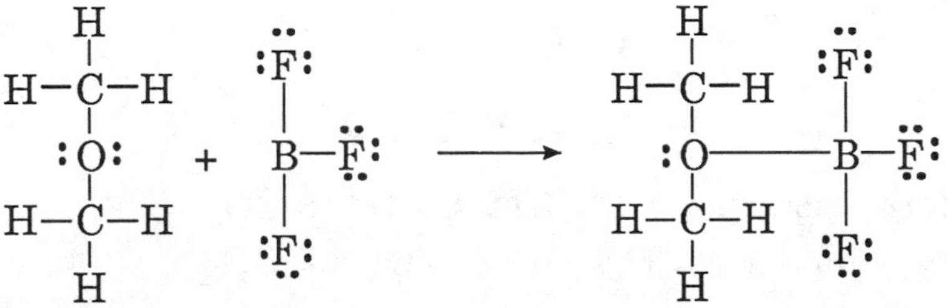 $(CH_3)_2\ddot{O}: + BF_3 \longrightarrow (CH_3)_2O\text{—}BF_3$

14. N—O < As—Cl < Li—I < Ba—F
15. Two electrons have to be in each bond. Since 2 × 4 = 8, more than four bonds means more than an octet.
16. (a) eight (b) eighteen (c) A phosphorus atom has *d* orbitals in its empty 3*d* subshell.
17. (a) 32 (b) 36 (c) 10 (d) 42 (e) 24 (f) 10
18. (a) $[PO_4]^{3-}$ (b) $[IF_4]^-$ (c) $[:N{\equiv}O:]^+$ (d) IF_5

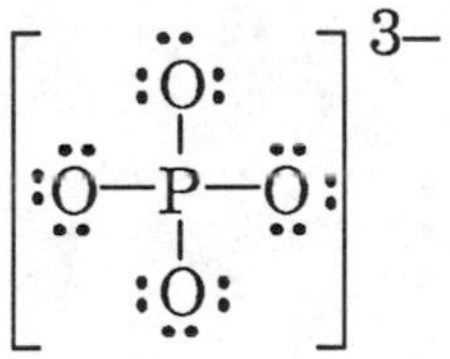

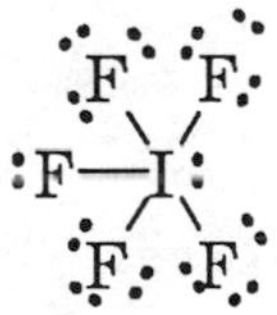

(e) $[CO_3]^{2-}$ (f) $[:C{\equiv}C:]^{2-}$

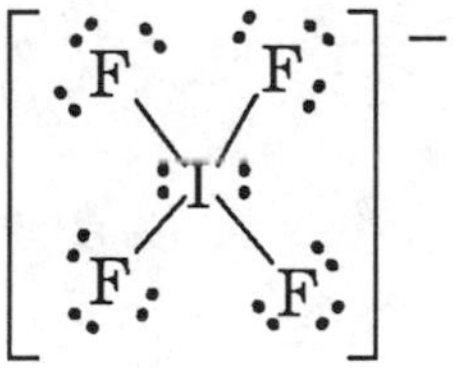

19. (a) H—O—Cl—O: (b) H—O—C(=O)—O—H

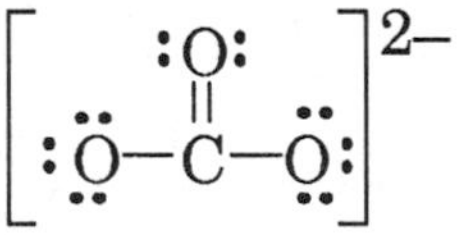

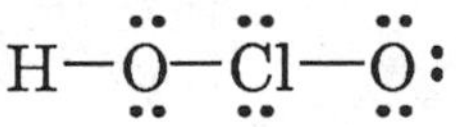

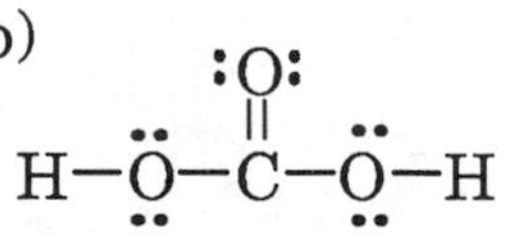

20. 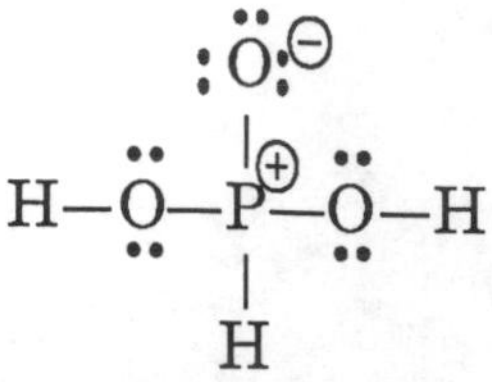 H—O—P(=O⁻)(+)—O—H with H on P. Preferred structure is ... H—O—P(=O)(H)—O—H

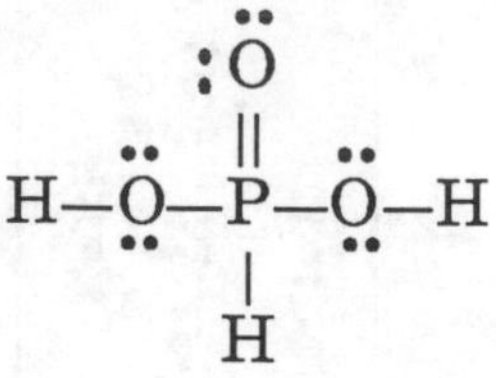

21. 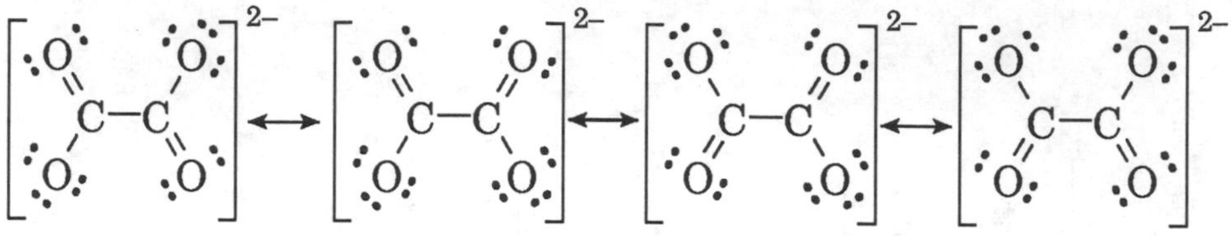 $[C_2O_4]^{2-}$ resonance structures: $\longleftrightarrow$ $\longleftrightarrow$ $\longleftrightarrow$

22. The C—O bond length decreases: $CH_3OH > C_2O_4^{2-} > CO_2$
The C—O bond energy increases: $CH_3OH < C_2O_4^{2-} < CO_2$

Tools you have learned

Remove this chart from the Study Guide and keep it handy when tackling homework problems.

Tool	*Function*
Rules for electron configurations of ions	To be able to write correct electron configurations for cations and anions.
Lewis symbols	A bookkeeping device for keeping track of valence electrons in atoms and ions. Learn to use the periodic table to construct the Lewis symbol
Periodic trends in electronegativity Electronegativity increases from left to right in a period and from bottom to top in a group.	Compare the relative polarities of chemical bonds and to identify the most (or least) electronegative element in a compound.
Method for drawing Lewis structures	A systematic method for constructing the Lewis structure for a molecule or polyatomic ion.
Correlation between bond properties and bond order As bond order increases, bond length decreases and bond energy increases	Compare experimental quantities related to covalent bonds with those predicted by theory.
Formal charges	Select of the best Lewis structure for a molecule or polyatomic ion.
Method for determining resonance structures The number of resonance structures equals the number of equivalent choices when forming double bonds in a Lewis structure.	Resonance structures provide a way of expressing the structures of molecules for which a single satisfactory Lewis structure cannot be drawn.

Summary of Useful Information

Rules for drawing Lewis structure (See also, page 232.)

1 Decide which atoms are bonded to each other.
(Remember, H cannot be a central atom.)

2 Count valence electrons. (Remember: add an electron for each negative charge on an ion or subtract an electron for each positive charge.)

3 Place two electrons in each bond.

4 Complete octets of atoms (except H) that surround the central atom.

5 Place any remaining electrons on central atom *in pairs*.

6 Form multiple bonds if central atom has less than an octet.
(Rule 6 is omitted for the elements Be and B.)

Calculation of formal charge

$$\text{Formal charge} = \begin{pmatrix} \text{number of electrons} \\ \text{an isolated atom} \\ \text{of the element has} \end{pmatrix} - \begin{pmatrix} \text{sum of the number of} \\ \text{bonds the atoms forms} \\ \text{plus the number of} \\ \text{unshared electrons} \end{pmatrix}$$

Chapter 7

CHEMICAL BONDING II

Now that you've learned a bit about the covalent bond and how to express the structures of molecules using Lewis structures, we turn our attention to the shapes of molecules and how to predict them. A knowledge of molecular shape is important because many of the physical properties of substances depend on the shapes of their molecules. In this chapter you will also obtain a more refined view of how and why covalent bonds form. You should also try to understand how modern theories *explain* covalent bonding in terms of the way the orbitals of the atoms interact.

Learning Objectives

Throughout your study of this chapter, keep in mind the following objectives:

1 To learn the basic molecular shapes found for various molecules and how to draw them. You should try very hard to visualize these molecular shapes as three-dimensional objects.

2 To learn a method for predicting the shapes of molecules by the use of Lewis structures.

3 To learn how to predict whether a molecule is polar based on its molecular shape.

4 To learn how a theory called Valence Bond Theory explains the formation of a covalent bond by the interactions of the orbitals of atoms forming the bond.

5 To learn how simple atomic orbitals are able to combine to produce hybrid orbitals which are better able to explain the shapes of molecules.

6 To learn how double and triple bonds are formed by the interaction of atomic orbitals.

7 To learn how bond energies can be determined from thermochemical data.

8 To learn how to use bond energies to estimate the heat of formation of a compound.

9 To learn how a theory called Molecular Orbital Theory views electron energy levels and orbitals in molecules.

10 To learn how molecular orbital theory is able to avoid the concept of resonance.

11 To learn about the electronic structures of solids and how substances can be electrical conductors, semiconductors, and nonconductors (insulators).

7.1 Some Common Molecular Shapes

Review

This section describes five basic geometric shapes that form the basis for most molecular structures you will encounter. The shapes are: *linear*, *planar triangular*, *tetrahedral*, *trigonal bipyramidal*, and *octahedral*. It is important that you develop the ability to visualize these shapes in three dimensions. It is also wise to learn to sketch them. Even if you don't consider yourself much of an artist, you should make the effort, because these three-dimensional concepts are likely to become important tools in other science courses you take.

The linear and planar triangular structures can be drawn on a two-dimensional surface, so you should have no difficulty with them. Instructions for drawing a tetrahedron are found in Figure 7.1, and simplified representations of the trigonal bipyramidal and octahedral structures are shown on pages 255 and 256, respectively. Instructions for drawing the trigonal bipyramidal and octahedral structures are given in Study Guide Figure 7.1 on the next page.

You should know the bond angles in the various structures. Notice that except for the trigonal bipyramid, all the other structures have equal angles between their bonds. Thus, in the tetrahedron, the bond angles are all 109.5°. In the octahedron, the bond angles are all 90°. In the trigonal bipyramid, the axial bonds make 90° angles with the equatorial bonds; the equatorial bonds are at 120° angles to each other.

Self-Test

1. On a separate sheet of paper, make sketches of the five basic molecular shapes described in this section. Indicate the bond angles on the drawings. (Continue to practice this until the structures you draw make three-dimensional sense to *you*.)

New Terms

Write the definitions of the following terms, which were introduced in this section. If necessary, refer to the Glossary at the end of the text.

linear molecule

planar triangular molecule

trigonal bipyramidal molecule

octahedron

(a) Sketching a simplified representation of a trigonal bipyramid

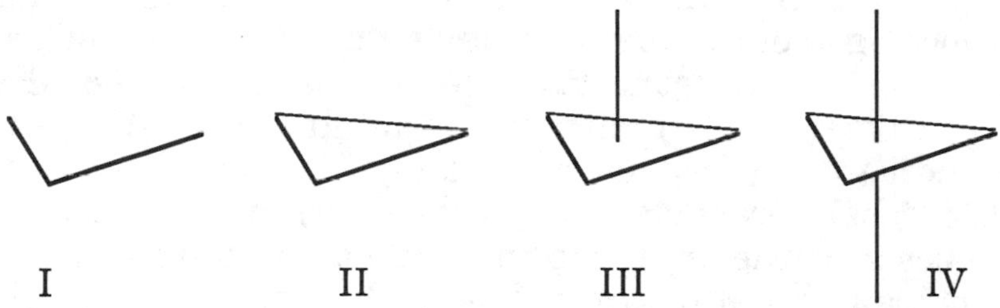

(b) Sketching a simplified representation of an octahedron

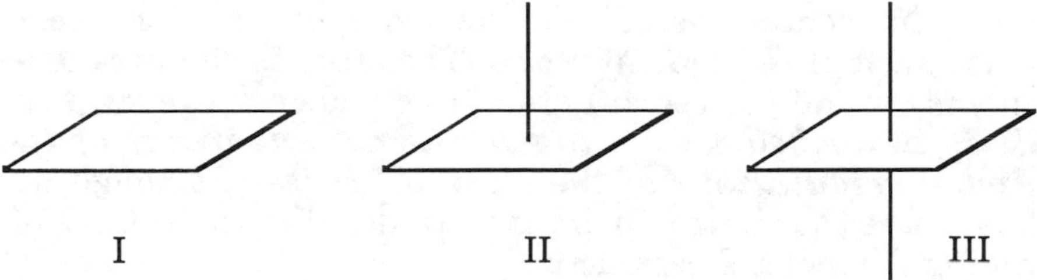

Figure 7.1

(a) To draw a trigonal bipyramid, begin by making a check mark (I), then draw a line across the top (II). Next, draw a line from the center of the triangle upward, representing an axial bond (III). Finally, draw a line downward (projecting from below the triangular plane), which represents the other axial bond (IV). (b) To draw an octahedron, begin with a parallelogram, representing a square viewed from the edge (I). Next, draw a line upward from the center (II), and finally, draw a line downward as if it comes out from behind the plane in the center (III).

tetrahedron
tetrahedral molecule
trigonal bipyramid
octahedral molecule
bond angle
axial and equatorial bonds

7.2 Predicting the Shapes of Molecules: VSEPR Theory

Review

The basic postulate of the VSEPR theory is very simple—*electron pairs in the valence shell of an atom stay as far apart as possible because they repel each other*. This allows us to predict how the electron pairs will arrange themselves around the central atom in a molecule or polyatomic ion. If all the electron pairs are used in bonds, the arrangement of electron pairs that minimizes repulsions also defines the shape of the molecule or ion.

When some electron pairs in the valence shell of a central atom are unshared (i.e., when they are *lone pairs*), the arrangement of the bonded atoms won't be the same as the arrangement of electron pairs. In ammonia, for example, the theory predicts that the electron pairs are arranged tetrahedrally, but there are only three hydrogen atoms to attach to the nitrogen. When we look at how the nitrogen and three hydrogens are arranged, we see a pyramid with the nitrogen at the top. Therefore, even though the electron pairs are distributed tetrahedrally, we describe the ammonia molecule as pyramid-shaped. (In fact, we say that NH_3 has a trigonal pyramidal shape because the pyramid has a three-sided base.)

The point of this is that when we apply the VSEPR theory to a molecule or ion, there are *two* shapes that concern us. One is the arrangement of the *electron pairs* in the valence shell of the central atom. The other is the arrangement of the *bonded atoms* around the central atom. Remember, however, ***the shape of a molecule is described according to the arrangement of its atoms, not the predicted orientations of the electron pairs***. Although we use the theory to tell us where the electron pairs are, we use the orientations of the *atoms* when describing the molecular structure.

You should be able to predict the shape of the molecule or ion if you are able to draw its Lewis structure following the simple rules in Figure 6.7 on page 232. The shape will be one of those shown in Figures 7.2, 7.3, 7.4, and 7.5. The procedure is as follows:

1. Draw the Lewis structure for the molecule or ion.
2. Count the number of electron pairs around the central atom. (Remember, however, that for the purposes of predicting molecular shapes, the electron pairs in double or triple bonds are viewed as groups that behave just like single pairs of electrons.)
3. Determine the arrangement of electron pairs around the central atom:

two pairs	linear
three pairs	planar triangular
four pairs	tetrahedral
five pairs	trigonal bipyramidal
six pairs	octahedral

   (Try to draw the appropriate structure so you can visualize the shape.)

4. Attach the necessary number of bonded atoms and then determine which shape from Figures 7.2 to 7.5 it corresponds to.

Study Examples 7.1 through 7.5 in the text.

Self-Test

2. Draw Lewis structures and, without referring to Figures 7.2 through 7.5, predict the geometry of each of the following:

(a) SO_4^{2-} (b) SF_2

(c) CS_2 (d) BrF_6^+

(e) PH_3 (f) SeF_4

(g) HCO_2^- (formate ion) (h) BrF_2^-

New Terms

Write the definitions of the following terms, which were introduced in this section. If necessary, refer to the Glossary at the end of the text.

VSEPR theory
lone pairs
nonlinear (bent) shape
square pyramidal
trigonal pyramidal
square planar

7.3 Molecular Shape and Molecular Polarity

Review

In this section we see that even when the bonds in a molecule are polar bonds, the effects of the individual bond dipoles can cancel if the molecule is symmetrical. If there are no lone pairs in the valence shell of an atom M in a molecule MX_n, and if all the atoms X that surround M are the same, then the molecule will have one of the symmetrical structures discussed in Section 7.2 and will be nonpolar.

Usually, molecules in which the central atom has lone pairs in its valence shell will be polar. The effects of their bond dipoles don't cancel. Two exceptions are linear molecules in which the central atom has three lone pairs, and square planar molecules in which the central atom has two lone pairs. These exceptions are illustrated in Figure 7.8 on page 266.

Self-Test

3. Which of the following molecules will be polar?

   (a) SCl_2 (b) SF_6 (c) ClF_3 (d) SO_3 ______________

4. The molecule CH_2Cl_2 has a tetrahedral shape, but is a polar molecule. Why?

   ____

New Terms

none

7.4 Wave Mechanics and Covalent Bonding: Valence Bond Theory

Review

The two main bonding theories based on the results of wave mechanics are valence bond theory (VB theory) and molecular orbital theory (MO theory). The VB theory retains the image of individual atoms coming together to form bonds. The MO theory, in its simplest form, considers the molecule after the nuclei are in their proper positions and examines the electron energy levels that extend over *all* the nuclei. Both theories ultimately give the same results when they are refined by eliminating simplifying assumptions.

Valence Bond Theory

According to VB theory, a pair of electrons can be shared between two atoms when orbitals (one from each atom) overlap. By *overlap* we mean that two orbitals from different atoms simultaneously share some region in space. This overlap provides the means by which the electrons spread themselves over both nuclei.

Two other main principles in VB theory are:

(1) Only two electrons, with their spins paired, can be shared between a pair of overlapping orbitals.

(2) The strength of a bond is proportional to the amount of overlap of the orbitals.

Atoms tend to form bonds that are as strong as possible because this lowers the energy of the atoms the most. This means that when several atoms surround some central atom, they tend to position themselves to give the maximum possible overlap of the orbitals that are used for bonding. According to VB theory, this is what determines molecular shapes and bond angles.

Self-Test (Use separate sheets of paper to answer these questions.)

5. Describe the formation of the chlorine-chlorine bond in Cl_2 using the principles of VB theory.

6. Hydrogen telluride, H_2Te, has a H—Te—H bond angle of 89.5°. Explain the bonding in H_2Te.

7. Arsine, AsH_3, has H—AS—H bond angles equal to 92°. Explain the bonding in AsH_3.

New Terms

Write the definitions of the following terms, which were introduced in this section. If necessary, refer to the Glossary at the end of the text.

valence bond theory (VB theory)	overlap
molecular orbital theory (MO theory)	bond angle

7.5 Hybrid Orbitals

Review

Because many molecules have bond angles that can't be explained by the directional properties of simple atomic orbitals, it is necessary to consider how these simple orbitals can "mix" or combine when bonds are formed. Mixing simple atomic orbitals gives *hybrid orbitals* with directional properties that differ from the basic atomic orbitals. In general, hybrid orbitals form stronger bonds than simple atomic orbitals because they give better overlap with orbitals of other atoms. Be sure you learn the directional properties of the hybrids described in Figures 7.17 and 7.22. These can be summarized as follows:

Hybrid type	Orbitals Mixed	Orientations of Orbitals
sp	$s + p$	linear
sp^2	$s + p + p$	planar triangular
sp^3	$s + p + p + p$	tetrahedral
sp^3d	$s + p + p + p + d$	trigonal bipyramidal
sp^3d^2	$s + p + p + p + d + d$	octahedral

Free rotation of groups of atoms around a single bond is possible because such rotation doesn't affect the overlap of the orbitals that form the bond, and therefore it does not affect appreciably the strength of the bond. Free rotation about bonds makes possible large numbers of different conformations. (In Section 7.6, you will see that quite a different situation exists with respect to rotation about a double bond.)

When unshared electron pairs exist in the valence shell of an atom, they are also found in hybrid orbitals. Study the explanations of the bonding for water and ammonia on page 281, as well as Example 7.12. Notice that the central atom needs one hybrid orbital for each attached atom *plus* one hybrid orbital for each lone pair.

Notice also that we can use the VSEPR theory to figure out which kind of hybrid orbitals are used by an atom when it forms bonds. The VSEPR theory predicts the orientations of the electron groups around the atom, which we then use to deduce the kinds of hybrid orbitals that are used.

Example 7.1 Using VSEPR Theory to Deduce Hybridization

Problem

Which kind of hybrid orbitals are used by iodine in the ICl_2^- ion?

Solution

First, let's draw the Lewis structure for the ion.

$$\left[:\ddot{\underset{\cdot\cdot}{Cl}}-\ddot{\underset{\cdot\cdot}{I}}-\ddot{\underset{\cdot\cdot}{Cl}}:\right]^-$$

There are five electron pairs around iodine, which the VSEPR theory predicts should be arranged in a trigonal bipyramid. The kind of hybrids that have this orientation is sp^3d, so we conclude that in this ion the iodine uses two sp^3d hybrids to form the bonds and the other three to house the three lone pairs.

Coordinate Covalent Bonds and Hybrid Orbitals

According to valence bond theory, a coordinate covalent bond is formed by the overlap of a filled orbital of one atom with an empty orbital of another atom. In the discussion describing the formation of BF_4^- from BF_3 and F^-, notice that sufficient hybrid orbitals are made available to hold all of the electrons in the bonds.

Self-Test

8. Why do atoms often use hybrid orbitals to form bonds instead of unhybridized atomic orbitals?

   ____

9. Use the VSEPR theory to predict which kinds of hybrid orbitals the central atom uses to form its bonds in each of the following molecules. To do this, use a separate sheet of paper to construct the Lewis structure for each.

   (a) $AsCl_5$ ______________

   (b) $SeCl_2$ ______________

   (c) $AlCl_3$ ______________

   (d) NF_3 ______________

10. Use orbital diagrams to give explanations using hybrid orbitals of the bonding and geometry of the following:

    (a) $GeCl_4$

    (b) $SeCl_4$

    (c) SCl_2

    (d) IF_5

11. Explain, in terms of valence bond theory, the reaction for the formation of an ammonium ion from an ammonia molecule and a hydrogen ion.

12. Which kind of hybrid orbitals are used by silicon in the $SiCl_6^{2-}$ ion? Give an orbital diagram for silicon that illustrates the bonding in this ion.

New Terms

Write the definitions of the following terms, which were introduced in this section. If necessary, refer to the Glossary at the end of the text.

- hybrid orbitals
- *sp* hybrid orbitals
- *sp*^{2} hybrid orbitals
- *sp*^{3} hybrid orbitals
- *sp*^{3}*d* hybrid orbitals
- *sp*^{3}*d*^{2} hybrid orbitals
- free rotation
- conformations

7.6 Double and Triple Bonds

Review

In this section we see that two kinds of bonds can be formed by the overlap of orbitals:

σ bonds in which the electron density is concentrated along an imaginary line joining the nuclei.

π bonds in which the electron density is divided into two regions that lie on opposite sides of an imaginary line joining the nuclei.

In every molecule that you will study, a single bond is a σ bond. A double bond consists of *one* σ and *one* π bond; a triple bond consists of *one* σ bond and *two* π bonds. Thus every bond is composed of at least a σ bond. In complex molecules, the molecular geometry is determined by this σ bond framework, and the kinds of hybrid orbitals that an atom uses is determined by the number of σ bonds it forms and the number of unshared pairs of electrons it holds. Review the brief summary at the bottom of page 289. The following example illustrates how we apply these concepts.

Example 7.2 **Determining the Kinds of Hybrids that Atoms Use in Molecules**

Problem

What kinds of hybrid orbitals do the carbon and nitrogen atoms use in the molecule

```
     H   H   H   H
     |   |   |   |
  H—C=C—C—N—H
             |   ••
             H
```

Solution

The carbon on the left and the one in the center each form three σ bonds and neither has any unshared electrons. Therefore, each of these carbons needs three hybrid orbitals, so sp^2 hybrids are used. The carbon at the right forms four σ bonds and therefore uses sp^3 hybrids. The nitrogen forms three σ bonds, which requires three hybrid orbitals, and it needs a fourth hybrid to house the lone pair; the total is four, so sp^3 hybrids are used by the nitrogen.

An important feature of double bonds is the restricted rotation around the axis of the bond. This is because such rotation misaligns the unhybridized *p* orbitals that form the π bond. Thus, rotation about a double bond axis involves bond breaking, which is very difficult.

Self-Test

13. Which kinds of hybrid orbitals are used by the carbon and oxygen atoms in acetic acid, which has the following structure?

```
     H  :O:
     |   ||   ••
  H—C—C—O—H
     |        ••
     H
```

14. Explain why it should be possible to isolate three compounds with the formula $C_2H_2Cl_2$ and in which there are carbon-carbon double bonds.

15. Why doesn't it matter whether or not rotation about a triple bond is restricted?

____

____

New Terms

Write the definitions of the following terms, which were introduced in this section. If necessary, refer to the Glossary at the end of the text.

sigma bond (σ bond)
pi bond (π bond)
restricted rotation

7.7 Bond Energies and Their Measurement

Review

In this section, you learn how values of ΔH_f° are used to calculate bond energies. The basis for these calculations is Hess's law and the fact that the enthalpy change is a state function; that is, the same enthalpy change takes place regardless of the path followed from the reactants to the products. Study Figure 7.30. The sum of the enthalpy changes corresponding to steps 1, 2, and 3 must be equal to ΔH_f° for the product.

In setting up an alternative path from the reactants (the elements in their standard states) to the product, we consider the following:

1 **The conversion of the elements on the reactant side into gaseous atoms.** The energy changes here are the standard heats of formation of the gaseous elements (Table 7.1). In the direction of the arrows in steps 1 and 2 in Figure 7.30, these changes are endothermic. We know this because it always takes

energy to vaporize a solid element, and it always takes energy to break bonds to give atoms.

2 **The formation of all the bonds in the product molecule.** In discussing this energy change, we define the atomization energy, ΔH_{atom}, which is the energy needed to *break* all the bonds in the molecule. Whether we are forming bonds or breaking them, the amount of energy involved is the same. Only the sign of the ΔH is different, positive (endothermic) for bond breaking and negative (exothermic) for bond making. We obtain the atomization energy by adding up all the bond energies for the bonds in the molecule. (Table 7.2).

Once an alternative path is established, we can use it to calculate bond energies if the value of ΔH_f° is known, or we can calculate the value of ΔH_f° if all the bond energies are known. This is illustrated in Example 7.13 on page 292.

Self-Test

16. Calculate the atomization energy of the molecule CH_3CN, in kJ/mol. The molecule has the structure

```
      H
      |
  H—C—C≡N:
      |
      H
```

____

17. Construct a figure similar to that in Figure 7.30 for the formation of $CH_3CN(g)$ from its elements. Be sure to show both the direct and alternative paths.

18. Estimate the standard heat of formation of CH_3CN vapor in kJ/mol using data in Table 7.1 and your answers to Questions 16 and 17 above.

____

New Terms

Write the definitions of the following terms, which were introduced in this section. If necessary, refer to the Glossary at the end of the text.

bond energy

bond dissociation energy

atomization energy

7.8 Molecular Orbital Theory

Review

According to molecular orbital theory (MO theory), when two orbitals overlap, their electron waves interact by constructive and destructive interference to give bonding and antibonding molecular orbitals. A bonding orbital is one that concentrates electron density between nuclei and leads to a lowering of the energy when occupied by electrons. An antibonding orbital concentrates electron density in regions that do not lie between nuclei. When occupied by electrons, an antibonding MO leads to a raising of the energy. Remember that compared to the atomic orbitals from which they are formed, bonding MOs are lower in energy and antibonding MOs are higher in energy.

The electronic structure of a molecule is obtained by feeding the appropriate number of electrons into the molecule's set of molecular orbitals. The rules that apply to the filling of molecular orbitals is the same as those for atomic orbitals.

1 An electron enters the lowest energy MO available.

2 No more than two electrons with spins paired can occupy any given MO.

3 When filling an energy level consisting of two or more MOs of equal energy, electrons are spread over the orbitals as much as possible with their spins in the same direction.

You should study the shapes of the σ and σ^* orbitals formed by the overlap of *s* orbitals as well as the shapes of the σ, σ^*, π, and π^* orbitals that arise from the overlap of *p* orbitals. Also study the order of filling of the MOs given by the energy level diagram in Figure 7.36 and Table 7.3.

Self-Test

19. On a separate piece of paper, construct the MO energy level diagrams and give the molecular orbital electron populations for (a) Li_2^- and (b) F_2^-

20. Which should be more stable, C_2^+ or C_2^-? Explain.

____

____

21. Which has the more stable bond, NO or NO^+? (See Practice Exercise 13 on page 298.). ____________________

New Terms

Write the definitions of the following terms, which were introduced in this section. If necessary, refer to the Glossary at the end of the text.

- molecular orbitals
- bonding molecular orbital
- antibonding molecular orbital
- bond order

7.9 Delocalized Molecular Orbitals

Review

Molecular orbital theory avoids resonance by allowing for the simultaneous overlap of more than two orbitals in such a way as to produce a large π-type cloud that extends over three or more atoms. This kind of an extended molecular orbital is said to be delocalized because the bond is not localized between just two atoms. One of the most important molecules having delocalized MOs is benzene.

Molecules in which there are delocalized bonds are more stable than they would be if the bonds were localized. The extra stability produced by delocalization is called the delocalization energy.

Self-Test

22. Draw resonance structures for the nitrate ion and show how the bonding would be represented using a delocalized MO.

23. How is the benzene molecule represented to show the delocalized molecular orbitals of the ring?

New Terms

Write the definitions of the following terms, which were introduced in this section. If necessary, refer to the Glossary at the end of the text.

delocalized bond
delocalization energy

7.10 Bonding in Solids

Review

In a solid, the energy levels of the atoms merge to form sets of energy bands each of which contains enormous numbers of closely spaced energy levels. The core electrons (electrons below the valence shell) of the atoms in a crystal are localized on the atoms. The valence shell energy levels combine to form a delocalized band called the valence band. This energy band contains the valence electrons of the atoms in the crystal. An energy band that is either filled or empty but which extends uninterrupted throughout the crystal is called a conduction band. (A conduction band is a delocalized band of energy levels.)

In a metal, the valence band is either partially filled or overlaps an empty conduction band, so metals are good electrical conductors. In a nonmetal, the valence band is filled and there is a large energy gap between the filled band and the lowest energy conduction band. At normal temperatures, the conduction band is empty and there is no means for the substance to conduct electricity, so it is an insulator. In a semiconductor, the band gap between the filled valence band and the conduction band is small. Thermal kinetic energy is able to raise some electrons into the conduction band, so semiconductors are weak electrical conductors.

p-type and n-type Semiconductors

If an element from Group IIIA is added as an impurity to a semiconductor from Group IVA, the valence band contains one electron vacancy for each atom of the Group IIIA element added. Electrical conduction begins by an electron filling this hole and leaving a positive hole elsewhere. Migration of such positive holes through the solid is the mode of electrical conduction in this p-type semiconductor.

If an element from Group VA is added as an impurity to a semiconductor from Group IVA, the valence band contains one extra electron for each atom of the Group VA element added. Electrical conduction takes place by the migration of these extra electrons through the conduction band in this n-type semiconductor.

Self-Test

24. Which kind of semiconductor is formed when
    (a) antimony is added to silicon? ______________
    (b) boron is added to germanium? ______________

New Terms

Write the definitions of the following terms, which were introduced in this section. If necessary, refer to the Glossary at the end of the text.

band theory	conduction band	n-type semiconductor
energy band	band gap	solar battery
valence band	p-type semiconductor	

Answers to Self-Test Questions

1. Check your answers by referring to the drawings of pages 254 through 256.
2. (a) 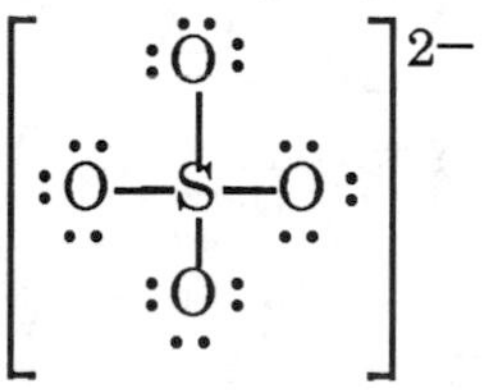 $[SO_4]^{2-}$ Lewis structure — tetrahedral
   (b) 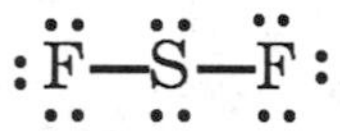 :F̤—S̤—F̤: — nonlinear
   (c) 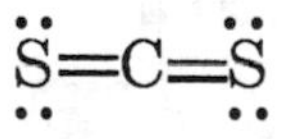 S=C=S — linear
   (d) 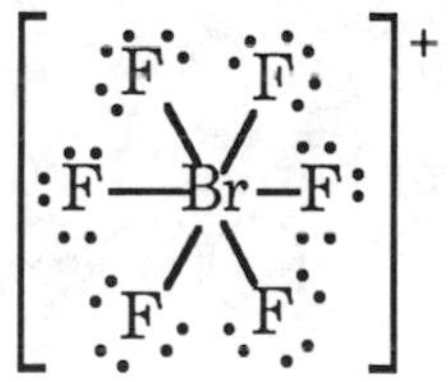 $[BrF_6]^+$ Lewis structure — octahedral
   (e) 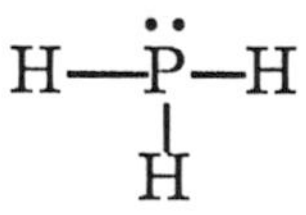 H—P—H with H below P — trigonal pyramidal
   (f) 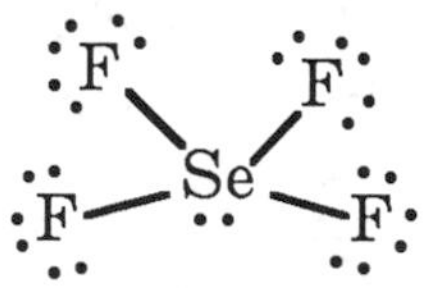 SeF_4 Lewis structure — unsymmetrical tetrahedral
   (g) 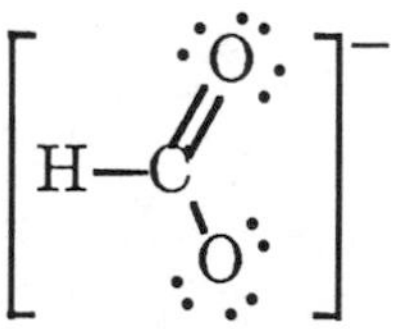 $[HCO_2]^-$ Lewis structure — planar triangular
   (h) 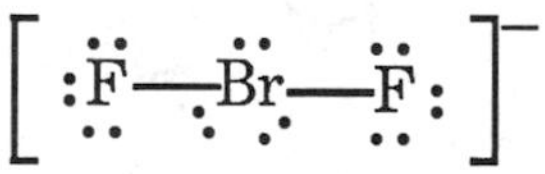 $[F—Br—F]^-$ — linear
3. SCl_2 and ClF_3 are polar. (Draw Lewis structures; note lone pairs.)

4. The C—Cl and C—H bonds differ in polarity, so the bond dipoles can't cancel to give a nonpolar molecule.
5. Overlap of the half-filled *p* orbitals of the chlorine atoms produces the bond. See Figure 7.14 for F_2.
6. Hydrogen 1*s* orbitals overlap two half-filled *p* orbitals of tellurium. Since *p* orbitals are at 90°, the bond angle is very close to 90°.

   Te (in H_2Te) (••) 5*s* (••)(•×)(•×) 5*p*   • = H electron; × = Te electron

7. Hydrogen 1*s* orbitals overlap 3 half-filled *p* orbitals of arsenic.

   As (in AsH_3) (••) 4*s* (•×)(•×)(•×) 4*p*   • = H electron; × = As electron

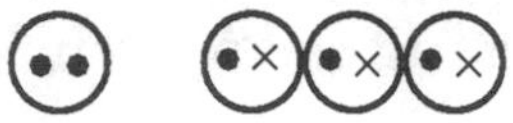

8. Hybrid orbitals overlap better with orbitals on neighboring atoms and form stronger bonds.
9. (a) sp^3d (b) sp^3 (c) sp^2 (d) sp^3
10. (a) Ge (in $GeCl_4$) (•×)(•×)(•×)(•×) sp^3 × = Cl electron

    molecule is tetrahedral

    (b) Se (in $SeCl_4$) (••)(•×)(•×)(•×)(•×) sp^3d ( )( )( )( ) *d*

    × = Cl electron

    molecule has unsymmetrical tetrahedral shape

    (c) S (in SCl_2) (••)(••)(•×)(•×) sp^3 × = Cl electron

    nonlinear or bent molecular shape

    (d) I (in IF) (••)(•×)(•×)(•×)(•×)(•×) sp^3d^2 ( )( )( ) *d*

    × = F electron

    square pyramidal molecular shape

11. N (in NH_4^+) (••)(•×)(•×)(•×) sp^3 × = H electron

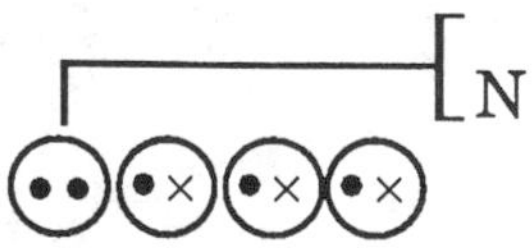

    [N to vacant 1*s* orbital of H]

12. sp^3d^2 hybrids

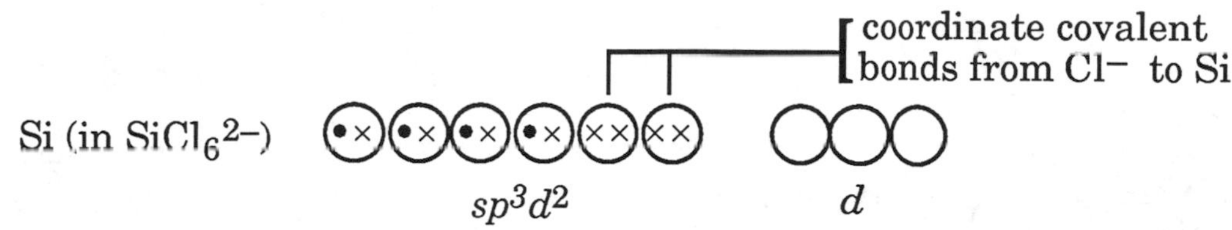

× = Cl electron

13. 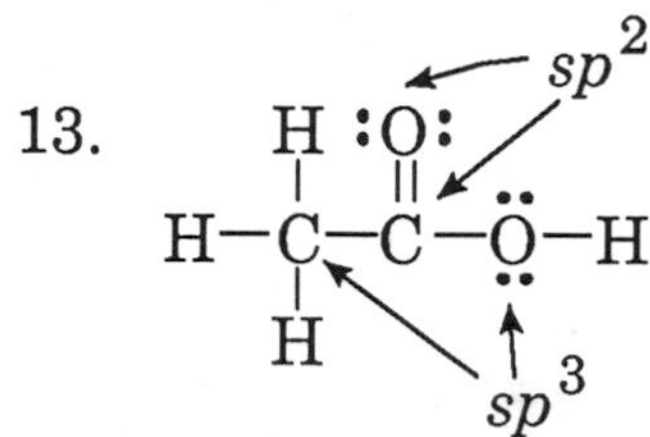

14. Restricted rotation about the double bond permits structures II and III (below) to be isolated as well as structure I.

I	II	III
$Cl_2C{=}CH_2$ (both Cl on one C, both H on the other)	HClC=CClH (Cl on opposite sides)	HClC=CClH (Cl on same side)

15. The atomic arrangement is linear.
16. ΔH_{atom} = 2478 kJ/mol
17. $2C(g) \ + \ 3H(g) \ + \ N(g)$

$$\begin{array}{ccccccc} 2C(g) & + & 3H(g) & + & N(g) & \longrightarrow & \downarrow \\ \uparrow & & \uparrow & & \uparrow & & \\ 2C(s) & + & \frac{3}{2}H_2(g) & + & \frac{1}{2}N_2(g) & \longrightarrow & CH_3CN(g) \end{array}$$

18. Estimated ΔH_f° = +79 kJ/mol (For comparison, the accepted value is +95 kJ/mol).
19.

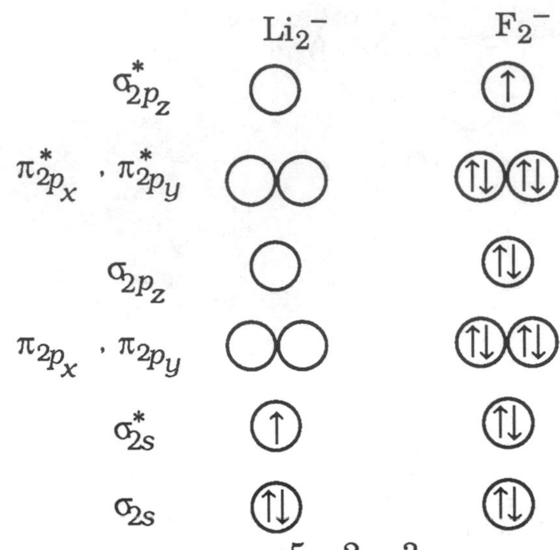

20. C_2^+, bond order $= \frac{5-2}{2} = \frac{3}{2}$

    C_2^-, bond order $= \frac{7-2}{2} = \frac{5}{2}$

    Therefore C_2^- is more stable than C_2^+.

21. NO, bond order is $\frac{8-3}{2} = \frac{5}{2}$

    NO^+, bond order is $\frac{8-2}{2} = \frac{6}{2}$

    NO^+ has a more stable bond than NO.

22.

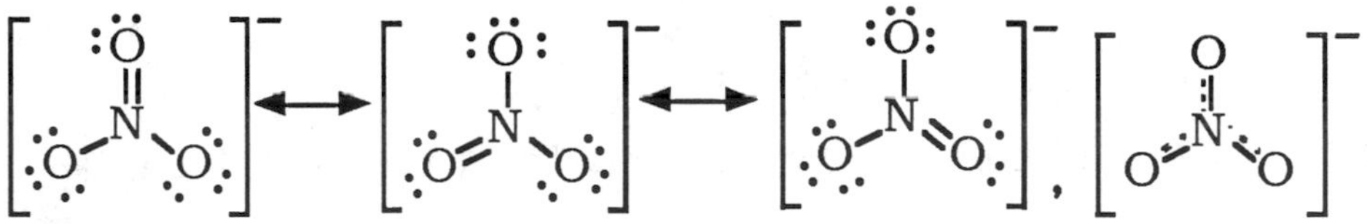

23.

24. (a) n-type, (b) p-type

Tools you have learned

Remove this chart from the Study Guide and keep it handy when tackling homework problems.

Tool	*Function*
Basic molecular shapes linear, planar triangular, tetrahedral, trigonal bipyramidal, octahedral	Knowing how to draw them enables you to sketch the shapes of most molecules.
VSEPR Theory	Enables you to determine the shape of a molecule or polyatomic ion when its Lewis structure is known. Also enables you to determine the kind of hybrid orbitals used by the central atom in a molecule or polyatomic ion.
Molecular shape	Enables you to determine whether a molecule is polar.

Summary of Useful Information

The various structures obtained by applying the VSEPR theory can be summarized as shown in the table below. In the formulas in the table, the symbol M stands for the central atom, X stands for an atom bonded to the central atom, and E stands for a lone pair of electrons in the valence shell of the central atom. Thus, MX_2E_3 represents a molecule in which the central atom is bonded to two other atoms and has three lone pairs in its valence shell. To use the information in the table, construct the Lewis structure for the substance. Then write its formula using the symbols as defined above. Find the formula in the table and read the name for the molecular shape alongside.

Formula	Structure
MX_2	linear
MX_3	planar triangular
MX_2E	nonlinear (bent)
MX_4	tetrahedral
MX_3E	trigonal pyramidal
MX_2E_2	nonlinear (bent)

Formula	Structure
MX_5	trigonal bipyramidal
MX_4E	unsymmetrical tetrahedral
MX_3E_2	T-shaped
MX_2E_3	linear
MX_6	octahedral
MX_5E	square pyramidal
MX_4E_2	square planar

Correlation between VSEPR theory and orbital hybridization.

Number of sets of electrons in valence shell of central atom	Geometric arrangement of electron pairs	Hybrid orbitals that give this geometry
2	linear	sp
3	planar triangular	sp^2
4	tetrahedral	sp^3
5	trigonal bipyramidal	sp^3d
6	octahedral	sp^3d^2

Chapter 8

PROPERTIES OF GASES

Of the four important variables of a sample of a gas—pressure, volume, temperature, and mass (or moles)—one cannot be changed without changing one or more of the others. Samples of liquids and solids do not display this feature. The purposes of this chapter are to study how P, V, T, and n are interrelated by the gas laws and to see how these laws can be explained in terms of a single theory, the kinetic theory of gases.

Learning Objectives

In this chapter, you should keep in mind the following goals.

1 To learn what pressure is, the units used to describe it, and ways to measure it.

2 To learn how the pressure and the volume of a fixed amount of gas are interrelated at constant temperature.

3 To learn how the volume of a fixed amount of gas changes with temperature if the pressure is kept constant.

4 To learn how the pressure of a fixed amount of gas changes with temperature if the volume is kept constant.

5 To learn to carry out gas law calculations using the combined gas law.

6 To learn the ideal gas law equation, the value of the universal gas constant, R, and to be able to do gas law calculations using the ideal gas law, including molecular mass calculations.

7 To learn how to carry out stoichiometric calculations when some or all of the substances are gases.

8 To learn what *partial pressure* means, what the relationship is between the total pressure of a gas mixture and the partial pressures of the mixture's components, and how to find what pressure a wet gas would have if made dry.

9 To learn how the rate of effusion of a gas is related to its density and molecular mass.

10 To learn the postulates of the kinetic theory of gases and the model for an ideal gas that these postulates describe.

11 To be able to explain the gas laws in terms of the kinetic model of gases.

12 To learn why real gases do not behave as an ideal gas and how van der Waals corrected the ideal gas law to make it fit real gases better.

8.1 Qualitative Facts About Gases

Review

Pressure, volume, temperature and mass (or number of moles) are the four physical properties of gases to be studied. Regardless of the size or shape of the container, a gas always fills it uniformly, so the volume of a gas, usually given either in liters or milliliters, is always the volume of its container. Temperature is always given in (or changed into) kelvins before calculations are made.

Self-Test

1. What four physical properties of gases are interrelated?

   ____

2. To keep the mass of a gas sample constant, what must be done experimentally? ______________________________________

New Terms

none

8.2 Pressure

Review

An unbalanced *force* can cause something to speed up, slow down, change direction, or crack apart. *Pressure* is the ratio of force to area. The earth exerts a gravitational force that makes the surrounding envelope of air—the atmosphere—press downward on the earth. The ratio of this force to area is the atmospheric pressure. The simplest device to measure it is the Torricelli barometer.

When the column of mercury in a Torricelli barometer has a temperature of 0 °C and stands 760 mm higher than the mercury level outside, the mercury in the column is exerting a pressure variously named as 1 *standard atmosphere*

(1 atm), or 760 torr, or 760 mm Hg. In the SI unit of pressure, the pascal (Pa), 1 atm = 101,325 Pa. The pressure unit that will be most often used in our study is the torr.

Manometers, either the open-end or the closed-end type, are used to measure the pressure of a confined gas. The separation between two mercury levels corresponds to the gas pressure when the closed-end manometer is used. With an open-end manometer, the separation (in mm) is added to the value of the atmospheric pressure (in mm). This then gives the pressure in torr, since 1 mm Hg = 1 torr.

Self-Test

3. A pressure of 745 torr has what value in atm? ____________
4. On a TV weather report, the pressure was reported to be 29.4 in., meaning "inches of mercury." What does this correspond to in torr?

   ________________ In atm? ________________
5. A pressure of 0.980 atm has what value in torr?____________
6. A pressure of 746 torr has what value in kilopascals (kPa)?________
7. On a day when the atmospheric pressure was 754 torr, the manometer reading on a gas sample was found to be 82 torr. What is the gas pressure in torr if the manometer is of the closed-end design?

   ____________ The open-end design? ________________

New Terms

Write the definitions of the following terms, which were introduced in this section. If necessary, refer to the Glossary at the end of the text.

atmospheric pressure
barometer
manometer
manometer, closed end
manometer, open end
pascal (Pa)
pressure
standard atmosphere (atm)
torr

8.3 Pressure—Volume—Temperature Relationships for a Fixed Amount of Gas

Review

Boyle's law (pressure-volume law). Gases generally obey the rule that their volumes are inversely proportional to their pressures, provided the temperature and mass of gas are kept constant. The best expression of this law—the pressure-volume law—is the equation

$$P_1V_1 = P_2V_2$$

The hypothetical gas that would obey this relationship exactly is called an ideal gas.

Charles's law (temperature-volume law). Provided we express the temperature of a gas in kelvins, the volume of a fixed quantity of gas is directly proportional to the temperature if the pressure is kept constant. One useful statement of this law—often called Charles' law—is

$$\frac{V_1}{T_1} = \frac{V_2}{T_2}$$

Gay-Lussac's law (pressure-temperature law). Provided that the volume is held constant, the pressure of a fixed quantity of gas is directly proportional to the kelvin temperature. One way to express this is the equation:

$$\frac{P_1}{T_1} = \frac{P_2}{T_2}$$

Combined gas law. For a fixed mass of gas the combined gas law tells us that

$$\frac{P_1V_1}{T_1} = \frac{P_2V_2}{T_2}$$

When $P_1 = P_2$, the equation reduces to that of Charles's law (the temperature-volume law). When $T_1 = T_2$, the combined gas law equation becomes that of Boyle's law (the pressure-volume law). When $V_1 = V_2$, the combined gas law equation reduces to the equation for Gay-Lussac's law (the pressure-temperature law).

Thinking It Through

1 Consider a sample of 2.15 g of krypton, one of the noble gases, occupying a volume of 640 mL at 25.0 °C under a pressure of 745 torr. It is desired to make the conditions of the final state of this amount of sample be a temperature of 35.0 °C and a pressure of 740 torr.

(a) Are enough data provided to enable another scientist to duplicate the sample in its initial state? If not, what additional data would be needed?

(b) For this specific quantity of sample to be in its final state, what other variable must change? To what value?

Self-Test

8. The pressure on 750 mL of a gas is changed at constant temperature from 720 torr to 760 torr. What is the new volume? ______

9. To change the volume of a gas at 755 torr from 400 mL to 350 mL at constant temperature, what new pressure must be provided? ______

10. What do we call the hypothetical gas that would obey the pressure-volume law exactly? ______

11. The volume of an ideal gas is inversely proportional to

    (a) $\frac{1}{P}$ (b) P (c) P^2 (d) $\sqrt{P}$ ______

12. What will be the new volume if 250 mL of oxygen is cooled from 25 °C to 15 °C at constant pressure? ______

13. A sample of 750 mL of nitrogen gas is heated from 20 °C to 100 °C. In order to keep its pressure constant, to what new volume should the gas be allowed to change? ______

14. A sample of 656 mL of oxygen gas at 740 torr is heated from 25.0°C to 55.0 °C in a sealed container. What gas variable changes and to what new value? ______

15. What will be the final volume of a 500-mL sample of helium if its pressure is changed from 740 torr to 780 torr and its temperature is changed from 30 °C to 50 °C? ______

16. In order for the pressure of a fixed mass of gas at 10 °C to change from 1 atm to 2 atm and its volume to go from 3 L to 4 L, what must the final temperature become? ______

17. What is the final pressure in torr exerted by 375 mL of argon at 755 torr and 25 °C if its temperature goes to 50 °C and its volume changes to 425 mL? ______

New Terms

Write the definitions of the following terms, which were introduced in this section. If necessary, refer to the Glossary at the end of the text.

absolute zero
combined gas law
ideal gas
pressure-temperature law (Boyle's law)
pressure-volume law (Gay-Lussac's law)
temperature-volume law (Charles's law)

8.4 The Ideal Gas Law

Review

From Avogadro's insights into the data on the combining volumes of gases came the knowledge—*Avogadro's principle*—that equal volumes of gases have equal numbers of moles (at the same temperature and pressure). The volume occupied by one mole of any gas at STP is 22.4 liters, the *standard molar volume.*

The *ideal gas law*, $PV = nRT$, gives us the relationship among all four physical variables of a gas—pressure, volume, temperature, and number of moles. The *universal gas constant*, R, has the value of 0.0821 L atm/mol K. Remember that when you use this value of R in the ideal gas law equation, the unit of pressure must be atm and that of volume *must* be liter (the temperature, of course, must be in kelvins). The ideal gas law equation can be used in the following kinds of calculations.

1. To find the *volume* that a given amount of gas will occupy at some value of pressure and temperature.
2. To find the *pressure* that a given amount of gas will exert if it is kept at a given temperature in some particular volume.
3. To find the *temperature* a given amount of a gas must have if it is confined in a specified volume and is to exert a certain pressure.
4. To use the measured values of P, V, and T to find the number of moles (n) of gas in the sample. If we then divide the mass of gas by the number of moles, we get the *molar mass*—which is numerically the same as the molecular mass.

Thinking It Through

2 A gaseous compound of sulfur and fluorine with an empirical formula of SF has a density of 4.102 g L^{-1} at 25.0 °C and 750 torr. What is the molecular mass and the molecular formula of this compound?

Self-Test

18. What is Avogadro's principle? ______________________

____

19. What do we call the reference conditions for gas temperature and pressure, and what are their values?

____

20. What is the standard molar volume (including its units)?

    ____

21. A sample of 11.2 L of a gas was trapped at STP. How many moles of the gas was this? ____________________

22. Write the equation for the ideal gas law. ____________________

23. Write the value of R, including the units. ____________________

24. What volume will 0.982 g of O_2 occupy if its temperature is 21 °C and its pressure is 748 torr? ____________________

25. If 0.156 g of N_2 is confined at 18.0 °C in a volume of 135 mL, what pressure will it have? ____________________

26. If a 0.138-gram sample of CO_2 is to have a pressure of 4.50 atm when kept in a container with a volume of 30.0 mL, what temperature in degrees Celsius must it have? ____________________

27. A 0.390-gram sample of a gas at 25 °C occupied a volume of 350 mL at a pressure of 740 torr. Calculate its molecular mass.

    ____

New Terms

Write the definitions of the following terms, which were introduced in this section. If necessary, refer to the Glossary at the end of the text.

Avogadro's principle
ideal gas law (equation of state of an ideal gas)
standard temperature and pressure (STP)
standard molar volume
universal gas constant (R)

8.5 The Stoichiometry of Reactions Between Gases

Review

For reactions involving gases, Avogadro's principle provides a new way to state stoichiometric equivalencies. For example, from the coefficients in the equation

$$N_2(g) + 3H_2(g) \rightarrow 2NH_3(g)$$

we know that in terms of *moles* 1 mol $N_2 \Leftrightarrow$ 3 mol H_2. We can now rephrase this in terms of gas *volumes* provide that the volumes are compared at identical pressures and temperatures.

$$1 \text{ vol } N_2 \Leftrightarrow 3 \text{ vol } H_2$$

Thinking It Through

3 Sodium hydroxide granules can remove carbon dioxide from an air stream by its reaction according to the following equation.

$$NaOH(s) + CO_2(g) \rightarrow NaHCO_3(s)$$

The air of a room measuring $6.00 \times 6.00 \times 3.00$ m containing CO_2 at a concentration of 1.00×10^{-6} mol/L is passed through a bed of NaOH granules. What is the minimum number of grams of NaOH needed to remove all of the CO_2 from this quantity of air?

Self-Test

28. In the following reaction, all substances are gases.

$$2NO + O_2 \rightarrow 2NO_2$$

Assuming that the gases are measured at the same values of P and T, how many liters each of NO and O_2 are needed to prepare 10 L of NO_2?

____

29. Nitrogen and hydrogen can be made to combine to give ammonia according to the following equation:

$$N_2 + 3H_2 \rightarrow 2NH_3$$

All the substances are gases. If one liter of nitrogen is used, what volume of hydrogen is needed and what volume of ammonia is made (assuming all volumes are measured under the same conditions of temperature and pressure)?

____

30. One step in the industrial synthesis of sulfuric acid is the following reaction:

$$2SO_2(g) + O_2(g) \rightarrow 2SO_3(g)$$

If 640 L of O_2, initially at 740 torr and 20.0 °C, is used to make SO_3, how many moles and how many grams of SO_2 are needed?

____

New Terms

none

8.6 Dalton's Law of Partial Pressures

Review

Gases in a mixture exert their own pressures, called partial pressures, independently of the other gases present. The total pressure of the mixture is a simple function of the partial pressures, P_a, P_b, P_c and so forth.

$$P_{total} = P_a + P_b + P_c + \ldots$$

This law of partial pressures is useful in doing calculations involving gases collected over water and that therefore have water vapor as one of the gases present.

$$P_{dry\ gas} = P_{total} - P_{water}$$

The partial pressure of water vapor when it is in the presence of liquid water is called the vapor pressure of the liquid water. It depends only on the temperature of the water. All liquids, including water, have their own vapor pressures. For any given liquid the value of its vapor pressure increases with increasing temperature of the liquid.

Thinking It Through

4 A 500-mL sample of nitrogen is collected over water at a temperature of 20 °C. The atmospheric pressure at the time of the experiment is 746 torr. What would be the volume of the nitrogen at this same temperature if it were free of water and at a pressure of 760 torr?

Self-Test

31. If the total pressure on a mixture of helium and neon is 745 torr and the partial pressure of the helium is 640 torr, what is the partial pressure of the argon? _______________

32. Suppose that all of the argon were removed from the sample in the previous question but the volume of the container was left the same and the temperature was not changed. What would a pressure gauge read for the remaining gas? ____________

33. How do we find out what is the partial pressure of water vapor in a sample of nitrogen gas collected over water as illustrated in Figure 8.8 in the text?____________

____

New Terms

Write the definitions of the following terms, which were introduced in this section. If necessary, refer to the Glossary at the end of the text.

law of partial pressures (Dalton's law of partial pressures)
partial pressure
vapor pressure

8.7 Graham's Law of Effusion

Review

The rate of effusion is inversely proportional to the square root of the density of a gas, or the square root of the molecular mass of the gas, provided that the pressure and temperature of the gas are kept constant. Equation 8.19 is the easiest equation to use in calculations.

Self-Test

34. State the law of gas effusion.

____

____

35. At room temperature and 760 torr the density of helium is 0.000160 g/mL. Under these conditions the density of hydrogen is 0.0000818 g/mL. How much more rapidly will hydrogen effuse than helium through the same pinhole?
(a) 1.95 times as rapidly
(b) 1.11 times as rapidly
(c) 1.22 times as rapidly
(d) 1.40 times as rapidly ____

New Terms

Write the definitions of the following terms, which were introduced in this section. If necessary, refer to the Glossary at the end of the text.

effusion

Graham's law (law of gas effusion)

8.8 Kinetic Theory and the Gas Laws

Review

The gas laws describe *observations*; they *explain* nothing. They would be true even if atoms and molecules were yet to be discovered. The kinetic theory answers the question, "What *must* be true about a gas at the microscopic level (or lower) to explain why the gas laws as we know them are observed and not some other gas laws?" This section takes each of the gas laws in turn and in a qualitative way uses the kinetic theory and its model of an ideal gas to explain them.

Kinetic Theory. The idea of matter in the gaseous state consisting of extremely tiny particles in chaotic, utterly random motions and collisions helps to explain *how* gases have both pressures and temperatures. The higher the average kinetic energy of gaseous particles, the higher the pressure and the higher the temperature.

Pressure-temperature law ($P \propto T$, at constant mass and V). If raising the temperature makes the average kinetic energy of the gas particles increase, they hit the walls harder giving a higher pressure.

Temperature-volume law ($V \propto T$, at constant mass and P). If gas particles are given higher average kinetic energies and velocities by giving the gas a higher temperature, the pressure tends to increase. To keep the pressure constant, the collisions with the walls have to be spread out over a larger area, which is accomplished by letting the particles take up a larger space or volume.

Pressure-volume law ($P \propto V$, at constant mass and T). If the volume is reduced, the gas particles hit the container's walls not more vigorously but more frequently—raising the pressure.

Effusion law (effusion rate $\propto \sqrt{\text{gas density}}$, at constant pressure and temperature). A gas having a low density has less massive particles than a gas with a high density. At the same temperature, the lighter particles move with a greater average velocity than the heavier particles. This means that particles of the less dense gas get from one place to another more quickly than those of the more dense gas, so the gas with the lower density must effuse more rapidly.

Law of partial pressures ($P_{total} = P_a + P_b + P_c + \ldots$). According to the model of an ideal gas its particles do not repel or attract neighboring particles.

They act independently, and therefore their partial pressures must add up to the total pressure, not some value less than or greater than the total.

Avogadro's principle ($V \propto n$, at constant P and T; or $P \propto n$, at constant V and T). At the same T, regardless of the gas, gas molecules have the same average kinetic energy and they exert the same average force per hit when they strike a unit area of the walls. The frequency of these hits per unit area is proportional to the molar concentration of the gas (moles per unit of volume), so the total force per unit area—the pressure—is proportional to the moles of the gas and not the molecular mass of the gas. Hence $P \propto n$, which is one way of stating the law.

Gas compressibility. Gases, but not liquids and solids, are relatively easily compressed because the volume occupied by a gas is mostly empty space.

Absolute zero. When all of the particles in a gas stop moving, the gas particles must have zero average kinetic energy, regardless of their mass. Since gas temperature is proportional to this average kinetic energy, a state of zero kinetic energy must be the coldest condition possible—absolute zero (0 K or −273.15 °C).

Self-Test

36. What are the postulates of the kinetic theory of gases?

____

____

____

37. The Kelvin temperature of a gas is proportional to what feature of the gas particles? ______________________

38. What is the fundamental difference between the gas laws—taken as a group— and the postulates of the kinetic theory?

____

____

____

39. In your own words explain each of the gas laws in terms of the model of the ideal gas.

____

____

____

New Terms

Write the definitions of the following terms, which were introduced in this section. If necessary, refer to the Glossary at the end of the text.

kinetic theory of gases

8.9 Real Gases: Deviations from the Ideal Gas Laws

Review

To simplify the theoretical treatment and its calculation, those who first postulated the kinetic theory simply ignored the fact that gas particles themselves do occupy some space and they assumed that ideal gas particles exert no forces on each other. For most practical purposes, these assumptions work very well, and we can use the ideal gas law ($PV = nRT$), in most ordinary situations, especially when the gas is at a relatively low pressure and relatively high temperature.

To correct for the real volume (the excluded volume) of just the gas particles, van der Waals subtracted a term, nb (where n = moles and b = a correction factor that differs for each gas), from the measured volume of the gas. In other words, he substituted the term $(V - nb)$ for V in the ideal gas law.

To correct for the real forces of attraction between gas particles, van der Waals added the term n^2a/V^2 (where n = moles, a = a correction factor that is different for each gas, and V = volume) to the measured pressure of the gas. In other words, he substituted the expression $(P + n^2a/V^2)$ for P in the ideal gas law.

The final result is called the van der Waals' equation of state:

$$(P + n^2a/V^2)(V - nb) = nRT$$

Self-Test

40. What are the two principal reasons why real gases deviate from ideal gas behavior?

    ____

41. When van der Waals suggested that the term $(V - nb)$ be substituted for V in the ideal gas law equation, he was saying that the measured volume, V, is actually a bit too__________ (large or small) to obtain a good fit of the data to the equation. When V is in liters, what are the units of b? __________

42. When van der Waals suggested that the term $(P + n^2a/V^2)$ be substituted for P in the ideal gas law equation, he was saying that the measured pressure, P, is actually a bit too ______________________

    (large or small) to obtain a good fit of the data to the equation. When P is in atm and V is in liters, what are the units of a?

    ____

43. If gas X has a larger van der Waals b constant than gas Y, what do we learn about the relative forces of attraction between the particles in these two gases?

    ____

44. The molecular mass of gas M is larger than that of gas Q. Which gas is likely to have the larger van der Waals' a constant?______________

New Terms

Write the definitions of the following terms, which were introduced in this section. If necessary, refer to the Glossary at the end of the text.

van der Waals' equation of state for a real gas
van der Waals' constants

Solutions to Thinking It Through

1 (a) No additional data are needed. (b) The volume must change. The combined gas law equation applies; $V_2 = 648$ mL.

2 Whenever you are stuck about what to do, go back to the basic definitions and let them lead you. In this problem, remember that a molecular mass is always the ratio of the grams of a substance to the moles of the substance—the grams per mole. We're told that 1 L of the gas has a mass of 4.102 g, so what we have to do next is find out how many moles this mass equals while being able to occupy 1 L at 750 torr and 25.0 °C. The universal gas law, $PV = nRT$ will help us because we know R, P (750 torr), T (25.0 °C), and V (for which we can assume at least 3 significant figures since "per liter" in the density's units can be taken as an exact number). Before we calculate n, we have to convert 750 torr into atm (0.987 atm) and 25.0 °C into kelvins (298 K) in order to use $R = 0.0821$ L atm mol^{-1} K^{-1}. Using these figures, the number of moles n is found to be 4.03×10^{-2} mol

The ratio of grams to moles (the molecular mass) is

$$\frac{4.102 \text{ g}}{4.03 \times 10^{-2} \text{ mol}} = 102 \text{ g mol}^{-1}$$

The formula mass for the empirical formula, SF, calculates to be 51.07. This is half the molecular mass (102/51.07), so each subscript in SF has to be multiplied by 2. The molecular formula of the compound is S_2F_2.

3 Don't let all the business about the room's dimensions throw you. With a 1:1 mole ratio of NaOH to CO_2, the stoichiometry is simple; it's getting the number of moles of CO_2 that requires work. Once we find the number of moles of CO_2 we automatically know the number of moles of NaOH. Then its a moles to grams conversion to find the number of grams of NaOH. We're told that the concentration of CO_2 in the room is 1.00×10^{-6} mol L^{-1}. So we need the volume of the room *in liters,* because

$$(\text{vol, in L}) \times 1.00 \times 10^{-6} \frac{\text{mol } CO_2}{\text{L}} = \text{mol } CO_2$$

The room's volume in cubic meters is: 6.00 m × 6.00 m × 3.00 m = 108 m^3

To convert 108 m^3 to liters is just a unit-conversion problem like those worked in Chapter 1. Since 1 m^3 = 1 × 10^3 L, 108 m^3 = 108 × 10^3 L. Now we can calculate the number of moles of CO_2 using our first equation above.

$$108 \times 10^3 \text{ L}) \times 1.00 \times 10^{-6} \frac{\text{mol } CO_2}{\text{L}} = 1.08 \times 10^{-1} \text{ mol NaOH}$$

Because the formula mass of NaOH is 40.00,

$$1.08 \times 10^{-1} \text{ mol NaOH} \times \frac{40.00 \text{ g NaOH}}{1 \text{ mol NaOH}} = 4.32 \text{ g NaOH}$$

The CO_2 in the room air requires a minimum of 4.32 g NaOH to be removed by the chemical reaction given.

4 The initial state of the gas is that it is wet at a total pressure of 746 torr in a volume of 500 mL at a temperature of 20 °C. The final state is at the same temperature, so we have ultimately a Boyle's law problem, only we need the initial pressure of the dry nitrogen, which Dalton's law lets us calculate. At 20 °C, the vapor pressure of water is 17.54 torr (from Table 8.2, page 333, of the text). Therefore, P_{total} = 746 torr = $P_{nitrogen}$ + 17.54 torr. This means that $P_{nitrogen}$ = 728 torr. So, using Boyle's law,

$$(728 \text{ torr})(500 \text{ mL}) = (760 \text{ torr})(V_2)$$

$$V_2 = 479 \text{ mL}$$

Answers to Self-Test Questions

1. Pressure, volume, temperature and mass (or moles)
2. Use a closed container.
3. 0.980 atm
4. 747 torr 0.983 atm
5. 745 torr
6. 99.5 kPa
7. 82 torr, 836 torr

8. 711 mL
9. 863 torr
10. An ideal gas
11. (b)
12. 242 mL
13. 955 mL
14. Pressure changes to 814 torr.
15. 506 mL
16. 755 K (482 °C)
17. 722 torr
18. Equal volumes of gases have equal numbers of moles when compared under identical conditions of pressure and temperature.
19. standard temperature and pressure, 273 K and 760 torr.
20. 22.4 L/mol at STP
21. 0.500 mol
22. $PV = nRT$
23. $R = 0.0821$ L atm/mol K
24. 0.753 L or 753 mL
25. $P = 0.986$ atm
26. 251 °C
27. 28.0 g/mol
28. 10 L of NO and 5 L of O_2
29. 3 L of H_2 and 2 L of NH_3
30. 51.8 mol of SO_2; 3.32×10^3 g of SO_2
31. 105 torr
32. 640 torr (This question gets at the fundamental meaning of partial pressure.)
33. Use a table giving the vapor pressures of water at various temperatures.
34. The rate of effusion of a gas is inversely proportional to the square root of the gas density.
35. d
36. Compare your answer with the three postulates of the kinetic theory given on page 336 of the text.
37. To the average kinetic energy of the gas particles.
38. The gas laws describe observations. The kinetic theory theorizes about the structure of a gas that is consistent with the observations.
39. Compare what you write with the explanations given in Section 8.8.
40. Individual gas particles have real volumes; individual particles in a gas sample do have small attractions for each other.
41. large; L/mol
42 small; (L^2 atm)/mol^2
43. The forces in X are greater than those in Y.
44. Gas M

Tools you have learned

Remove this chart from the Study Guide and keep it handy when tackling homework problems.

Tool	*Function*
Pressure-Volume law $P_1V_2 = P_2V_2$	Pressure-volume calculations when the temperature is constant for a fixed mass of gas.
Temperature-Volume law $\frac{V_1}{T_1} = \frac{V_2}{T_2}$	Temperture-volume calculations when the pressure is constant for a fixed mass of gas.
Pressure-Temperature law $\frac{P_1}{T_1} = \frac{P_2}{T_2}$	Pessure-temperature calculations when the volume is constant for a fixed mass of gas.
Combined gas law $\frac{P_1V_2}{T_1} = \frac{P_2V_2}{T_2}$	Gas law calculations for a fixed mass of gas when three gas variables change.
Ideal gas law PV = nRT	Gas law calculations involving possible changes in any of the four gas variables.
Dalton's law of partial pressures $P_{total} = P_a + P_b + P_c + \ldots$	Partial pressure calculations for mixtures of gases
Graham's law $\frac{\text{effusion rate } (A)}{\text{effusion rate } (B)} = \sqrt{\frac{d_B}{d_A}} = \sqrt{\frac{FM_B}{FM_A}}$	Calculation of relative rates of effusions of two gases either from their densities or their molecular masses.

Summary of Important Equations

Boyle's law equation

$$P_1V_2 = P_2V_2$$

Charles's law equation

$$\frac{V_1}{T_1} = \frac{V_2}{T_2}$$

Gay-Lussac's law equation

$$\frac{P_1}{T_1} = \frac{P_2}{T_2}$$

Combined gas law equation

$$\frac{P_1V_2}{T_1} = \frac{P_2V_2}{T_2}$$

Ideal gas law equation

$$PV = nRT$$

Dalton's law of partial pressures

$$P_{\text{total}} = P_a + P_b + P_c + \ldots$$

Graham's Law equation

$$\frac{\text{effusion rate }(A)}{\text{effusion rate }(B)} = \sqrt{\frac{d_B}{d_A}} = \sqrt{\frac{FM_B}{FM_A}}$$

Chapter 9

INTERMOLECULAR ATTRACTIONS AND THE PROPERTIES OF LIQUIDS AND SOLIDS

In Chapter 8 you learned about the physical properties of gases, and you saw that gas behavior is essentially independent of the chemical composition of the gas molecules. In this chapter we turn our attention to the other two states of matter, where attractions between molecules play a dominant role, and where physical properties are strongly affected by the chemical makeup of the particles.

Learning Objectives

Throughout your study of this chapter, keep in mind the following objectives:

1 To learn why the physical properties of liquids and solids depend so heavily on their chemical composition while the properties of gases do not.

2 To learn the nature and relative strengths of the principal kinds of intermolecular attractions. You should also learn the factors that influence the strengths of intermolecular attractions.

3 To learn some general properties of liquids and solids, and how these properties are related to the closeness of packing of the molecules and to the intermolecular attractive forces.

4 To learn about the kinds of changes of state and to learn about the concept of dynamic equilibrium.

5 To learn about the factors that control the pressure that a vapor exerts when it is in equilibrium with a liquid or a solid.

6 To learn how boiling point is defined in scientific terms and the factors that influence the boiling point.

7 To examine the kinds of energy changes that accompany changes of state.

8 To learn how we can predict the way that the composition of an equilibrium system is affected by outside influences that are

able to upset the equilibrium.

9 To learn how the pressure-temperature relationships between the states of substances can be represented graphically.

10 To learn how atoms, molecules, or ions are arranged in crystalline solids and how we are able to describe their structures in simple ways.

11 To see how data obtained by X-ray diffraction experiments are used to obtain structural information about crystals.

12 To learn how the physical properties of crystalline solids can be related to the kinds of particles at lattice positions and the kinds of attractive forces between the particles.

13 To learn how some substances solidify without forming a crystalline solid.

9.1 Why Gases Differ From Liquids and Solids

Review

Intermolecular attractions are attractions that exist between molecules. They are most significant when molecules are close together and are practically insignificant when they are far apart. In gases, the molecules hardly feel the intermolecular attractions at all because the particles are so widely spaced. Differences in the strengths of these attractions caused by differences in chemical makeup are so small that all gases behave in nearly the same way. This is why we are able to have gas laws. But in liquids and solids, where the molecules are practically touching, the intermolecular attractions are very strong and differences in their strengths are significant and are reflected in differences in physical properties.

Self-Test

1. Under what conditions do the properties of real gases deviate *most* from the predicted properties of an ideal gas?

___

New Terms

Write the definition of the following term, which was introduced in this section. If necessary, refer to the Glossary at the end of the text.

intermolecular attractions

9.2 Intermolecular Attractions

Review

The attractions between neighboring molecules are always much weaker than the chemical bonds within molecules. Whereas the strengths of chemical bonds determine the chemical properties of substances, it is the strengths of the intermolecular attractions that determine many of the physical properties of substances. The three principal kinds of intermolecular attractions are dipole-dipole attractions, hydrogen bonds, and London forces.

Dipole-dipole attractions occur between polar molecules and are about 1% as strong as normal covalent bonds.

Hydrogen bonding is an especially strong type of dipole-dipole attraction (about 5 times the strength of the usual dipole-dipole attraction). It occurs between molecules in which hydrogen is covalently attached to nitrogen, oxygen, and fluorine. Therefore, molecules that contain O—H and N—H bonds experience hydrogen bonding.

London forces result from attractions between instantaneous dipoles and induced dipoles in neighboring molecules. London forces flicker on and off and their average strength is usually less than most dipole-dipole attractions. Remember that London forces are present in *all* substances.

Large molecules and atoms have large, more easily distorted electron clouds than small molecules and atoms, so London forces are greater between the larger particles. When atoms are of about the same size, the longer the chain of atoms, the greater is the *total* strength of the London forces of attraction.

Ions are also able to induce dipoles in neighboring particles, so ion-dipole attractions are present between ions and molecules and other ions.

Self-Test

2. How do the strengths of the different kinds of intermolecular attractive forces compare?

   ____

   ____

3. What kinds of attractive forces (both intermolecular and chemical bonding) are present between the particles in each of the following substances?

   (a) helium ____________________

   (b) sodium nitrate, $NaNO_3$ ____________________

(c) ethyl alcohol, CH_3CH_2OH ________________

(d) sulfur dioxide ________________

(e) butane, $CH_3CH_2CH_2CH_3$ ________________

4. For each pair of substances below, choose the one in which there will be the stronger intermolecular attractions.

(a) Cl_2 or Br_2 ______ (d) CH_4 or SiH_4 ______

(b) HF or HCl ______ (e) PF_3 or PCl_3 ______

(c) C_2H_6 or C_6H_{14} ______ (f) CH_4 or CH_3Cl ______

New Terms

Write the definitions of the following terms, which were introduced in this section. If necessary, refer to the Glossary at the end of the text.

dipole-dipole attractions
London forces
instantaneous dipole
induced dipole
hydrogen bond

9.3 General Properties of Liquids and Solids

Review

The physical properties of liquids and solids depend on both the tightness of packing of their particles as well as on the strengths of the intermolecular attractions. However, some properties depend more on one of these factors than the other.

Properties that Depend Mostly on Tightness of Packing

The closeness of packing is the reason that liquids and solids resist compression when pressure is applied to them, and it is also primarily responsible of the slow rates of diffusion in these states of matter.

Liquids and solids are nearly incompressible because there is almost no empty space into which to squeeze the molecules when pressure is applied. Diffusion is slow in liquids because to move from one place to another the particles must work their way around and past so many other near neighbors. In human terms, it's like attempting to cross a crowded room; you must squeeze past so many other people that movement is slow. Diffusion in solids is nearly nonexistent; the particles are not free to move about as they are in liquids.

Properties that Depend Mostly on Intermolecular Attractions

Properties mentioned in this section that are controlled mostly by the strengths of intermolecular attractions are retention of shape and volume, surface tension, wetting of a surface by a liquid, and evaporation.

Attractive forces hold the particles close together in liquids and solids, so their volume doesn't change when transferred from one container to another. In solids, the attractive forces are even greater and the particles cannot easily move away from their equilibrium positions. This rigidity allows solids to keep their shapes when transferred from one container to another.

To understand surface tension, as well as many of the other physical properties of liquids and solids, it is necessary to understand the relationship between attractive forces and potential energy, so let's review these important concepts. First, *whenever something feels an attractive force, it has potential energy.* A large reservoir of water high in the mountains has potential energy because the water feels the earth's gravitational attraction. Allowing the water to flow to a lower altitude releases some of this potential energy, which we can use to generate electricity or for some other energy-consuming activity. If there were no attractive forces between the earth and the water, it would not flow downhill and it would not release energy that we could use. Second, *the greater the sum-total of the attractive forces felt by something, the lower will be its potential energy.* Consider the water, for example. You probably know that the earth's gravitational attraction decreases with increasing distance from the earth's center. When the water flows to a lower altitude, its comes closer to the center of the earth, so it feels a *greater* gravitational attraction at the same time that its potential energy decreases.

Now let's consider the phenomenon known as surface tension. This property of liquids arises because molecules at the surface feel fewer attractions than the molecules within the liquid. This is because the molecules at the surface are only partially surrounded by other molecules, whereas those inside the liquid are completely surrounded. This causes the molecules within the liquid to experience greater *total* attractive forces than molecules at the surface, and as a result, the molecules in the interior of the liquid have lower potential energies than those at the surface. The *surface tension* is a quantity that's related to the energy difference between a molecule at the surface and a molecule inside the liquid. We can also say that this is the amount of energy needed to bring a molecule from the interior to the surface. When the intermolecular attractions are large, the surface tension is large. The tendency of a liquid to achieve a minimum potential energy causes the liquid to minimize its surface area and thereby minimize the number of higher-energy surface molecules.

For a liquid to wet a surface, the attractive forces between the liquid and the surface must be about as strong as between molecules of the liquid. Substances with low surface tensions easily wet solid surfaces because the attractions in the liquid are weak. Water has a large surface tension and can wet a glass surface because the surface contains oxygen atoms to which water molecules can hydrogen bond. Water doesn't wet waxy or greasy surfaces because the water molecules are only weakly attracted to hydrocarbon molecules by London forces. The London forces are much weaker than the hydrogen bonding between water molecules in the liquid.

Evaporation of a liquid or sublimation of a solid occurs when molecules leave their surfaces and enter the vapor space that surrounds them. Remember that it is always the molecules with very large kinetic energies that escape, so the average kinetic energy of molecules left behind is lowered, and this means that the temperature is also lowered. (In other words, evaporation leads to a cooling effect, and so is endothermic.)

Factors Affecting the Rate of Evaporation

Two factors control the rate of evaporation. One is the temperature and the other is the strengths of the intermolecular attractions. Increasing the temperature increases the rate of evaporation. Study Figure 9.11 to be sure you understand why this is so. Substances with weak intermolecular attractive forces evaporate more rapidly than those that have strong intermolecular attractive forces. Study Figure 9.12.

Thinking It Through

1 For the following pairs of substances, choose the one that will have the greater rate of evaporation at a given temperature.

(a) $CH_3CH_2CH_3$ or $CH_3CH_2CH_2CH_2CH_3$

(b)

$$\begin{array}{ccccccccccccccc} & H & & H & & H & & H & & & & & & & \\ & | & & | & & | & & | & & & & & & & \\ H- & C & - & C & - & C & - & C & -O-H & \quad \text{or} \quad & H- & C & - & C & -O- \\ & | & & | & & | & & | & & & & | & & | & \\ & H & & H & & H & & H & & & & H & & H & \end{array}\begin{array}{ccccc} H & & H & & \\ | & & | & & \\ C & - & C & -H & \\ | & & | & & \\ H & & H & & \end{array}$$

Self-Test

5. Which of the properties discussed in this section are primarily the result of the tightness of packing of molecules in liquids and solids?

____

6. In ethylene glycol (antifreeze), the intermolecular attractive forces are much greater than between molecules in gasoline. Molecules that evaporate from ethylene glycol therefore carry with them much more kinetic energy than most of the molecules that evaporate from gasoline. Yet, if you spill these two liquids on yourself, the gasoline gives a much greater

cooling effect than the ethylene glycol. Why? (Think!)

____

____

New Terms

Write the definitions of the following terms, which were introduced in this section. If necessary, refer to the Glossary at the end of the text.

incompressibility	surfactant
surface tension	evaporation
wetting of a surface	sublimation

10.5 Changes of State and Dynamic Equilibrium

Review

Changes of state include:

solid $\rightarrow$ liquid	liquid $\rightarrow$ vapor	vapor $\rightarrow$ solid
solid $\rightarrow$ vapor	liquid $\rightarrow$ solid	vapor $\rightarrow$ liquid

Changes of state take place under conditions of dynamic equilibrium. This is one of the most important concepts for you to grasp in this chapter. In a dynamic equilibrium, two opposing events are taking place at equal rates. For example, when a liquid evaporates into an enclosed space, the concentration of molecules in the vapor rises until the rate at which molecules condense becomes equal to the rate at which they evaporate. Once this happens, the *number* of molecules in the vapor remains constant with time; in the time it takes for a hundred molecules to evaporate into the vapor, another hundred molecules condense and leave the vapor. Thus a dynamic equilibrium is a condition in which an unending sequence of balanced events leads to a constant composition within the equilibrium system.

Dynamic equilibria can exist between liquid and vapor, solid and vapor, and between solid and liquid.

Self-Test

7. In some high-altitude cities, such as Denver, Colorado, snow sometimes disappears gradually without ever melting. What happens to it?

____

8. What is another term that has the same meaning as "change of state?"

____

9. What changes are constantly taking place during the equilibrium involved in sublimation?

____

New Terms

Write the definitions of the following terms, which were introduced in this section. If necessary, refer to the Glossary at the end of the text.

change of state
phase change
melting point
dynamic equilibrium
condensation

9.4 Vapor Pressures of Liquids and Solids

Review

When the rates of evaporation and condensation are equal, the vapor above a liquid exerts a pressure called the equilibrium vapor pressure, or simply the vapor pressure. The vapor pressure of a particular liquid is independent of the size of the container or the amount of liquid that is present, as long as some liquid remains at equilibrium. A similar situation exists for solids. The vapor that is in equilibrium with a solid also exerts a pressure that is independent of the size of the container or the amount of solid that remains at equilibrium.

The vapor pressure differs for different substances and reflects the strengths of the intermolecular attractions. When these attractions are strong, the vapor pressure is low.

The vapor pressure also depends on the temperature. For any given liquid, the vapor pressure increases as the temperature rises. This is because at the higher temperature a larger fraction of the molecules have sufficient energy to escape from the liquid's surface. For the same reason, the equilibrium vapor pressure of a solid also rises with increasing temperature.

Self-Test

10. The following are the vapor pressures of some common solvents at a temperature of 20 °C. Arrange these substances in order of increasing strengths of their intermolecular attractive forces.

carbon disulfide, CS_2	294 torr

ethyl alcohol, C_2H_5OH	44 torr
methyl alcohol, CH_3OH	96 torr
acetone, $(CH_3)_2CO$	186 torr

____

11. On a separate sheet of paper, sketch a graph that shows how vapor pressure varies with temperature for a typical liquid.

New Terms

Write the definitions of the following terms, which were introduced in this section. If necessary, refer to the Glossary at the end of the text.

vapor pressure equilibrium vapor pressure

9.6 Boiling Points of Liquids

Review

Boiling occurs when large bubbles of vapor form below the surface of a liquid. The boiling point is the *temperature* at which a liquid boils, and it is the temperature at which the liquid's vapor pressure becomes equal to the atmospheric pressure exerted on the liquid's surface. The temperature at which the vapor pressure equals 1 atm is the liquid's normal boiling point. Substances with strong intermolecular attractions generally have high boiling points.

Thinking It Through

2 Arrange the following in order of increasing normal boiling point?

$(CH_3)_2CNH_2$, $(CH_3)_2CH_2$, $(CH_3)_2CO$

Self-Test

12. Why does a pressure cooker cook foods more rapidly than when they are boiled in an open pot? ________________

____

13. Arrange the substances in Question 11 in increasing order according to their expected boiling points.

____

14. What happens to the temperature if heat is added more rapidly to an already boiling liquid? Explain.____________________________

____

15. Why is boiling point a useful physical property for the purposes of identifying liquids? ____

____

16. Which would be expected to have a higher boiling point, argon or krypton? ____________________

17. Which would be expected to have a higher boiling point, hydrogen sulfide or hydrogen chloride?

____

18. Which would be expected to have the higher vapor pressure at 20 °C, C_3H_8 or C_5H_{12}? ____________________

New Terms

Write the definitions of the following terms, which were introduced in this section. If necessary, refer to the Glossary at the end of the text.

boiling point

normal boiling point

9.7 Energy Changes During Changes of State

Review

Figure 9.20 describes the shapes of heating and cooling curves. The flat portions of these graphs represent potential energy changes associated with solid ↔ liquid and liquid ↔ vapor phase changes. You should know the general shapes of such graphs and what kinds of energy changes take place along the various line segments.

Every change of state is accompanied by an energy change. They are enthalpy changes because phase changes take place under conditions of constant pressure. Study the definitions of molar heat of fusion, molar heat of vaporization, and molar heat of sublimation given on page 375. Melting of a solid, evaporation of a liquid, and sublimation are changes that absorb energy (they are endothermic), so their values of ΔH are positive. Phase changes in the opposite direction—freezing of a liquid and condensation of a vapor to either a liquid or a solid—release energy and are exothermic. Their values of ΔH are equal in magnitude to, but opposite in sign to the heats of fusion, vaporization, and sublimation.

The Clausius-Clapeyron equation (Equation 9.1 on page 376, and its alternative form, Equation 9.2 on page 377) relates the heat of vaporization to the variation of vapor pressure with temperature. Study Example 9.1 to see how this equation can be used to calculate the vapor pressure of a liquid at a particular temperature when the heat of vaporization and the vapor pressure at some other temperature are known.

The size of the enthalpy change associated with a change of state depends on the strengths of the intermolecular attractions in the substance and also by how much these forces are altered during the change of state. The $\Delta H_{\text{vaporization}}$ is larger than ΔH_{fusion}, and the $\Delta H_{\text{sublimation}}$ is larger than $\Delta H_{\text{vaporization}}$.

Thinking It Through

3 Butane, C_4H_{10}, is the liquid fuel in disposable cigarette lighters. The normal boiling point of butane is –0.5 °C and its heat of vaporization is 24.27 kJ/mol. What is the approximate pressure of butane gas over liquid butane in a cigarette lighter at room temperature (25 °C)?

Self-Test

19. If you know the values of the molar heat of fusion and the molar heat of vaporization, you can calculate the value of the molar heat of sublimation. Why can this be done, and how can it be done?

____

20. The molar heat of vaporization of benzene (C_6H_6) is 30.8 kJ/mol and water has a molar heat of vaporization of 40.7 kJ/mol. Which of these substances has the larger intermolecular attractive forces?

____

21. Referring to the data in Question 20, which substance, water or benzene, should have the larger molar heat of sublimation?

____

22. How many kilojoules are required to vaporize 10.0 g of liquid benzene? (See Question 20.)

____

23. Arrange the following in increasing order of their expected molar heats of vaporization: HCl, NH_3, CH_4, HBr, He.

____

24. According to Figure 9.19, what would the boiling point of water be if it were not for hydrogen bonding? Could life as we know it exist without hydrogen bonding? ______________________

____

25. Which compound should have the higher heat of vaporization,

(a) $CH_3CH_2CH_3$ or $CH_3CH_2CH_2CH_2CH_3$ ______________________

(b)

```
   H  H  H  H                 H  H     H  H
   |  |  |  |                 |  |     |  |
H—C—C—C—C—O—H      or    H—C—C—O—C—C—H
   |  |  |  |                 |  |     |  |
   H  H  H  H                 H  H     H  H
```

____

New Terms

Write the definitions of the following terms, which were introduced in this section. If necessary, refer to the Glossary at the end of the text.

heating curve	molar heat of fusion
cooling curve	molar heat of vaporization
supercooling	molar heat of sublimation
fusion	Clausius-Clapeyron Equation

9.8 Dynamic Equilibrium and Le Châtelier's Principle

Review

Learn Le Châtelier's principle, which is set off between colored bars on page 379. We will use this principle frequently when studying both physical and chemical equilibria. For physical changes, the following apply:

1. An increase in temperature shifts an equilibrium in the direction of an endothermic change.
2. An increase in pressure shifts an equilibrium in a direction that will favor a decrease in volume.

Self-Test

26. State Le Châtelier's principle. ______________________________

____

____

27. The equilibrium between a liquid and a vapor can be written

$$\text{liquid} \rightleftharpoons \text{vapor}$$

If the volume of a container in which this equilibrium is established is decreased while the temperature is kept the same, what happens to the relative amounts of liquid and vapor in the container?

____

____

____

New Terms

Write the definitions of the following terms, which were introduced in this section. If necessary, refer to the Glossary at the end of the text.

Le Châtelier's principle

position of equilibrium

9.9 Phase Diagrams

Review

A typical phase diagram has three lines that serve as temperature-pressure boundaries for the three states of the substance—solid, liquid, and gas. Any point on one of these lines represents a temperature and pressure in which there is an equilibrium between two phases. The three equilibrium lines intersect at a common point called the triple point. Each substance has a unique triple point corresponding to the temperature and pressure at which solid, liquid, and gas are simultaneously in equilibrium with each other.

The liquid-vapor line terminates at the critical point, corresponding to the critical temperature and pressure. Below the critical temperature, a substance can be liquefied by the application of pressure. Above the critical temperature, no amount of pressure can create a distinct liquid phase. Above the critical temperature the substance is said to be a supercritical fluid.

Self-Test

28. According to the phase diagram for water (Figure 9.22), what phase will exist at

    (a) –5 °C and 330 torr ____________________

    (b) 15 °C and 3.00 torr ____________________

    (c) 15 °C and 330 torr ____________________

    (d) 0 °C and 850 torr ____________________

    (e) 100 °C and 330 torr ____________________

29. What changes will be observed if water at –15 °C and 20 torr is gradually warmed at a constant pressure?

    ____

30. What changes will be observed if water is heated from –15 °C to 100 °C at a pressure of 4.58 torr?

    ____

31. At what temperature does liquid and gaseous water exist in equilibrium at 1 atm? What do we call this temperature?

    ____________ °C ____________________

New Terms

Write the definitions of the following terms, which were introduced in this section. If necessary, refer to the Glossary at the end of the text.

phase diagram
triple point
critical point
critical temperature
critical pressure
meniscus
supercritical fluid

9.10 Crystalline Solids

Review

A crystalline solid is characterized by a highly organized, regular, repeating pattern of particles. The overall pattern, called the crystal lattice, can be described by its smallest repeating unit—the unit cell. The number of kinds of lattices (or unit cells) is very small, and all substances can be described in terms of this limited set.

There are three kinds of cubic unit cells—simple cubic, face-centered cubic, and body-centered cubic. Figure 9.29 illustrates how different substances can have the same kind of unit cell but with different cell dimensions.

Self-Test

32. Why do crystals have such regular surface features?

    ____

33. On a separate sheet of paper, make sketches of simple cubic, face-centered cubic and body-centered cubic unit cells. Compare your drawings to Figures 9.27, 9.28, and 9.30.

34. How do the unit cells of KCl and NaCl compare? ______________

    ____

35. Which kind of unit cell is depicted in each of the following

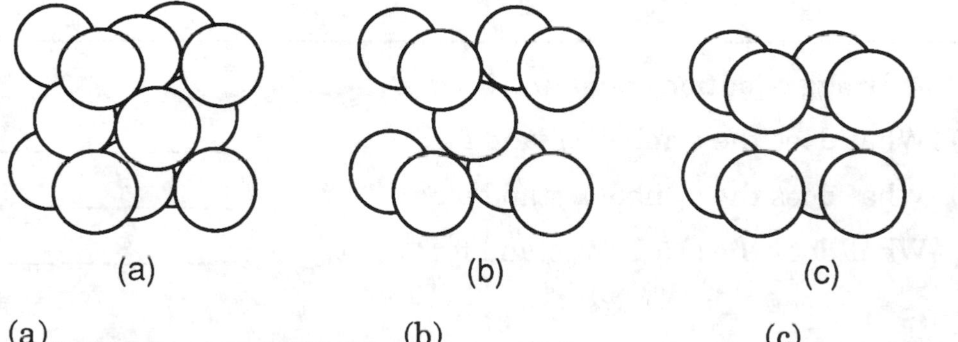

(a) ______________ (b) ______________ (c) ______________

36. The radius of a K^+ ion is 133 pm and the radius of a Cl^- ion is 181 pm. The salt KCl has the same kind of unit cell that NaCl has (Figure 9.31). Calculate the length of an edge of this unit cell in units of picometers.

    ____

New Terms

Write the definitions of the following terms, which were introduced in this section. If necessary, refer to the Glossary at the end of the text.

crystal lattice
unit cell
simple cubic unit cell
face-centered cubic (fcc) unit cell
body-centered cubic (bcc) unit cell

9.11 X-Ray Diffraction

Review

When a crystal is bathed in X rays, it produces a diffraction pattern that can be recorded on film or by other detection devices. From the angles at which the diffracted beams emerge from the crystal and the wavelength of the X rays, the distances between planes of atoms can be computed using the Bragg equation. By a complex procedure, scientists can use these interplanar distances to figure out the structure of the crystal.

Self-Test

37. Why do diffracted X-ray beams occur in only certain directions when a crystal is bathed in X rays?______________________________

    ____

38. Draw three additional sets of parallel lines through the points in Figure 9.35. Do all the sets of lines have the same spacing?

    ____

39. In the Bragg equation (Equation 9.4):
    (a) What does the symbol θ stand for? ________________
    (b) What does the symbol λ stand for? ________________
    (c) What does the symbol d stand for? ________________

New Terms

Write the definitions of the following terms, which were introduced in this section. If necessary, refer to the Glossary at the end of the text.

diffraction pattern
Bragg equation

9.12 Physical Properties and Crystal Types

Review

In this section we divide crystalline solids into four types: ionic, molecular, covalent, and metallic. For each type you should learn the kinds of particles found at the lattice sites and the kinds of attractive forces that exist between them. You should be able to relate these forces to the physical properties of the solid. Study Table 9.5 and Example 9.4.

Self-Test

40. What is the electron-sea model of a metallic crystal?

 ____

 ____

41. One of the elements was once named columbium, and that name is still used by some people. It is shiny, soft, ductile, and melts at 2468 °C. It also conducts electricity. What type of crystal does columbium form?

 ____

42. Antimony pentachloride forms white crystals that do not conduct electricity and that melt at 2.8 °C to give a nonconducting liquid. What kind of solid are $SbCl_5$ crystals?

 ____

43. Strontium fluoride forms brittle, nonconducting crystals that melt at 1450 °C and give an electrically conducting liquid. The solid itself is not an electrical conductor. What type of solid is SrF_2?

 ____

New Terms

Write the definitions of the following terms, which were introduced in this section. If necessary, refer to the Glossary at the end of the text.

ionic crystal
covalent crystal
molecular crystal
metallic crystal

9.13 Noncrystalline Solids

Review

Amorphous solids lack the order found within crystalline solids. They can be thought of as liquids that have lost their ability to flow, so they are sometimes called supercooled liquids. The general term glass is often used to refer to these kinds of solids. Note the difference between the cooling curve for a substance that forms a crystal and one that forms a glass.

Self-Test

44. Why do substances that consist of very long chainlike molecules often form amorphous solids?

____

____

New Terms

Write the definitions of the following terms, which were introduced in this section. If necessary, refer to the Glossary at the end of the text.

amorphous solid
supercooled liquid
supercooling

Answers to Thinking It Through

1 The "tool" we use to answer this question is the relationship between the rate of evaporation and the strengths of intermolecular attractions: the weaker the intermolecular attractions, the faster the rate of evaporation. This then raises the question, which of these has the weaker intermolecular attractions? Three kinds of attractions were discussed: dipole-dipole, hydrogen bonding, and London forces.

In (a), we have hydrocarbon molecules, which are not particularly polar molecules. They don't have N—H or O—H bonds, so hydrogen bonding isn't important either. The principal attractions are therefore London forces. Then we ask, what factors affect the strengths of London forces. For molecules formed by the same elements, the longer the molecular chain, the stronger the attractions, so $CH_3CH_2CH_3$ has the weaker attractions and the higher vapor pressure.

In (b) we have one molecule with an O–H bond, so this molecule experiences hydrogen bonding, which is so much stronger than other intermolecular attraction we can safely assume the molecule with the O–H bond will have the stronger intermolecular attractions. Therefore, it is the second molecule, $C_2H_5–O–C_2H_5$, that will have the higher vapor pressure.

2 Boiling point increases with increasing strengths of the intermolecular attactive forces. The compound $(CH_3)_2CNH_2$ has N–H bonds and so experiences hydrogen bonding, a very strong kind of intermolecular attraction. The compound $(CH_3)_2CH_2$ is a hydrocarbon, so we only expect to find London forces here. The compound $(CH_3)_2CO$ has a polar carbon-oxygen bond, so it will have dipole-dipole attractions in addition to London forces. All the molecules are about the same size, so the London forces in each liquid should be about the same strength. Therefore, the order of overall intermolecular attractions are $(CH_3)_2CH_2 < (CH_3)_2CO < (CH_3)_2CNH_2$. The

boiling points should therefore increase from $(CH_3)_2CH_2$ to $(CH_3)_2CNH_2$.

3 To calculate the vapor pressure of the butane at room temperature, our tool is the Clausius Clapeyron equation. The form of the equation on page 377 is appropriate if we have both an initial pressure and an initial temperature. By definition, the normal boiling point is the boiling point at a pressure of 1 atm, so the initial conditions are P = 1 atm and t = –0.5 °C. The final temperature is 25 °C. We take R = 8.314 J mol^{-1} K^{-1}, convert temperatures to kelvins, change ΔH_{vap} to joules, and then substitute into the equation to solve for P at the higher temperature. If you do this, you should obtain a vapor pressure of 2.5 atm.

Answers to Self-Test Questions

1. high pressure and low temperature
2. Hydrogen bonding is stronger than ordinary dipole-dipole attractions, which are usually stronger than London forces.
3. (a) London forces, (b) ionic bonding, covalent bonding, London forces, (c) covalent bonding, hydrogen bonding, London forces, (d) dipole-dipole forces, London forces, (e) London forces
4. (a) Br_2, (b) HF, (c) C_6H_{14}, (d) SiH_4, (e) PF_3, (f) $CHCl_3$
5. retention of volume, incompressibility, rate of diffusion
6. At room temperature gasoline has many molecules that have enough energy to escape the liquid and it evaporates very quickly. This removes heat very quickly.
7. It sublimes.
8. Phase change
9. Evaporation from the solid; condensation.
10. $CS_2 < (CH_3)_2CO < CH_3OH < C_2H_5OH$
11. See Figure 9.17 on page 369.
12. The higher pressure causes the water to boil at a higher temperature.
13. $CS_2 < CH_3COCH_3 < CH_3OH < CH_3CH_2OH$
14. Temperature remains constant. Adding heat just makes the liquid boil more rapidly.
15. It is easily measured.
16. Krypton
17. HCl (more polar)
18. C_3H_8 (weaker total London forces)
19. Enthalpy is a state function, so $\Delta H_{sublimation} \cong \Delta H_{fusion} + \Delta H_{vaporization}$
20. Water
21. Water
22. 3.95 kJ
23. $He < CH_4 < HBr < HCl < NH_3$
24. Water would boil at about –80 °C. Life as we know it couldn't exist because it would be too cold.
25. (a) $CH_3CH_2CH_2CH_2CH_3$, (b) $CH_3CH_2CH_2CH_2OH$
26. Check your answer on page 379.

27. Decreasing the volume increases the pressure above the equilibrium value. The pressure can be returned to the equilibrium value by decreasing the amount of vapor and increasing the amount of liquid. Le Châtelier's principle says that raising the pressure shifts the equilibrium toward the lower-volume liquid.
28. (a) solid, (b) gas, (c) liquid, (d) liquid, (e) gas
29. solid $\rightarrow$ liquid $\rightarrow$ gas
30. It would begin as a solid. At 0.01 °C, all three phases would appear. Above 0.01 °C it would all be gas.
31. 100 °C; the boiling point.
32. Because they have such ordered internal structures.
33. See Figures 9.27, 9.28, and 9.30.
34. Both are fcc, but the edge length is greater for KCl than for NaCl because K^+ is larger than Na^+.
35. (a) face centered cubic, (b) body centered cubic, (c) simple cubic
36. Edge length = 2(133 pm) + 2(181 pm) = 628 pm
37. Constructive and destructive interference
38. More sets can be drawn. Their spacings differ.
39. (a) The angle at which X rays are reflected from a set of planes of atoms in a crystal. (b) The wavelength of the X rays. (c) The distance between planes of atoms that are giving the reflection.
40. Positive ions of the metal at lattice sites surrounded by a sea of mobile electrons.
41. Metallic (the element is niobium, Nb)
42. Molecular
43. Ionic
44. The molecules become intertwined and cannot find their way into the proper crystalline pattern.

Tools you have learned

Remove this chart from the Study Guide and keep it handy when tackling homework problems.

Tool	*Function*
Intermolecular attractions dipole-dipole, H-bonding, London forces	Knowing the different kinds of intermolecular attractions and the factors that affect their strengths enables you to compare the attractive forces between molecules.
Factors that control rate of evaporation; Factors the control the vapor pressure	Comparison of rates of evaporation or vapor pressures at a given temperature allows you to estimate the relative strengths of intermolecular attractive forces.
Boiling points of liquids Boiling points are proportional to strengths of intermolecular attractions.	Compare relative strengths of intermolecular attractions in liquids.
Heats of vaporization, fusion, sublimation	Compare relative strengths of intermolecular attractions
Clausius-Clapeyron equation	Determine ΔH_{vap} from variation of vapor pressure with temperature; calculate vapor pressure from ΔH_{vap} and known vapor pressure at a known temperature
Le Châtelier's Principle	Predict the effect of a disturbance on the position of equilibrium.
Phase diagram	Determine the phase(s) present at a given temperature and pressure for a substance.
Physical properties and crystal type	Determine the likely crystal type for a substance based on the observed properties of the solid.

Summary of Important Equations and Other Information

Equations

Clausius-Clapeyron equation:

$$\ln P = \frac{-\Delta H_{\text{vap}}}{RT} + C$$

$$\ln \frac{P_1}{P_2} = \frac{\Delta H_{\text{vap}}}{R}\left(\frac{1}{T_2} - \frac{1}{T_1}\right)$$

Bragg equation:

$$n\lambda = 2d \sin \theta$$

Intermolecular attractions and physical properties

As the strengths of intermolecular attractions *increase*, there is an *increase* in

- surface tension
- molar heat of vaporization
- molar heat of sublimation
- normal boiling point
- critical temperature

and a *decrease* in

- rate of evaporation
- vapor pressure

Chapter 10

SOLUTIONS

The emphasis in this chapter is on the *physical* properties common to all solutions, not on their chemical properties. The latter depend on the chemicals themselves.

Learning Objectives

In this chapter, you should keep in mind the following goals.

1 To learn the importance of randomness and of attractive forces—ion-dipole attractions and hydrogen bonds—to the ability of a solute to dissolve in a solvent.

2 To learn how the like-dissolves-like rule helps us qualitatively predict whether a solution can be made.

3 To learn how the heat of solution is the net effect of lattice energies and heats of solvation.

4 To learn how the effect of temperature on solubility correlates with the sign of the heat of solution.

5 To learn the general response to pressure of any equilibrium between a saturated solution of a gas in a liquid.

6 To learn how to work problems involving mole fractions, mole percents, percentage concentrations, and molal concentration.

7 To learn how to convert a concentration in one set of units to another set.

8 To learn how solutes affect the vapor pressure of a solution and to use the Raoult's law equation.

9 To learn how a nonvolatile solute elevates the boiling point and depresses the freezing point of a solution and how the data involved in these changes can be used to calculate a molecular mass.

10 To learn how osmotic pressure data can be used to determine molecular masses.

11 To learn that the colligative properties of solutions of ionic compounds depend not just on the molality in terms of the formula

units of the compound but on the number of ions into which the compound breaks up in solution.

12 To learn the chief physical characteristics of colloidal dispersions and to see how particle size is the key factor in understanding the differences among solutions, colloidal dispersions, and suspensions.

10.1 The Formation of Solutions

Review

In the mixing of gases we have probably the clearest illustration of how nature's drive toward randomness alone can strongly influence a physical change. There is little if any intermolecular attraction or heat of solution, yet the solution spontaneously forms.

When a solid or a liquid dissolves in a liquid, randomness is still a factor, but now intermolecular attractions become very important. *Like-dissolves-like* becomes an operating rule, where "like" refers to similarities in polarity.

The ability of water to *solvate* (*hydrate*) ions largely explains why ionic compounds dissolve better in water than in any other solvent.

Self-Test

1. What are the driving forces behind the formation of the following solutions at room temperature and pressure? (Go beyond the like-dissolves-like rule to the fundamental reasons.)

   (a) Ethyl alcohol in water. ______________________

   ____

   (b) Potassium bromide in water. ______________________

   ____

   (c) The fragrance of a flower bouquet spreading in a room's air.

   ____

2. Wax consists of nonpolar molecules. In terms of intermolecular attractions, explain why wax does not dissolve in water.

   ____

   ____

3. Why should the hydration of the ions of NaCl help this compound to dis-

solve in water?

____

____

____

4. Liquids *X* and *Y* consist of nonpolar molecules. Will *X* dissolve in *Y*?

   __________ Explain ______________________________

   ____

5. $NaNO_3$, an ionic solid, does not dissolve in gasoline. Explain.

   ____

New Terms

Write the definitions of the following terms, which were introduced in this section. If necessary, refer to the Glossary at the end of the text.

hydration
solvation
like-dissolves-like rule

10.2 Heats of Solution

Review

It costs energy to separate solute particles from each other and to separate solvent molecules from each other, because forces of attraction have to be overcome. But when new forces of attraction become established between solute and solvent molecules, energy—*solvation energy*—is liberated. The *heat of solution* is the net energy change, and enthalpy diagrams are helpful to show these relationships.

For salts, we have information about *lattice energies* and *hydration energies* from independent sources, and the net values of these energies are roughly of the same magnitude as measured heats of solution.

Two liquids form an *ideal solution* when the net heat of solution is zero, indicting there is no change in forces of attraction between molecules in the separate components as compared to the forces between them in the solution.

Self-Test

6. What is another technical name for "heat of solution?"

   ____

7. To what does the term "lattice energy" refer in connection with a solid solute?

____

8. Consider a sample of NaCl and a separate beaker of water. The potential energy of this system ______________________ (increases or decreases)

   when the NaCl is converted to its gaseous ions. Why?

   ____

   The separated ions now enter the water. Each becomes surrounded by water molecules, which is a phenomenon called ______________. The potential energy of the system ______________ (increases or decreases) as a result of this.

   Why? ____

9. For a particular salt, the true lattice energy is –400 kJ/mol and the hydration energy is –500 kJ/mol. If the heat of solution for the salt is a function just of these two energies, what is its value? ________ The formation of the solution is therefore ______________________ (exothermic or endothermic).

10. When one liquid is dissolved in another, we can envision a 3-step process. Which step or steps is endothermic and why?

    ____

    ____

    Which step or steps is exothermic and why? ______________________

    ____

    ____

11. What is true about two liquids and their solution if the solution is properly described as ideal?

    ____

    ____

12. Briefly explain why all gases dissolve in liquids exothermically.

    ____

    ____

New Terms

Write definitions of the following terms, which were introduced in this section. If necessary, refer to the Glossary at the end of the text.

heat of solution
hydration energy
ideal solution
solvation energy

10.3 The Effect of Temperature on Solubility

Review

We can include the heat of solution as a product (when it is negative) or as a reactant (when it is positive) in the equilibrium expression for a saturated solution. When we do this, it is easy to apply Le Châtelier's principle to the prediction of how the equilibrium will shift if the solution is heated or cooled.

When two solutes differ substantially in solubility in a solvent, their separation by *fractional crystallization* becomes possible.

Self-Test

13. Suppose a saturated solution involves the following equilibrium.

$$\text{solute}_{\text{undissolved}} + \text{heat} \rightleftharpoons \text{solute}_{\text{dissolved}}$$

(a) Is the formation of the solution endothermic or exothermic?

____

(b) If the saturated solution is cooled, more solute will come out of solution or will go into solution? ________________________

14. Write the equilibrium expression for a saturated solution of a gas in a liquid, and include the heat of solution either as a product or as a reactant.

____

15. In order for a solid to be purified by fractional crystallization, what must happen to the impurities as the boiling, saturated solution of the sample is cooled?

____

____

New Term

Write the definition of the following term, which was introduced in this section.

If necessary, refer to the Glossary at the end of the text.

fractional crystallization

10.4 The Effect of Pressure on the Solubilities of Gases

Review

The relevant equilibrium is

$$\text{gas} + \text{solvent} \rightleftharpoons \text{solution}$$

As the gas dissolves, a large decrease in the volume of this system occurs, so an increase in pressure—a volume-reducing action—will shift the equilibrium to the right. *Henry's law*—gas solubility is proportional to pressure—gives us the quantitative relationship between pressure and gas solubility.

Some important gases—CO_2, SO_2, and NH_3, for example—are helped into aqueous solution by reacting to some extent with the solvent.

Self-Test

16. At 20 °C, the solubility of oxygen in water is 43.0 mg/L at a pressure of 760 torr. Calculate its solubility in these units at a pressure of 540 torr.

____

New Term

Write the definition of the following term, which was introduced in this section. If necessary, refer to the Glossary at the end of the text.

Henry's law (pressure-solubility law)

10.5 Concentrations of Solutions

Review

All of the new concentration expressions in this Section give us, directly or indirectly, information about the ratio of particles of solute and solvent. The values of concentrations defined here are not affected by changes in the temperature of the solution. Be sure that you thoroughly learn the definitions of these expressions, which means that you can automatically write down the actual units (even when they cancel each other) that are implied or directly stated in the definitions. Those who do not learn these units invariably flounder when trying to work problems involving concentrations. You must not fail to learn the following.

The units implied in a *mole fraction* are

$$\frac{\text{number of moles of one component}}{\text{sum of moles of all components}}$$

Thus if a solution of sugar in water is described as having 0.10 mole fraction of sugar, the mole fraction of the solvent must be 0.90 because 1.00 – 0.10 = 0.90. Either of the following conversion factors are thus available.

$$\frac{0.10 \text{ mol sugar}}{0.90 \text{ mol water}} \qquad \frac{0.90 \text{ mol water}}{0.10 \text{ mol sugar}}$$

A mole percent, of course, is simply a mole fraction multiplied by 100.

When working with a mixture of gases, we can use partial pressure data to calculate the mole fraction of any component, because the mole fraction of one gas is its partial pressure divided by the total pressure.

The units for *molality* are number of moles of solute per kilogram of solvent. Usually, it's easier to substitute 1000 g of solvent for 1 kilogram, so most commonly you would be able to chose from the following conversion factors when you know the molality of a solution.

$$\frac{\text{mol of solute}}{1000 \text{ g of solvent}} \qquad \frac{1000 \text{ g of solvent}}{\text{mol of solute}}$$

The most commonly used *percentage concentration* is the weight/weight (w/w) percent. In the lab, if weight/weight percentage is not specified, you can assume that this is what is meant. When you have such a percentage, you can choose from the following kinds of conversion factors.

$$\frac{\text{g of solute}}{100 \text{ g of solution}} \qquad \frac{100 \text{ g of solution}}{\text{g of solute}}$$

(Actually, you don't have to use the gram unit. You can use any mass unit you wish provided you use the same one in both the numerator and the denominator.)

Only with solutions of a liquid in a liquid or a gas in a gas is the volume/volume (v/v) percent useful, and when you know a value you can choose from the following conversion factors.

$$\frac{\text{vol. of solute}}{100 \text{ vol. of solution}} \qquad \frac{100 \text{ vol. of solution}}{\text{vol. of solute}}$$

It does not matter what unit of volume you use, so long as you can use the same unit in both the numerator and denominator.

In converting among concentration units, remember that when you want to convert from a molar concentration to another unit, or from another unit to a molarity, you also have to know the density of the solution. Density provides the link between a solution's mass and its volume, because density is mass per unit volume.

Thinking It Through

1 What are the molarity and the molality of a solution of NaBr with a concentration of 5.00% (w/w)? The density of the solution is 1.04 g/mL.

Self-Test

17. A solution for disinfecting clinical thermometers was made by dissolving together 160 g of ethyl alcohol and 40.0 g of water. Calculate the mole fraction and the mole percent of each component. (For the molecular masses, use the following values: ethyl alcohol, 46.0; water, 18.0.)

    ____

18. The average composition of the air we exhale in terms of the partial pressures of the gases is as follows: N_2, 569 torr; O_2, 116 torr; CO_2, 28 torr; and $H_2O(g)$, 47 torr. Calculate the composition of the air in mole percents.

    ____

19. A bottle bears the label: "0.250 molal sodium chloride." What are the two conversion factors made possible by this information?

    ____

20. If the bottle in Question 19 actually holds only 0.100 mol NaCl, what mass of *water* is present as the solvent? ____________________

21. What mass of solute in grams would be needed to make a 0.100 molal glucose solution if you intended to use 250 g of water as the solvent? Use 180 as the molecular mass of glucose.

    ____

22. How would you prepare 500 g of 0.500% (w/w) sugar?

    ____

    ____

23. For an experiment you need 8.42 g of H_2SO_4, and it is available as 10.0% (w/w) H_2SO_4. How many grams of this solution should you take?

    ____

24. How many milliliters of ethyl alcohol are present in 48.0 mL of 40.0% (v/v) aqueous ethyl alcohol? ____________

25. How would you prepare 250 mL of 10.0% (v/v) acetone in water?

____

26. What is the molality of a 15.0% (w/w) LiCl solution? ____________

27. What are the mole fractions and the mole percents of the components in 15.0% (w/w) LiCl (the same solution in Question 26)?

____

28. In a two-component solution of sugar in water, the mole fraction of sugar is 0.0150. Calculate the percent by weight of the sugar. Use 342 as the molecular mass of sugar.

____

29. A solution of NaCl has a concentration of 1.15 *m*. What are the mole fractions and mole percents of NaCl in this solution?

____

30. What is the percent concentration (w/w) of a sodium carbonate solution with a concentration of 1.04 mol/L? The density of the solution is 1.105 g/mL?

____

New Terms

Write the definitions of the following terms, which were introduced in this section. If necessary, refer to the Glossary at the end of the text.

mole fraction	molality	weight/weight (w/w)
mole percent	percent concentration	volume/volume (v/v)
molal concentration		

10.6 The Effect of a Solute on the Vapor Pressure of a Solution

Review

The vapor pressure of a liquid is lowered when a nonvolatile solute is dissolved in it because the solute particles interfere with the escape of molecules of the liquid into the vapor state but not their return to the liquid solution. When the solute is molecular, not ionic, a particularly simple relationship exists between the vapor pressure of the solution, $P_{solution}$, the vapor pressure of the pure solvent, $P°_{solvent}$, and the mole fraction of the *solvent* (not of the solute). This is the vapor-pressure concentration law or *Raoult's law.*

$$P_{solution} = X_{solvent} \times P°_{solvent}$$

(If the solute breaks up into ions, we have to modify this, but that is a subject for Section 10.9.)

When the solute is also volatile, like another liquid, the presence of any other component, whether it is also volatile or not, interferes with the escape of the solute molecules. So the vapor pressure of such a solute is lowered by other components in proportion to the extent to which they make the mole fraction of the solute less. We now use a variation of the Raoult's law equation to calculate the *partial pressure* of the vapor of each such volatile solute.

$$P_A = P_A°X_A$$

P_A is the partial vapor pressure of volatile component A; $P_A°$ is the vapor pressure of A when it is a pure liquid; and X_A is the mole fraction of A in the solution.

The total vapor pressure exerted by a mixture of volatile substances is—as indicated by Dalton's law of partial pressures—the sum of the individual partial vapor pressures. Notice in Figures 10.17 and 10.18 in the text how the top lines of the plots actually represent the sums of the partial vapor pressures given by the lower lines.

Only *ideal solutions* obey Raoult's law (in the form of either of the above equations) exactly. In many real two-component solutions of volatile liquids, the plot of the total vapor pressure bows upward—a positive deviation from Raoult's law. This happens when intermolecular forces in the solution are less than they are in the separated components. Such solutions form endothermically because it costs more energy to overcome forces of intermolecular attraction in the separate components than is recovered when new forces of attraction operate in the solution. When the plot bows downward—a negative deviation—intermolecular forces in the solution are greater than they are in the separated components. And such solutions form exothermically because less energy is needed to overcome attractive forces in the separate components than is recovered when attractive forces start to operate in the newly forming solution.

Self-Test

31. What is the vapor pressure of a 1.00 molal sugar solution at 25 °C? Sugar, $C_{12}H_{22}O_{11}$, is a nonvolatile, non-ionizing solute and the vapor pressure of water at 25 °C is 23.8 torr.

____

32. At 20 °C the vapor pressure of pure toluene is 21.1 torr and of pure cyclohexane is 66.9 torr. What is the vapor pressure of a solution made of 25.0 g of cyclohexane and 25.0 g of toluene? (Cyclohexane is C_6H_{12} and toluene is C_7H_8.)

New Terms

Write the definitions of the following terms, which were introduced in this section. If necessary, refer to the Glossary at the end of the text.

colligative properties

Raoult's law (vapor pressure-concentration law)

10.7 The Effects of a Nonvolatile Solute on the Freezing Point and the Boiling Point of a Solution

Review

To overcome the lowering of the vapor pressure by a nonvolatile solute, we have to raise the solution's temperature a small amount to get it to boil. To make the solution freeze, we must decrease the solution's temperature below the solvent's freezing point. The value of the change in temperature, Δt, is directly proportional to the solution's molal concentration. The proportionality constant is different for each solvent and is different for boiling and freezing. Depending on the actual physical change, the constant is called the *molal boiling point elevation constant,* k_b, or the *molal freezing point depression constant,* k_f. If we measure Δt, and can find k_b or k_f for the solvent in a table, then we can calculate m, our symbol here for molal concentration:

$$\Delta t = k_b m.$$

$$\Delta t = k_f m.$$

When we know m by this procedure, we can calculate the moles of solute present in the mass of solute used for the solution. When we know both mass (in grams) and moles, it's simply a matter of taking the ratio of grams to moles to find the molecular mass. This is the principal application of the phenomenon of boiling point elevation or freezing point depression. To review the procedure:

Step 1. Select a solvent, and dissolve a measured mass of solute in a measured mass of the solvent.

Step 2. Measure the boiling point of this solution or its freezing point. (Unusually precise thermometers have to be used because Δt is nearly always small.)

Step 3. Use Δt and k_b or k_f to calculate m.

Step 4. Multiply m by the number of kilograms of solvent actually used to find the moles of solute. Notice how the units work out to give "mol solute":

$$\underbrace{\frac{\text{mol solute}}{\text{kg solvent}}}_{\text{molality}} \times \text{kg solvent} = \text{mol solute}$$

Step 5. Divide the number of grams of solute by the number of moles of solute to find the grams per mole, the molar mass. This numerically equals the molecular mass.

Restudy Example 10.12 in the text to see how these steps were used.

Thinking It Through

2 Vapona is the ingredient in flea collars and pesticide "strips." A sample of 0.347 g of vapona was melted together with 35.0 g of camphor, and this mixture was cooled to give a solid. The solid was pulverized and its melting point was found to be 35.99 °C. Using the identical thermometer, pure camphor was found to melt at 37.68 °C. For camphor,

$$k_f = 37.7\ \frac{°\text{C}}{m}.$$

What is the molecular mass of vapona?____________________

Self-Test

33. A solution of 8.32 g of PABA, a common sunscreen agent, in 150 g of chloroform boiled at 62.62 °C. At the same pressure and with the same thermometer, pure chloroform boiled at 61.15 °C. What is the molecular mass of PABA? For chloroform,

$$k_b = 3.63\ \frac{°\text{C}}{m}$$

____

New Terms

Write the definitions of the following terms, which were introduced in this section. If necessary, refer to the Glossary at the end of the text.

boiling point elevation	freezing point depression
boiling point elevation constant	freezing point depression constant

10.8 Dialysis and Osmosis; Osmotic Pressure

Review

Whether the movement of a fluid through a semipermeable membrane is *osmosis* or *dialysis* depends on the kind of membrane separating the solutions or dispersions of unequal concentration. As this membrane becomes less and less permeable, we approach osmosis as the limiting case of dialysis in general.

With the help of Equation 10.6 in the text—$\Pi = MRT$—we can use data on temperature (T) and *osmotic pressure* (Π) to find moles per liter (M). From the data obtained when the solution was prepared, grams/liter, we can take the ratio of g/L to mol/L to get g/mol, the molar mass.

$$\frac{\text{g/L}}{\text{mol/L}} = \frac{\text{g}}{\text{mol}} = \text{molar mass}$$

The technique of using osmotic pressure to determine a molecular mass is particularly useful with high-molecular-mass substances, because even though a given mass gives a very low concentration in units of molarity, the osmotic pressure is still sizeable enough for an accurate measurement.

Self-Test

34. In some sciences, a body fluid might be described as having a "high osmotic pressure." How should we translate this—as meaning a high or a low concentration?____________________

35. When raisins or prunes are placed in warm water they soon swell up and the water becomes slightly sweet to the taste. What phenomenon is occurring, dialysis or osmosis? ____________________

36. If a steak on the grill is heavily salted as it grills, barbecue experts say that the "salt draws the juices." Explain how that happens.

____

____

37. An aqueous solution of a protein with a concentration of 1.30 g/L at 25 °C had an osmotic pressure of 0.0160 torr. What was its molecular mass? ______________________

New Terms

Write the definitions of the following terms, which were introduced in this section. If necessary, refer to the Glossary at the end of the text.

dialysis
osmosis
osmotic pressure

10.9 Colligative Properties of Solutions of Electrolytes

Review

If you know that a solute breaks up into ions when it dissolves then, as an estimate of the effect of this breakup on colligative properties, multiply the molality (or molarity) by the number of ions each formula unit gives as it dissociates. This works best for very dilute solutions.

As solutions of electrolytes become increasingly concentrated, their ions behave less and less well as fully independent particles. To compare the degrees of dissociation at different concentrations, the *van't Hoff factor* is determined. It is the ratio of the degrees of freezing point depression actually observed for the solution to the freezing point that would be calculated assuming that the electrolyte does not dissociate at all.

One way to become comfortable about this concept is to study Table 10.5 in the text. Look at the first row of data for NaCl. The last column tells us that if NaCl were 100% dissociated in solution, at any concentration its van't Hoff factor would be 2.00. This is because two ions form from each NaCl unit. And in a very dilute solution (0.001 *m*), the van't Hoff factor is 1.97 or very close to 2.00. In this quite dilute solution, the electrolyte behaves as if it were almost 100% dissociated. In the more concentrated solution of 0.1 *m*, the van't Hoff factor is less—1.87. But even this is not far from 2.00, so the solute behaves as if it were still mostly dissociated. Notice in this table that the salts that give the largest deviations are those made of ions with more than one charge. These are able to attract each other strongly, and so they are less able to act independently in a solution.

This section also describes how a *percent dissociation* can be estimated from freezing point depression data, but the method is not highly accurate.

The section also alerts you to the existence of some solutes that give weaker colligative properties than expected because they are *associated* in solution.

Self-Test

38. Calculate the freezing point of aqueous 0.250 m NaCl on the assumption that it is 100% dissociated. ________________

    Now do this calculation on the assumption that it is not dissociated at all. ________________

39. The van't Hoff factor for a salt at a concentration of 0.01 m is 2.70. At a concentration of 0.05 m, its van't Hoff factor would be greater or less?

    __________ Explain. ________________

    ____

New Terms

Write the definitions of the following terms, which were introduced in this section. If necessary, refer to the Glossary at the end of the text.

van't Hoff factor

association

10.10 Colloidal Dispersions

Review

In solutions, the solute particles are the smallest chemical species we know—atoms, ions or molecules of ordinary size (0.1 to 1 nm in average diameter). These are too small to scatter light and too easily buffeted about by solvent molecules to settle under the influence of gravity.

Particles that are in a *colloidal dispersion* are next up in size—1 to 1000 nm in average diameter—and whether they settle under the influence of gravity depends much on the presence of stabilizing agents—e.g., *emulsifying agents*. Colloidally dispersed particles are large enough to scatter light—*Tyndall effect*—and to be observed (with the aid of a microscope) to be buffeted about—*Brownian movement*. *Sols*, gels, smokes, *emulsions* and foams are among the common types of colloidal dispersions. If colloidal particles collect and merge to sizes larger than 1000 nm, the system becomes a *suspension* that must be continuously agitated to be kept somewhat (but imperfectly) homogeneous.

Soaps and detergents are good emulsifying agents for making an emulsion of oily material in water. The molecules (or ions) of these agents have water-attracting or *hydrophilic* portions and water-avoiding or *hydrophobic* sections. In the absence of oily or greasy material with which to form an emulsion, detergent molecules (or ions) in water spontaneously form colloidal sized particles called *micelles.*

Self-Test

40. What law of chemical combination do compounds obey that mixtures do not obey? ____________________

41. The three chief kinds of mixtures are named

    ____

42. Suspensions are mixtures in which one kind of intermixed particles have what dimensions? ____________________

43. Finely divided starch, a white solid, dispersed in water easily passes through a filter paper. Is this mixture a suspension?

    ____

44. A headlight set on high beam cutting into an early morning fog makes a strong glare. What effect is this?

    ____

45. Ions and molecules of ordinary size have mass and therefore are influenced by gravity. However, they do not settle under that influence when they are in a solution in a liquid or a gas. Why not?

    ____

46. What are the important differences between solutions and colloidal dispersions that are *observed* (as opposed to the explanations for the observations)?

    ____

47. What natural force is at work when a colloidal dispersion spontaneously separates?

    ____

48. Classify these substances as a foam, aerosol, smoke, emulsion, sol, or gel.

    (a) marshmallow ____________________

    (b) milk ____________________

(c) grape jelly ________________

(d) soap suds ________________

(e) cheese ________________

(f) clouds ________________

(g) cream ________________

(h) mayonnaise ________________

(i) the black discharge from a power plant stack ________________

49. What one fact about the particles in a colloidal dispersion gives rise to the Tyndall effect?

____

50. A typical synthetic detergent consists of the following ions.

$$CH_3CH_2CH_2CH_2CH_2CH_2CH_2CH_2CH_2CH_2CH_2{-}O{-}\overset{\overset{\displaystyle O}{\|}}{\underset{\underset{\displaystyle O}{\|}}{S}}{-}O^- \quad \text{and} \quad Na^+$$

Separately circle and label its hydrophilic and its hydrophobic parts. Explain in your own words how this substance can emulsify oily materials.

____

____

____

New Terms

Write the definitions of the following terms, which were introduced in this section. If necessary, refer to the Glossary at the end of the text.

- Brownian movement
- colloid
- colloidal dispersion
- emulsifying agent
- hydrophilic
- hydrophobic
- micelle
- sol
- suspension
- Tyndall effect

Answers to Thinking It Through

1 The ability to solve a problem like this develops as you remember *the basic definitions* of the concentration expressions and to *express them with all of their units displayed.* To go from 5.00% (w/w) NaBr to the molarity of this solution, we have to think of converting the quantities in the given weight percentage into the equivalent amounts of substances expressed as a molarity:

$$\frac{5.00 \text{ g NaBr}}{100 \text{ g NaBr soln}} \Rightarrow \frac{? \text{ mol NaBr}}{1 \text{ L NaBr soln}}$$

In other words, for the numerator, we have to do a grams-to-moles conversion on 5.00 g NaBr, so we'll need the formula mass of NaBr (which calculates to be 102.89). This calculates to 4.86×10^{-2} mol NaBr. We also need to change $\text{mass}_{\text{soln}}$ to $\text{volume}_{\text{soln}}$: 100 g NaBr soln $\Leftrightarrow$? L NaBr soln

The mass-to-volume tool is provided by the solution's density, which is in units of g/mL.

$$100 \text{ g NaBr soln} \times \frac{1 \text{ mL NaBr soln}}{1.04 \text{ g NaBr soln}} = 96.2 \times 10^{-3} \text{ L NaBr soln}$$

Notice that we've changed mL to L. We now have "mol NaBr" and "L NaBr solution," so we take their ratio to calculate the molarity, which is 0.505 *M*

To calculate the molality, we also begin with the basic definitions and proceed methodically to convert quantities from one set of units into the units we want. As above, to go from 5.00% (w/w) NaBr to the molality of this solution, we have to think of converting the quantities in the given weight percentage into the following units for molality:

$$\frac{5.00 \text{ g NaBr}}{100 \text{ g NaBr soln}} \Rightarrow \frac{? \text{ mole NaBr}}{1 \text{ kg water}} = \text{molality}$$

We already know the number of moles of NaBr (4.86×10^{-2} mol NaBr). So what's left is convert 100 g NaBr *solution* into kilograms of just the *solvent*, water.

$$100 \text{ g NaBr solution} \Leftrightarrow ? \text{ kg } H_2O$$

But the 100 g of the *solution* contains 5.00 g of *solute,* so the net mass of the solvent, H_2O, is 95.0 g. Because 1 g = 10^{-3} kg, we now have the mass of the solvent in kilograms: 95.0×10^{-3} kg. All we have left to do is take the ratio of the number of moles of NaBr to the number of kilograms of water.

$$\text{molality} = \frac{4.86 \times 10^{-2} \text{ mol NaBr}}{95.0 \times 10^{-3} \text{ kg water}} = 0.512\ m$$

2 Remember that the molecular mass is the ratio of grams to moles. We've

been given the number of grams: 0.347 g vapona. We have to use the freezing point depression data to calculate how many moles this corresponds to. The equation that relates Δt to the molality of the solution is

$$\Delta t = k_f m.$$

The value of Δt is (37.68 – 35.99) °C = 1.69 °C. Knowing that k_f = 37.7 °C/m, we can write:

$$1.69\ °\text{C} = 37.7\ \frac{°\text{C}}{m} \times \text{molality of solution}$$

$$\text{molality} = \frac{1.69\ °\text{C}}{37.7\ °\text{C}/m} = 4.48 \times 10^{-2}\ \text{mol vapona/kg camphor}$$

But the solution did not involve a whole kilogram of camphor, only 35.0 g or 35.0×10^{-3} kg of camphor. So the actual number of moles of vapona is found by

$$4.48 \times 10^{-2}\ \frac{\text{mol vapona}}{\text{kg camphor}} \times 35.0 \times 10^{-3}\ \text{kg camphor} = 1.57 \times 10^{-3}\ \text{mol}$$

Now we can take the ratio of grams of vapona to moles of vapona:

$$\text{molecular mass} = \frac{0.347\ \text{g vapona}}{1.57 \times 10^{-3}\ \text{mol vapona}} = 221\ \text{g/mol}$$

Thus the molecular mass of vapona is 221

Answers to Self-Test Question

1. (a) Both the tendency toward randomness and the intermolecular attractions between the molecules of these two polar liquids assist the formation of their solution.
   (b) The hydration of the K+ and Br- ions is the major factor, but the tendency toward randomness is also important.
   (c) When one gas intermixes in another, only the tendency toward randomness is operating.
2. Water molecules are attracted to each other far more strongly than they can be attracted to wax molecules, so the molecules of water remained separated in their own phase.
3. The very strong attractions between Na^+ and Cl^- ions within crystalline NaCl are sharply reduced when each kind of ion becomes surrounded and hydrated by water molecules.
4. Yes, X will dissolve in Y. There are no forces of attraction to inhibit the operation of the tendency toward randomness.
5. The strong forces of attraction between the Na^+ and NO_3^- ions in the crystalline salt can find no substitutes in new forces of attraction between these ions and the nonpolar molecules of gasoline.

6. Enthalpy of solution
7. True lattice energy is the energy released when separated solute ions or molecules attract each other and come together to form the crystalline lattice. Therefore, true lattice energy carries a minus sign indicating that it refers to an exothermic event. When lattice energy is used in discussing how a solution forms, the reverse event—the separation of the ions or molecules—is envisioned, so lattice energy must be taken as positive, indicating an endothermic event.
8. Increases. Because it must receive outside energy to separate the ions. Hydration. Decreases. Because now energy is released as forces of attraction operate.
9. –100 kJ/mol, exothermic. (It costs 400 kJ/mol to break up the crystal and 500 kJ/mol is returned, so the net is 100 kJ/mol released by the system.)
10. The endothermic steps are the separation of the molecules of the individual liquids from each other. The exothermic step is the intermingling of these separated molecules to form the solution.
11. The forces of attraction in the separated liquids are identical to those in the solution and there is no heat of solution.
12. The forces of attraction between gas particles are zero or extremely small, so when a gas goes into solution only the energy of solvation (always exothermic) is a factor. It becomes the heat of solution.
13. (a) Endothermic, (b) Come out of solution
14. $\text{Gas}_{\text{undissolved}} \rightleftharpoons \text{gas}_{\text{dissolved}} + \text{heat}$
15. The impurities must remain in solution even as the system is cooled to very low temperatures.
16. 30.6 mg/L
17. $X_{\text{ethyl alcohol}} = 0.610 = 61.0$ mol %, $X_{\text{water}} = 0.390 = 39.0$ mol %
18. 74.9 mol % N_2, 15.3 mol % O_2, 3.7 mol % CO_2, 6.2 mol % H_2O
19. $\dfrac{0.250 \text{ mol NaCl}}{1000 \text{ g } H_2O}$ or $\dfrac{1000 \text{ g } H_2O}{0.250 \text{ mol NaCl}}$ (One could replace "1000 g H_2O" by 1 kg H_2O.)
20. 400 g of H_2O
21. 4.50 g of glucose
22. Dissolve 2.50 g of sugar in water to make the final mass of the solution equal 500 g by adding water.
23. Weigh out 84.2 g of 10% (w/w) H_2SO_4 solution.
24. 19.2 mL of ethyl alcohol
25. Dissolve 25.0 mL of acetone in water and make the final volume of the solution equal 250 mL by adding water.
26. 4.23 *m*
27. $X_{\text{LiCl}} = 0.0710$ or 7.10 mol %, $X_{\text{water}} = 0.929$ or 92.9 mol %
28. 22.4% (w/w) sugar
29. $X_{\text{NaCl}} = 0.0203$ or 2.03 mol %
30. 9.98% (w/w)

31. P_{soln} = 23.4 torr
32. P_{total} = 45.1 torr
33. 137
34. High concentration
35. Dialysis (Both sugar and water pass across the membrane, not just water, so it's dialysis, not osmosis.)
36. The salt forms a very concentrated solution on the surface of the meat and this draws water by dialysis from the less concentrated solution inside the meat and the meat cells.
37. 1.51×10^6
38. −0.93 °C. −0.47 °C
39. Less. At the higher concentration, the ions would interact with each other more and so the solution would behave as if the solute were even less dissociated.
40. law of definite proportions
41. solutions, colloids, suspensions
42. larger than 1000 nm
43. no
44. Tyndall effect
45. They are buffeted too much by the solvent molecules.
46. Colloids show the Tyndall effect and ultimately settle.
47. force of gravity
48. (a) foam, (b) emulsion, (c) gel, (d) sol, (e) emulsion, (f) aerosol, (g) emulsion, (h) emulsion, (i) smoke.
49. their size
50.

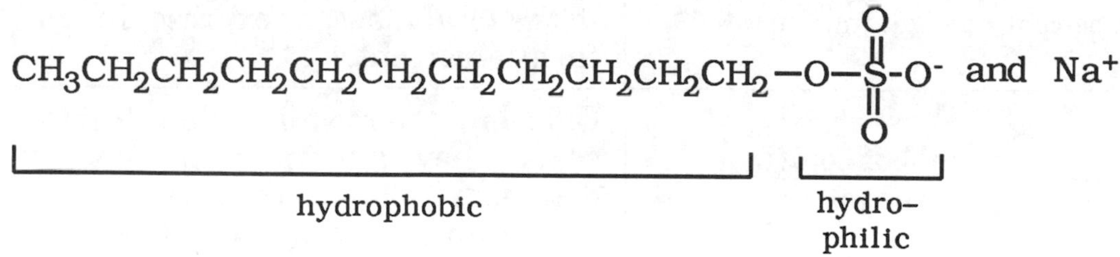

The hydrophobic tails of the anions embed themselves in the oily material, leaving the hydrophilic (ionic) heads exposed to the water. As the pincushioned oily matter breaks up into droplets, the droplets all bear the same kind of charge. Since they repel each other, the droplets are emulsified.

Tools you have learned

Remove this chart from the Study Guide and keep it handy when tackling homework problems.

Tool	*Function*
Henry's Law $\frac{C_1}{P_1} = \frac{C_2}{P_2}$	Calculate the solubility of a gas at some pressure from its solubility at a different pressure.
Mole Fraction $X_A = \frac{n_A}{n_A + n_B + n_C + ... + n_Z}$	Calculate the mole fraction of one component in a mixture so as to express the ratio of the formula units of this component to the total of all formula units.
Mole Percent mole percent = mole fraction $\times$ 100%	To express a mole fraction as a percentage.
Mole Fraction (gaseous mixture) $X_A = \frac{P_A}{P_A + P_B + P_C + ... + P_Z}$	Calculate the mole fraction of a gas in a mixture of gases from the partial pressures of the components.
Weight fraction (weight percent) weight fraction = $\frac{\text{mass of component}}{\text{mass of solution}}$ weight percent = weight fraction $\times$ 100%	Calculate the (temperature-independent) ratio of the mass of one component of a solution (or other mixture) to the total mass, or express this ratio as a percent.
Molality molality = $m = \frac{\text{mol of solute}}{\text{kg of solvent}}$	Calculate the molality of a solution so as to have a temperature-independent expression for the ratio of formula units of the solute(s) to a given mass of the solvent.
Raoult's law $P_{\text{solution}} = X_{\text{solvent}} P^{\circ}_{\text{solvent}}$	Calculate the vapor pressure of a solution of a nonvolatile solute from the mole fraction of the solvent and its vapor pressure.
Freezing point depression $\Delta t_f = K_f m$	Calculate an expected freezing point of a solution from the molality. Calculate a molecular mass from a freezing point depression.

Boiling point elevation $\Delta t_b = K_b m$	Calculate an expected boiling point of a solution from the molality. Calculate a molecular mass from a boiling point elevation.
Osmotic pressure $\Pi V = nRT$, or $\Pi = MRT$	Calculate an expected osmotic pressure from a molarity. Calculate a molecular mass from osmotic pressure.
van't Hoff factor $i = \frac{(\Delta t)\text{measured}}{(\Delta t)\text{calcd as nonelectrolyte}}$	Estimate the degree of dissociation of an electrolyte.

Summary of Important Equations

Henry's Law

$$\frac{C_1}{P_1} = \frac{C_2}{P_2}$$

Mole fraction

$$X_A = \frac{n_A}{n_A + n_B + n_C + \ldots + n_Z}$$

Mole fraction for a gaseous mixture

$$X_A = \frac{P_A}{P_A + P_B + P_C + \ldots + P_Z}$$

Mole percent

$$\text{mole percent} = \text{mole fraction} \times 100\%$$

Weight fraction or weight percent

$$\text{weight fraction} = \frac{\text{mass of component}}{\text{mass of solution}}$$

$$\text{weight percent} = \text{weight fraction} \times 100\%$$

Molality

$$\text{molality} = m = \frac{\text{mol of solute}}{\text{kg of solvent}}$$

Raoult's law

$$P_{\text{solution}} = X_{\text{solvent}} P^{\circ}_{\text{solvent}}$$

Freezing point depression

$$\Delta t_f = K_f m$$

Boiling point elevation

$$\Delta t_b = K_b m$$

Osmotic pressure

$$\Pi V = nRT, \text{ or}$$

$$\Pi = MRT$$

van't Hoff factor

$$i = \frac{(\Delta t)_{\text{measured}}}{(\Delta t)_{\text{calcd as nonelectrolyte}}}$$

Chapter 11

ACID-BASE AND IONIC REACTIONS

All *electrolytes* can produce ions in solution, and *nonelectrolytes* cannot. Some electrolytes consist in the pure state of already existing ions that simply *dissociate* from each other as the solution forms. Other electrolytes generate ions—they *ionize*—by chemical reactions with water. The *net ionic equation* is a particularly important tool in discussing reactions of ions.

Arrhenius said that acids provide *hydrogen ions (hydronium ions)* in solution and bases provide *hydroxide ions.* Later, Brønsted extended the concept of acids and bases to say that *acids are proton-donors* and *bases are proton-acceptors.* Then Lewis broadened the picture further by saying that *acids are electron-pair acceptors* and *bases are electron-pair donors. Complex ions* are seen as special examples of the products of Lewis acid-base neutralization reactions.

The common feature is that acids and bases, however you might wish to define them, have the property of being able to react with each other and destroy their acidic and basic characters through a process called *neutralization.* Such a reaction can be carried out by a procedure known as *titration.*

Acids and bases vary widely in their strengths, and we learn that an equilibrium equation with oppositely pointing arrows can be useful in discussing relative strengths. The periodic table and the concept of electronegativity help us sort out trends in these strengths, too.

Learning Objectives

In this chapter, you should keep in mind the following goals.

1 To learn what happens to ionic solutes when they dissolve in water and to learn how to write chemical equations to represent the changes that take place.

2 To learn how molecular acids and bases form ions by reacting with water and how a dynamic equilibrium is able to account for the low concentrations of ions in solutions of weak acids and bases.

3 To learn how to write chemical equations for the reactions of electrolytes in aqueous solutions.

4 To learn which factors cause a metathesis reaction and how to predict the products of metathesis reactions.

5 To learn how to work problems that deal with the stoichiometry of reactions in solution, especially those that deal with the experimental technique of titration.

6 To learn the Brønsted concept of acids and bases.

7 To use the periodic table to organize information about the relative strengths of acids and to use the concept of electronegativity to explain trends in acid strengths.

8 To learn how to define acids and bases so that they are independent both of solvents and the transfer of protons.

9 To learn the terminology of complex ions, the kinds of molecules and anions that form complex ions with metals, and how to write the formulas of complex ions.

11.1 Electrolytes

Review

Solutions of ionic compounds contain hydrated ions, which is why they conduct electricity and why they are called *electrolytes*. You should be sure that you can write equations for the *dissociation* of an ionic compound in water. To do this, it is necessary that you know the formulas and the charges of the ions. If the formulas of polyatomic ions are still difficult, review them in Table 2.8 on page 68 of the text.

In studying this section, notice how the formula of the salt determines the number of each kind of ion that is found in the solution. For example, the dissociation of chromium(III) sulfate, $Cr_2(SO_4)_3$, in water produces two Cr^{3+} ions and three SO_4^{2-} ions for each formula unit of the salt.

$$Cr_2(SO_4)_3(s) \rightarrow 2Cr^{3+}(aq) + 3SO_4^{2-}(aq)$$

Self-Test

1. What are the formulas of the ions that would be found in aqueous solutions of the following salts?

   (a) $AgC_2H_3O_2$ ____________________

   (b) $(NH_4)_2Cr_2O_7$ ____________________

(c) $Ba(OH)_2$ ______________________

2. Write chemical equations for the dissociation of the compounds in Question 1 when they are dissolved in water.

   (a) ______________________

   (b) ______________________

   (c) ______________________

New Terms

Write the definitions of the following terms, which were introduced in this section. If necessary, refer to the Glossary at the end of the text.

dissociation

electrolyte

nonelectrolyte

11.2 Acids and Bases as Electrolytes

Review

The modern version of the Arrhenius description of acids and bases defines an *acid* as a substance that yields H_3O^+ by transferring H^+ ions to water molecules. We call this an *ionization reaction* because ions are formed from neutral molecules—before the reaction there were no ions and afterwards there are. For *strong acids,* the ionization reaction is complete and the acid is 100% ionized. All of the molecules of the strong acid react with the solvent. For a *weak acid,* only a small fraction of the molecules are ionized. Most of the weak acid is present in the solution as nonionized molecules, and there is an equilibrium between the ions of the acid and its molecules.

Only a small handful of acids are strong—most are weak. The strong ones are hydrochloric, hydrobromic, hydriodic, nitric, sulfuric, perchloric, and chloric acids. Be sure to learn their names and formulas. It is very useful to know the strong acids, because if you encounter a substance that's an acid and it isn't among the small group of strong ones, then you can assume that it is weak. A few oxides of nonmetals react with water to give hydrogen ions and so are called *acidic anhydrides.*

There are two kinds of Arrhenius bases. One consists of the metal hydroxides—compounds such as NaOH and $Ca(OH)_2$. These are ionic and simply dissociate in water to give the metal ion and the hydroxide ion. Since they are completely dissociated, metal hydroxides are *strong bases.* This applies even to metal hydroxides that have a very low solubility in water. Even though very little of the solute is dissolved, all of it that does dissolve is completely dissociated.

The second kind of base consists of molecules that react with water and release OH^- ions. These include molecules such as ammonia. Molecular bases generally are *weak bases.*

Because strong acids and bases as well as all soluble salts are 100% broken up into ions, they are all strong electrolytes and their solutions are good conductors. A few oxides of metals react with water to give hydroxide ions and are therefore called *basic anhydrides.*

When describing the breakup into ions of weak acids or bases, an equilibrium equation with oppositely pointing arrows is usually used. The forward reaction (read left to right) is the ionization and the reverse reaction (read right to left) is the reconversion of the ions to the un-ionized or undissociated species. The double arrows symbolize the dynamic nature of an equilibrium—both forward and reverse reactions occur in the solution continuously and at equal rates when there is equilibrium. But there is no net change. For a weak acid or base, the extent of completion of the forward reaction in the equilibrium is small and we say that the reactants are favored, not the products.

Thinking It Through

1 A white solid readily dissolves in water to form a solution that turns red litmus paper blue. Which of the following compounds could this solid be?

$$CO_2,\ Na_2O,\ HNO_3,\ KOH,\ SO_3,\ HC_2H_3O_2$$

2 Is butyric acid a strong or a weak acid? How can you tell without even knowing its formula and without having an access to a table?

3 Consider the oxide of an element in group IIA of the periodic table. Let Z be the symbol of this element. (a) Write the formula of this compound using Z. (b) Will a solution of this oxide in water give a blue or a red color to litmus paper? (c)Is this compound very soluble in water or sparingly soluble?

Self-Test

3. Hydrogen cyanide, HCN, is a weak acid when dissolved in water. Write the chemical equation that illustrates the equilibrium that exists in the aqueous solution.

____

Which reaction, the forward or the reverse, is far from completion?

____

4. Chloric acid, $HClO_3$ is a strong acid. Write a chemical equation that shows the reaction to this acid with water.

____

5. Aniline, $C_6H_5NH_2$, is a weak base in water. Write a chemical equation that illustrates the equilibrium that exists in aqueous aniline solutions.

____

6. Why is Na_2O called a *basic anhydride*? Write the equation for its reaction with water.

____

____

7. Why is SO_3 called an *acidic anhydride*? Write the equation for its reaction with water.

____

____

8. From memory,write the names and the formulas of the six strong acids.

____

____

New Terms

Write the definitions of the following terms, which were introduced in this section. If necessary, refer to the Glossary at the end of the text.

acid-base neutralization
acidic anhydride
acids
bases
basic anhydride
dynamic equilibrium
salts
strong acid
strong base
strong electrolyte
weak acid
weak base
weak electrolyte

11.3 Ionic Reactions in Aqueous Solutions

Review

In this section you learned that there are three ways of writing chemical equations for reactions between solutes that give ions in solution—so-called *metathesis* or double *replacement reactions.* When whole formulas of the compounds are used, the equation is called a *molecular equation.* (Keep in mind, though, that we do not mean that the solutes are actually molecular. The formulas are just written as if they were.) When you write an *ionic equation,* the formulas of any soluble strong electrolytes are written in dissociated form. You learned how to divide a solute into its ions in Section 11.1. When all the spectator ions are deleted from an ionic equation, we get the *net ionic equation.* Remember that an ionic or net ionic equation must be balanced both in terms of atoms as well as charge. If both criteria are not met, the equation isn't balanced.

Thinking It Through

4 What specifically is wrong, *if anything,* about the following net ionic equations? (The *states* such as *aq, l, g,* or *s,* have been intentionally omitted.)

(a) $Zn^{+} + Ag_2O + H_2O \rightarrow Zn(OH)_2 + 2Ag^{+}$

(b) $PbO_2 + P + 4H^{+} + 2SO_4^{2-} \rightarrow 2PbSO_4 + 2H_2O$

(c) $6H^{+} + 11ClO_3^{-} \rightarrow 5ClO_4^{-} + 2ClO_2 + 2Cl_2 + 3O_2 + 3H_2O$

Self-Test

9. Balance the following molecular equations and then write their ionic and net ionic equations:

(a) __ $Cu(NO_3)_2(aq)$ + __ $KOH(aq)$ → __ $Cu(OH)_2(s)$ + __ $KNO_3(aq)$

Ionic equation:

____

Net ionic equation:

____

(b) __ $NiCl_2(aq)$ + __ $AgNO_3(aq)$ → __ $Ni(NO_3)_2(aq)$ + __ $AgCl(s)$

Ionic equation:

____

Net ionic equation:

____

(c) __ $Cr_2(SO_4)_3(aq)$ + __ $BaCl_2(aq)$ → __ $CrCl_3(aq)$ + __ $BaSO_4(s)$

Ionic equation:

____

Net ionic equation:

____

New Terms

Write the definitions of the following terms, which were introduced in this section. If necessary, refer to the Glossary at the end of the text.

double replacement reaction
ionic equation
ionic reaction
metathesis reaction
molecular equation
net ionic equation
spectator ions

11.4 Predicting When Metathesis Reactions Will Occur

Review

In this section we learn how to predict whether a *metathesis reaction* will occur between a pair of reactants. The first step is to write an equation for what the reaction could be. Then we examine this equation to determine whether, in fact, the reaction does occur.

To write the equation, it is necessary to determine what the products are. This is done by exchanging the anions between the two cations. However, we must be very careful to write the correct formulas of the products—that is, we must be sure that the formulas represent electrically neutral formula units. For example, in the reaction between $Cr_2(SO_4)_3$ and $BaCl_2$, we have to be sure to check the changes of the anions and cations before we write the products. The anions are SO_4^{2-} and Cl^-; the cations are Cr^{3+} and Ba^{2+}. When we exchange the anions, the Cl^- goes with the Cr^{3+} to give $CrCl_3$ and the SO_4^{2-} goes with the Ba^{2+} to give $BaSO_4$. Therefore, the equation before balancing is

$$Cr_2(SO_4)_3 + BaCl_2 \rightarrow CrCl_3 + BaSO_4$$

The next step is to write the ionic equation for the reaction. To do this we must know whether the reactants and products are soluble in water, and if they are soluble, whether they are strong or weak electrolytes. We must also know if one of the products is a gas. For you to accomplish this step, therefore, you must learn the solubility rules on page 471 and the contents of Table 11.1 on page 478. You must also know the formulas of the strong acids, because then you will know that if an acid is a product of a reaction, and if it is not among the list of strong acids, it must be a weak acid, and therefore a weak electrolyte. If necessary, review the names and formulas of the strong acids.

A metathesis reaction occurs if there is anything left after deleting spectator ions from the ionic equation. A net ionic equation will remain when a precipitate forms in a solution of soluble reactants, when a weak electrolyte forms from ionic reactants, and when a gas is produced. A particularly important class of reactions involves the reaction of a metal oxide or a metal hydroxide with either a weak or strong acid. With each, one of the products is water and another is the metal salt of the acid.

The last part of this section discusses writing molecular equations from net ionic equations. This is something you would do, for example, if you wished to carry out a reaction in the lab. The net ionic equation tells you what the net reaction should be, but to carry out the reaction, you must choose the proper chemicals so that all the ions except those actually involved in the net ionic equation are spectator ions. Often, there is more than one set of reactants that will satisfy these requirements, so there is generally no one correct "answer." Working backwards from the net ionic equation is not always easy, and requires a very thorough knowledge of the solubility rules and Table 11.1.

Thinking It Through

5 What *general* chemical facts must be recalled in order to predict if the given reactants actual undergo a reaction, and what is the net ionic equation (including the physical states) for the reactions that are predicted on the basis of these facts? Assume that *solutions* of the reactants are used where the reactants are quite soluble in water.

(a) $HC_2H_3O_2 + KOH \rightarrow$ ____________________

(b) $HI + KOH \rightarrow$ ____________________

(c) $KHCO_3 + HCl \rightarrow$ ____________________

(d) $AgNO_3 + LiI \rightarrow$ ____________________

(e) $NaHSO_3 + HCl \rightarrow$ ____________________

(f) $Mg(OH)_2 + HBr \rightarrow$ ____________________

(g) $Na_2S + H_2SO_4 \rightarrow$ ____________________

Self-Test

10. Test your knowledge of the solubility rules by *circling* the formulas of the following salts that are soluble in water.

    Na_2SO_4 $CuCO_3$ $Ni(NO_3)_2$ Hg_2Cl_2 PbI_2 Cr_2O_3 $Ca_3(PO_4)_2$

    $(NH_4)_2CO_3$ $Ba(ClO_4)_2$ AgI ZnS $MgSO_3$ $Sr(C_2H_3O_2)_2$ $(NH_4)_2S$

11. Write molecular, ionic, and net ionic equations for the reactions, if any, that occur between the following compounds. If there is no reaction, state so on the line provided for a net ionic equation.

    (a) $AgC_2H_3O_2 + (NH_4)_2S \rightarrow$

    Molecular equation:

    ____

    Ionic equation:

    ____

    Net ionic equation:

    ____

    (b) $Cr(ClO_4)_3 + MgSO_4 \rightarrow$

    Molecular equation:

    ____

    Ionic equation:

    ____

    Net ionic equation:

    ____

    (c) $Fe(NO_3)_3 + KOH \rightarrow$

    Molecular equation:

    ____

    Ionic equation:

    ____

    Net ionic equation:

    ____

(d) $Cr_2O_3 + HClO_4 \rightarrow$

Molecular equation:

____

Ionic equation:

____

Net ionic equation

____

(e) $BaSO_3 + HClO_4 \rightarrow$

Molecular equation:

____

Ionic equation:

____

Net ionic equation:

____

(f) $(NH_4)_2SO_4 + Ba(OH)_2 \rightarrow$

Molecular equation:

____

Ionic equation:

____

Net ionic equation:

____

12. What nitrate salt and ammonium salt would be used to carry out the following net ionic reaction?

$$Ba^{2+}(aq) + CrO_4^{2-}(aq) \rightarrow BaCrO_4(s)$$

____

New Terms

none

11.5 Stoichiometry of Ionic Reactions: Acid-Base Titrations

Review

The concentration unit used for dealing with the stoichiometry of reactions in solution is molarity. You should be familiar with how it is used as a conversion factor—that is, it should be second nature to you to be able to translate the molarity on the label of a bottle into its equivalent form as a conversion factor. Thus the label on a solution of 0.10 *M* NaOH gives the conversion factors. (Be sure that the units are *complete.*)

$$\frac{0.10 \text{ mol NaOH}}{1 \text{ L NaOH soln}} \qquad \frac{1 \text{ L NaOH soln}}{0.10 \text{ mol NaOH}}$$

Since the volumes of the solutions are usually expressed in milliliters, equivalent forms of these expressions are

$$\frac{0.10 \text{ mol NaOH}}{1000 \text{ mL NaOH soln}} \qquad \frac{1000 \text{ mL NaOH soln}}{0.10 \text{ mol NaOH}}$$

Be sure to review Example 11.6 and work Practice Exercises 9 and 10 to be certain that you can use molarity correctly.

Example 11.7 illustrates how we can keep track of each of the ions in an ionic reaction. Notice that we calculate the number of moles of each ion in the reaction mixture before and after the reaction, taking into account any reaction that takes place *as well as a change in total volume.*

Example 11.8 illustrates an important principle often used in chemical analyses. We begin with a mixture with an unknown composition—the *relative amounts* of the various components are unknown. A reaction is then carried out that transfers one component of the mixture into a different compound, but one having a known composition. By measuring the quantity of this compound, the amount of the component in the original mixture can be calculated. By means of procedures of this sort, repeated for as many components as necessary, the composition of the entire mixture is determined.

Titrations. Measuring the volumes of solutions needed to give complete reaction is easily accomplished if an appropriate indicator can be found that is able to signal the end of the reaction. This is the technique known as titration. Be sure to learn the boldface terms on page 485.

Thinking It Through

6 You are hired to take charge of an analytical laboratory and one task is to analyze samples of a commercial product used to remove iron stains.

It consists of $NaHSO_4$, an acid salt that reacts with sodium hydroxide as follows:

$$NaHSO_4(aq) + NaOH(aq) \rightarrow Na_2SO_4(aq) + H_2O$$

The intent of the analysis is to ensure that the bottle labels show the actual percentage of $NaHSO_4$. The legal standard is pure $NaHSO_4$. You propose to analyze the samples by titrating 0.300 g samples of the product (dissolved in water) with 0.100 *M* NaOH. How many milliliters of the NaOH solution will be needed if the sample is pure $NaHSO_4$?

Self-Test

13. A solution is labeled 0.50 *M* HNO_3. Translate this into two conversion factors that can be used for calculations.

    ____

14. How many moles of each ion are in 50.0 mL of 0.250 *M* $(NH_4)_2SO_4$?

    ____

15. A student mixed 30.0 mL of 0.25 *M* NaI solution with 45.0 mL of 0.10 *M* $Pb(NO_3)_2$ solution.

    (a) Write the molecular equation for the reaction that occurred.

    ____

    (b) How many moles of each of the ions were in the mixture before any reaction took place?

    Na^+ ____________ I^- ____________

    Pb^{2+} ____________ NO_3^- ____________

    (c) How many moles of which ions reacted?

    ____

    (d) How many moles of what compound were formed?

    ____

    (e) How many moles of each of the ions were present in the solution after the reaction was over?

    Na^+ ____________ I^- ____________

    Pb^{2+} ____________ NO_3^- ____________

(f) What were the concentrations of each of the ions in the solution after the reaction was over?

Na^+ ______________ I^- ______________

Pb^{2+} ______________ NO_3^- ______________

16. A 0.250-g sample of a mixture known to contain some $CaCO_3$ was treated with some dilute HCl solution and warmed to expel all the CO_2 that formed. The CO_2 was collected and found to occupy a volume of 35.0 mL at a temperature of 25 °C and a pressure of 756 torr.

   (a) Write a molecular equation for the reaction that occurs when HCl is allowed to react with $CaCO_3$.

   ____

   (b) Use the ideal gas law to calculate the number of moles of CO_2 that were collected. ______________

   (c) If it is assumed that no CO_2 was lost during the experiment, how many moles of $CaCO_3$ were in the original sample of the mixture?

   ____

   (d) What was the percentage by weight of $CaCO_3$ in the mixture?

   ____

17. A sample of a weak acid weighing 0.256 g was dissolved in water and titrated with 0.200 *M* NaOH solution. The titration required 17.80 mL of the base.

   (a) How many moles of H^+ were neutralized in the titration?

   ____

   (b) If the acid is monoprotic, what is its molecular mass?

   ____

New Terms

Write the definitions of the following terms, which were introduced in this section. If necessary, refer to the Glossary at the end of the text.

acid-base indicator
buret
end point
standard solution
stopcock
titrant
titration

11.6 Brønsted Concept of Acids and Bases

Review

The need to have water as a solvent for acid-base reactions can be eliminated by defining an acid as a proton (H^+) donor and a base as a proton acceptor. Acids and bases defined this way are called Brønsted acids and bases, after the scientist who invented the definition. A Brønsted acid-base reaction simply involves the transfer of a hydrogen ion (which is the same as a proton) from the acid to the base.

Brønsted acid-base reactions can always be considered to be reversible, so in both the forward and reverse directions there is an acid and a base. The acid on the left becomes the base on the right; the base on the left becomes the acid on the right. This is illustrated for the reaction of HCN with H_2O on page 488.

Two substances that are related to each other by just the gain or loss of a single proton are called an acid-base conjugate pair. In any Brønsted acid-base reaction, there are two conjugate pairs. Be sure to study Examples 11.11 and 11.12 so that you are able to identify the conjugate pairs in a reaction.

Two examples of acid-base conjugate pairs are H_3O^+ and H_2O, and H_2O and OH^-. Hydronium ion is the conjugate acid of water, and water is the conjugate base of the hydronium ion. Similarly, water is the conjugate acid of hydroxide ion, and hydroxide ion is the conjugate base of water. Water is the base in one pair and the acid in the other. Such substances, which can be either an acid or a base depending on the circumstances, are said to be amphoteric or amphiprotic.

It is important to remember the inverse relationship between the strengths of an acid and its conjugate base. The stronger is the acid, the weaker is its conjugate base. This means, for example, that very strong acids, such as HCl or $HClO_4$, have very weak conjugate bases.

Thinking It Through

7 The reactants NO(*g*) and O_2(*g*), although they do react, cannot give a Brønsted acid-base reaction. Why not?

8 Consider the *possibility* that H_2O and S^{2-} might give a Brønsted acid-base reaction.

(a) *If they do give a Brønsted acid-base reaction,* what would be the likeliest products? Identify each of the Brønsted acids and each of the Brønsted bases.

(b) What chemical fact would you have to know about the individual Brønsted acids in order to predict if the reaction would actually occur?

Self-Test

22. Write the formula of the conjugate acid of each of the following:

    (a) CH_3NH_2 ____________

    (b) NH_2OH ____________

    (c) ClO_3^- ____________

    (d) $HC_2O_4^-$ ____________

23. Write the formula of the conjugate base of each of the following:

    (a) $HC_2O_4^-$ ____________

    (b) $H_2AsO_4^-$ ____________

    (c) $HCHO_2$ ____________

    (d) H_2S ____________

24. Identify the conjugate acid-base pairs in the following reaction. In each pair, identify the acid and the base.

    $H_2SO_4 + Cl^- \rightarrow HSO_4^- + HCl$ ____________

    ____

25. Identify the acid-base conjugate pairs in the following reaction. In each pair, identify the acid and the base.

    $NH_3 + NH_3 \rightarrow NH_4^+ + NH_2^-$ ____________

26. The ion NH_2^- is a stronger Brønsted base than OH^-. Which is the stronger acid, NH_3 or H_2O? ____________

27. Acetic acid, $HC_2H_3O_2$, is a base in concentrated sulfuric acid and it is an acid in water. Write chemical equations that illustrate these reactions.

    ____

    ____

    Why is acetic acid said to be *amphiprotic* in these reaction?

    ____

New Terms

Write the definitions of the following terms, which were introduced in this section. If necessary, refer to the Glossary at the end of the text.

- Brønsted acid
- Brønsted base
- conjugate acid
- conjugate base
- conjugate acid-base pair
- amphiprotic
- amphoteric

11.7 Trends in the Strengths of Brønsted Acids

Review

Oxoacids have one or more OH groups, and sometimes extra O atoms (lone oxygens), bound by covalent bonds to a central atom, usually a nonmetal. The OH groups are polar because O is more electronegative than H, and when the OH group is joined to another electronegative atom, like Cl, S, or N, the group is made even more polar. Now the $\delta+$ on the H atom of the OH group is sufficiently positive to make it relatively easy for H to be transferred as H^+ to an acceptor, like H_2O. In other words, the substance is an acid. All of the correlations in this Section between acidity and structure or location in the periodic table ultimately come down to the magnitude of $\delta+$ on H. Whatever makes this $\delta+$ greater, makes the acid stronger.

Acids whose central atoms are in the same group and hold identical numbers of oxygen atoms increase in strength from bottom to top in the group. This is the same trend in electronegativities of central atoms in the same group. (Example: H_2SO_4 is a stronger acid than H_2SeO_4. S stands above Se in its group and is more electronegative than Se.)

Acids whose central atoms are in the same period and hold identical numbers of oxygen atoms increase in strength from left to right in the period. This is the same trend in electronegativities of central atoms in the same period. (Example: H_2SO_4 is a stronger acid than H_3PO_4. S stands to the right of P in the same period.)

Acids with the same central atom increase in strength with increasing numbers of O atoms bound to this atom. By increasing the number of O atoms, each with electron-withdrawing ability, the O-H bond is made more polar and the $\delta+$ on H becomes greater. (Example: H_2SO_4 is a stronger acid than H_2SO_3.)

The presence of lone oxygens, those without attached H atoms, on the central atom of an oxoacid makes the acid stronger not only because these extra oxygens (by their electronegativity) increase the polarizations of adjacent O-H bonds. They also help to distribute the negative charge in the anion left when the acid ionizes.

In binary acids, H is joined directly to some central atom. These acids become stronger as the location of the central atom in the periodic table moves from left to right in a period or from top to bottom in a group. The left to right trend reflects a left-to-right trending increase in electronegativity. The top to bottom trend seems to defy the opposite trend in electronegativity, but another factor overpowers it, a top-to-bottom weakening of the bond holding H.

Thinking It Through

9 An oxoacid of the general formula H_xZO_y has been discovered and found to be a stronger acid than H_2SeO_4.

(a) If $y = 4$, which location in the periodic table is more likely for Z, to the *left* of Se or to the *right* of Z in the same row?

(b) If $y = 4$ and $x = 2$, where does Z most likely lie in the periodic table, *above* or *below* Z in the same group?

Self-Test

24. Which is the stronger acid, H_3PO_3 or H_3PO_4? __________

    Explain. __________

25. Which is the stronger acid, H_3AsO_4 or H_3PO_4? __________

    Explain. __________

26. Which is the stronger acid, H_2Te or H_2Se? __________

    Explain. __________

27. Which is the stronger acid, H_2Te or HI? __________

    Explain. __________

New Terms

Write the definitions of the following terms, which were introduced in this section. If necessary, refer to the Glossary at the end of the text.

oxoacid

binary acid

11.8 Lewis Acids and Bases

Review

According to the *Lewis definition,* an *acid* is a substance that accepts the share of a pair of electrons during the formation of a coordinate covalent bond. A *base* is a substance that donates a pair of electrons in the formation of a coordinate covalent bond. The principal benefit of this definition is that acids and bases are made free of the need of a solvent, or even the need of protons. Study the reactions that are shown using Lewis structures on pages 498-499. Notice that the attachment of a proton to a base, which is a necessary step in the reaction of Brønsted acids and bases, is just a special case of a much more general phenomenon.

Compounds that contain *complex ions* are called *coordination compounds* because coordinate covalent bonds generally bind the pieces that make up the complex ion. One piece is an acceptor atom—a central metal ion—and the other is a *ligand*—a Lewis base. Many ligands are anions, like halide ions, S^{2-}, CN^-, OH^-, SCN^-, $S_2O_3^{2-}$, and NO_2^-. Other ligands are electrically neutral, like NH_3 and H_2O. All these are *monodentate ligands* because they "bite" or are joined to the donor atom by one coordinate covalent bond. *Bidentate ligands*, like $NH_2CH_2CH_2NH_2$ or the oxalate ion are held by two coordinate covalent bonds to the donor. EDTA is a *polydentate ligand.*

In the formula of a complex, the *acceptor atom* is always written first followed by the ligands. When several ligands are held by the same acceptor atom, as in $[Cr(H_2O)_5Cl]^{2+}$, the net charge—the algebraic sum of the charges on the acceptor and the ligands—is placed outside the squared brackets.

Coordination compounds are found throughout nature, and many are important in analytical chemistry.

Thinking It Through

10 The zinc ion forms a complex with four OH^- ions.

(a) What is the formula of this complex ion.

(b) Could the complex ion be isolated as a potassium salt or a chloride salt? Write the formula.

Self-Test

28. The aluminum chloride molecule has the Lewis structure

```
        ..
       :Cl:
    ..  |  ..
   :Cl–Al–Cl:
    ..     ..
```

Explain why $AlCl_3$ is a strong Lewis acid.

____

29. The oxide ion, O^{2-}, is able to function well as a Lewis base, but not as a Lewis acid. Explain why.

____

30. The formula of boric acid is usually written H_3BO_3, although a better representation of its structure is $B(OH)_3$. Boric acid is not a Brønsted acid—that is, it is not a proton donor. It is, however, a Lewis acid. Use the Lewis structure of boric acid to explain this behavior.

```
       OH
       |
  HO—B—OH   (boric acid)
```

____

____

31. Fe^{3+} and SCN^- form a complex ion with a net charge of 3–. Write the formula of this complex ion.

____

32. Ni^{2+} and NH_3 form a complex ion with a net charge of 2+. The ratio of donor atom to ligand is 1 to 6. Write the formula.

____

New Terms

Write the definitions of the following terms, which were introduced in this section. If necessary, refer to the Glossary at the end of the text.

acceptor ion

Lewis acid

bidentate ligand
complex ion (complex)
coordination compound
donor atom
Lewis base
ligand
monodentate ligand
polydentate ligand

Answers to Thinking It Through

1 Go through the list and identify the *kinds* of substances and recall their general acid-base properties. Carbon dioxide is a gas and a *nonmetal oxide* that dissolves in water to make a weakly acidic solution. Sodium oxide is the oxide of a group IA element *all of which oxides react with water to give solutions of the hydroxides*. HNO_3 is nitric acid, a strong acid. KOH is the hydroxide of a group IA metal *all of which are strongly basic*. SO_3 is a *nonmetal oxide (and a gas) which gives an acid in water* (H_2SO_4). $HC_2H_3O_2$ is acetic acid, a weak acid. Thus only the metal oxide and the metal hydroxide are both solids and give a basic solution in water.

2 We can tell that it is a *weak acid* because it is not on the list of strong acids.

3 (a) ZO, because group IIA metals have charges of 2+, which balances the 2– charge on the O.
(b) Blue; oxides of group IA and IIA metals all do this.
(c) Sparingly soluble, like all of the group IIA oxides.

4 (a) There is no electrical balance. Moreover, the zinc ion is Zn^{2+}, not Zn^+.
(b) There is no material balance. (The P on the left should be Pb.)
(c) This is balanced both electrically and materially.

5 (a) Facts needed: What is $HC_2H_3O_2$? (It's acetic acid, a *weak* water-soluble acid.) What is KOH? (A strong, water-soluble base.) The two react but the weak acid is shown un-ionized.

$$HC_2H_3O_2(aq) + OH^-(aq) \rightarrow C_2H_3O_2^-(aq) + H_2O$$

(b) Facts needed: HI is a strong, water-soluble acid; KOH is a strong water-soluble base; water is an un-ionized compound.

$$H^+(aq) + OH^-(aq) \rightarrow H_2O$$

(c) $KHCO_3$ is a bicarbonate of a group IA metal and so is water-soluble. HCl is a strong, water-soluble acid. Bicarbonates react with strong acids to give CO_2 and H_2O.

$$HCO_3^-(aq) + H^+(aq) \rightarrow CO_2(g) + H_2O$$

(d) $AgNO_3$ is a nitrate and so is water-soluble; LiI is the iodide of a group IA metal and so is water-soluble. Ag^+ and I^- together form a water-insoluble salt, AgI.

$$Ag^+(aq) + I^-(aq) \rightarrow AgI(s)$$

(e) $NaHSO_3$ is a sodium salt and so is water-soluble; moreover, it is a hydrogen sulfite salt and so will react with strong acid (HCl) to give SO_2 and H_2O.

$$HSO_3^-(aq) + H^+(aq) \rightarrow SO_2(g) + H_2O$$

(f) $Mg(OH)_2$ is a water-insoluble hydroxide of a group IIA metal and so will neutralize strong acid (HBr) to give a water-soluble salt, $MgBr_2$, and H_2O.

$$Mg(OH)_2(s) + 2H^+(aq) \rightarrow Mg^{2+}(aq) + 2H_2O$$

(g) Na_2S is a sodium salt and so is water-soluble. It is also a sulfide salt and so will react with strong acid (H_2SO_4) to give H_2S and a different (but water-soluble) sodium salt, Na_2SO_4.

$$S^{2-}(aq) + 2H^+(aq) \rightarrow H_2S(g)$$

6 The important fact is that the stoichiometric ratio is 1:1, meaning

$$1 \text{ mol } NaHSO_4 \Leftrightarrow 1 \text{ mol } NaOH$$

So we must convert 0.300 g $NaHSO_4$ into moles using the grams to moles tool (a conversion factor made from the formula mass of $NaHSO_4$, 120.07).

$$0.300 \text{ g } NaHSO_4 \times \frac{1 \text{ mol } NaHSO_4}{120.07 \text{ g } NaHSO_4} = 2.50 \times 10^{-3} \text{mol } NaHSO_4$$

Therefore, we must take as many milliliters of 0.100 *M* NaOH to provide 2.50×10^{-3}mol NaOH (because 1 mol $NaHSO_4 \Leftrightarrow$ 1 mol NaOH).

$$\text{volume of NaOH solution} \times \frac{0.100 \text{ mol NaOH}}{1000 \text{ mL NaOH solution}} =$$

$$2.50 \times 10^{-3} \text{mol NaOH}$$

Thus, the volume of the NaOH solution = 25.0 mL.

7 In the Brønsted theory, an acid-base reaction is the transfer of a proton, and neither reactant is able to donate H^+.

8 (a) The likeliest products are OH^- and HS^-, the results of a proton transfer from H_2O to S^{2-}.

$$H_2O + S^{2-}(aq) \rightarrow OH^-(aq) + HS^-(aq)$$

In this particular equation, the two Brønsted acids are H_2O and HS^-. The two Brønsted bases are S^{2-} and OH^-.

(b) You would have to know which of the two acids is stronger, H_2O or HS^-. (Actually H_2O is the stronger acid, so the reaction occurs.)

9 (a) To the right. For oxoacids with the same number of oxygens, acid strength increases from left to right in the same row.

(b) Above. For oxoacids of elements in the same group and so having the same general formula, acid strength increases from bottom to top within the group.

10 (a) The zinc ion is Zn^{2+}, so its complex with 4 OH^- ions must be $Zn(OH)_4^{2-}$.

(b) Because the complex ion is negatively charged, it has to form a salt with a cation, K^+. The salt is $K_2[Zn(OH)_4]$.

Answers to Self-Test Questions

1. Ag^+, $C_2H_3O_2^-$ (b) NH_4^+, $Cr_2O_7^{2-}$ (c) Ba^{2+}, OH^-
2. (a) $AgC_2H_3O_2(s) \rightarrow Ag^+(aq) + C_2H_3O_2^-(aq)$
   (b) $(NH_4)_2Cr_2O_7(s) \rightarrow 2NH_4^+(aq) + Cr_2O_7^{2-}(aq)$
   (c) $Ba(OH)_2(s) \rightarrow Ba^{2+}(aq) + 2OH^-(aq)$
3. $HCN(aq) + H_2O \rightleftharpoons H_3O^+(aq) + CN^-(aq)$; forward reaction
4. $HClO_3(aq) + H_2O \rightarrow H_3O^+(aq) + ClO_3^-(aq)$
5. $C_6H_5NH_2(aq) + H_2O \rightleftharpoons C_6H_5NH_3^+(aq) + OH^-(aq)$
6. Na_2O reacts with water to give sodium hydroxide.
   $Na_2O(s) + H_2O \rightarrow 2NaOH(aq)$
7. SO_3 reacts with water to give sulfuric acid
   $SO_3(g) + H_2O \rightarrow H_2SO_4(aq)$
8. Hydrochloric acid (HCl), hydrobromic acid (HBr), hydroiodic acid (HI), nitric acid (HNO_3), perchloric acid ($HClO_4$), and sulfuric acid (H_2SO_4)
9. (a) 1, 2, 1, 2
   $$Cu^{2+}(aq) + 2NO_3^-(aq) + 2K^+(aq) + 2OH^-(aq) \rightarrow Cu(OH)_2(s) + 2K^+(aq) + 2NO_3^-(aq)$$
   $$Cu^{2+}(aq) + 2OH^-(aq) \rightarrow Cu(OH)_2(s)$$
   (b) 1, 2, 1, 2
   $$Ni^{2+}(aq) + 2Cl^-(aq) + 2Ag^+(aq) + 2NO_3^-(aq) \rightarrow Ni^{2+}(aq) + 2NO_3^-(aq) + 2AgCl(s)$$
   $$Cl^-(aq) + Ag^+(aq) \rightarrow AgCl(s)$$
   (c) 1, 3, 2, 3
   $$2Cr^{3+}(aq) + 3SO_4^{2-}(aq) + 3Ba^{2+}(aq) + 6Cl^-(aq) \rightarrow 2Cr^{3+}(aq) + 6Cl^-(aq) + 3BaSO_4(s)$$
   $$SO_4^{2-}(aq) + Ba^{2+}(aq) \rightarrow BaSO_4(s)$$
10. Soluble: Na_2SO_4, $Ni(NO_3)_2$, $(NH_4)_2CO_3$, $Ba(ClO_4)_2$, $Sr(C_2H_3O_2)_2$, $(NH_4)_2S$
11. (a) $2AgC_2H_3O_2 + (NH_4)_2S \rightarrow Ag_2S + 2NH_4C_2H_3O_2$
   $$2Ag^+(aq) + 2C_2H_3O_2^-(aq) + 2NH_4^+(aq) + S^{2-}(aq) \rightarrow Ag_2S(s) + 2NH_4^+(aq) + 2C_2H_3O_2^-(aq)$$
   $$2Ag^+(aq) + S^{2-}(aq) \rightarrow Ag_2S(s)$$
   (b) $2Cr(ClO_4)_3 + 3MgSO_4 \rightarrow Cr_2(SO_4)_3 + 3Mg(ClO_4)_2$
   $$2Cr^{3+}(aq) + 6ClO_4^-(aq) + 3Mg^{2+}(aq) + 3SO_4^{2-}(aq) \rightarrow 2Cr^{3+}(aq) + 3SO_4^{2-}(aq) + 3Mg^{2+}(aq) + 6ClO_4^-(aq)$$
   No net reaction
   (c) $Fe(NO_3)_3 + 3KOH \rightarrow Fe(OH)_3 + 3KNO_3$
   $$Fe^{3+}(aq) + 3NO_3^-(aq) + 3K^+(aq) + 3OH^-(aq) \rightarrow Fe(OH)_3(s) + 3K^+(aq) + 3NO_3^-(aq)$$
   $$Fe^{3+}(aq) + 3OH^-(aq) \rightarrow Fe(OH)_3(s)$$
   (d) $Cr_2O_3 + 6HClO_4 \rightarrow 2Cr(ClO_4)_3 + 3H_2O$

$Cr_2O_3(s) + 6H^+(aq) + 6ClO_4^-(aq) \rightarrow$
$2Cr^{3+}(aq) + 6ClO_4^-(aq) + 3H_2O$
$Cr_2O_3(s) + 6H^+(aq) \rightarrow 2Cr^{3+}(aq) + 3H_2O$

(e) $BaSO_3 + 2HClO_4 \rightarrow Ba(ClO_4)_2 + SO_2 + H_2O$
$BaSO_3(s) + 2H^+(aq) + 2ClO_4^-(aq) \rightarrow$
$Ba^{2+}(aq) + 2ClO_4^-(aq) + SO_2(g) + H_2O$
$BaSO_3(s) + 2H^+(aq) \rightarrow Ba^{2+}(aq) + SO_2(g) + H_2O$

(f) $(NH_4)_2SO_4 + Ba(OH)_2 \rightarrow 2NH_3 + 2H_2O + BaSO_4$
$2NH_4^+(aq) + SO_4^{2-}(aq) + Ba^{2+}(aq) + 2OH^-(aq) \rightarrow$
$2NH_3(g) + 2H_2O + BaSO_4(s)$
Net ionic equation is the same as the ionic equation.

12. $Ba(NO_3)_2$ and $(NH_4)_2CrO_4$
13. $\frac{0.50 \text{ mol } HNO_3}{1000 \text{ mL } HNO_3 \text{ soln}}$ and $\frac{1000 \text{ mL } HNO_3 \text{ soln}}{0.50 \text{ mol } HNO_3}$. (Note. 1000 mL can be replaced by 1L.)
14. 0.0250 mol NH_4^+, 0.0125 mol SO_4^{2-}
15. (a) $Pb(NO_3)_2(aq) + 2NaI(aq) \rightarrow PbI_2(s) + 2NaNO_3(aq)$
(b) 0.0075 mol Na^+, 0.0075 mol I^-, 0.0045 mol Pb^{2+}, 0.0090 mol NO_3^-
(c) 0.0075 mol I^- and 0.0038 mol Pb^{2+}
(d) 0.0038 mol PbI_2
(e) 0.0075 mol Na^+, 0.0 mol I^-, 0.0007 mol Pb^{2+}, 0.0090 mol NO_3^-
(f) 0.10 *M* Na^+, 0.0 *M* I^-, 0.009 *M* Pb^{2+}, 0.12 *M* NO_3^-
16. (a) $CaCO_3(s) + 2HCl(aq) \rightarrow CaCl_2(aq) + CO_2(g) + H_2O$
(b) 1.42×10^{-3} mol CO_2, (c) 1.42×10^{-3} mol $CaCO_3$, (d) 56.8%
17. (a) 3.56×10^{-3} mol H^+, (b) 71.9 g/mol
18. (a) $CH_3NH_3^+$, (b) NH_3OH^+, (c) $HClO_3$, (d) $H_2C_2O_4$
19. (a) $C_2O_4^{2-}$, (b) $HAsO_4^{2-}$, (c) CHO_2^-, (d) HS^-
20. $\mathbf{H_2SO_4}$, HSO_4^-; Cl^-, **HCl** (acid in bold type)
21. NH_3 (base), NH_4^+(acid); NH_3(acid), NH_2^-(base)
22. H_2O
23. $HC_2H_3O_2 + H_2SO_4 \rightarrow H_2C_2H_3O_2^+ + HSO_4^-$
$HC_2H_3O_2 + H_2O \rightarrow H_3O^+ + C_2H_3O_2^-$
Acetic acid can be either an acid or a base.
24. H_3PO_4 It has one more O.
25. H_3PO_4 P is more electronegative than As, standing above As in the table.
26. H_2Te The H–Te bond is weaker, because Te is below Se in the same group in the table.
27. HI I is more electronegative than Te, standing to the right of Te in the same period in the table.

28. The Al in $AlCl_3$ has less than an octet, so it can easily accept another pair of electrons.
29. The oxygen in O^{2-} has a completed valence shell and cannot accept additional electrons.
30. The boron in $B(OH)_3$ has less than an octet and can accept a pair of electrons from a Lewis base.
31. $[Fe(SCN)_6]^{3-}$
32. $[Ni(NH_3)_6]^{2+}$

Tools you have learned

Remove this chart from the Study Guide and keep it handy when tackling homework problems.

Tool	*Function*
List of strong acids	To enable the recognition of any other acids as weak. To decide which formula should represent the acid in a net ionic equation
Criteria for balanced ionic equations	To balance equations
Solubility rules	To tell if a given salt is soluble in water and dissociates fully. To enable the prediction of chemical reactions.
Types of substances that react to give gases either with acids or bases	To predict when to expect CO_2, SO_2, HCN, H_2S, or NH_3 to form from a reaction.
Brønsted definitions	To recognize possible proton donors or proton acceptors To tell from a given Brønsted acid-base reaction which member of a conjugate pair is stronger.

Use of the periodic table to predict acid strengths When the central atoms of oxoacids are in the same *group* and they hold the same number of oxygen atoms, the acid strength increases from bottom to top within the group. When the central atoms of oxoacids are in the same *period* and they hold the same number of oxygen atoms, the acid strength increases from left to right within the period. For a given central atom, the acid strength of an oxoacid increases with the number of oxygens it holds. The strengths of binary acids increase from left to right in a period and from top to bottom in a group.	To enable a better use of the periodic table to "store" knowledge about acid strengths among the oxoacids or the binary acids

Chapter 12

OXIDATION–REDUCTION REACTIONS

In Chapter 11 we discussed various kinds of chemical reactions, including metathesis and acid/base reactions. These reactions have certain features that make discussing them together convenient. Now we turn our attention to another class of chemical reactions. These are often characterized by the transfer of electrons from one species to another, and they include many of our most important chemical changes, as you will learn in this chapter.

Learning Objectives

As you study of this chapter, keep in mind the following goals:

1 To learn the criteria that we use in classifying reactions as oxidation-reduction reactions.

2. To learn a bookkeeping method called oxidation numbers which we use to follow the course of redox reactions.

3 To learn how to balance redox reactions in aqueous solution by dividing the reaction into parts, balancing the parts separately, and then recombining them to give the balanced equation.

4 To learn how acids are able to attack metals, and to learn what the products are in such reactions.

5 To learn how to predict reactions in which one metal displaces another from its compounds.

6 To learn how the ease with which a metal loses electrons varies according to its location in the periodic table.

7 To learn how the tendency that nonmetals have to gain electrons varies with the position of the nonmetal in the periodic table.

8 To learn about the reactions that molecular oxygen undergoes with organic compounds, with metals, and with nonmetals.

12.1 Oxidation-Reduction Reactions

Review

Oxidation-reduction reactions (often called *redox* reactions) are very common, so it is important that you learn how to identify them and know the terminology used in discussing them. These reactions can be thought of as involving the transfer of electrons from one substance to another. (Sometimes an electron transfer actually takes place, as when Na and Cl_2 react to form ions, but for many reactions it is just a convenience to think of them as involving electron transfer.)

Remember these definitions:

Oxidation can be viewed as a loss of electrons.

Reduction can be viewed as a gain of electrons.

Oxidation and reduction always occur simultaneously during a chemical reaction. If one substance loses electrons and is oxidized, then another substance must gain electrons and be reduced.

The terms *oxidizing agent* and *reducing agent* often cause confusion. The reason for assigning these terms is given on page 516, but the simplest way to remember how to apply the terms properly is to find which substance is oxidized and which is reduced. Then, switch words—if a certain substance is oxidized, then it's the reducing *agent*; if the substance is reduced, then it's the oxidizing *agent*.

When you write separate equations showing electron gain or loss, as in Example 12.1, keep in mind that electrons are negatively charged particles. That way, you will be sure to write the electrons on the correct side. For instance, in Example 12.1 we find that calcium atoms become calcium ions.

$$Ca \rightarrow Ca^{2+}$$

The calcium is becoming more positive, so it must be losing electrons. They are written on the product side of the equation to show that they have been separated from the calcium atom, which has become a calcium ion.

$$Ca \rightarrow Ca^{2+} + 2e^-$$

Self-Test

1. Consider the reaction of magnesium with fluorine to give the ionic compound MgF_2.

$$Mg + F_2 \rightarrow MgF_2$$

(a) Write separate equations showing the gain and loss of electrons.

____

____

(b) Which substance is oxidized? ______________

(c) Which substance is reduced? ______________

(d) Which reactant is the oxidizing agent? ______________

(e) Which reactant is the reducing agent? ______________

New Terms

Write the definitions of the following terms, which were introduced in this section. If necessary, refer to the Glossary at the end of the text.

oxidation	oxidation-reduction reaction
reduction	oxidizing agent
redox reaction	reducing agent

12.2 Oxidation Numbers

Review

Oxidation numbers are a bookkeeping device. We assign them following the rules given on page 518, which you should study carefully. The best way to learn how to apply the rules is to practice assigning oxidation numbers, so study Example 12.2 through 12.6 on pages 518 to 520, work the Practice Exercises, and then work the questions in the Self-Test here in the Study Guide. As you apply these rules, remember that if you encounter a conflict between two rules, the one with the lower number is the one that applies and we ignore the rule with the higher number. This is illustrated in Example 12.4.

For binary compounds of a metal and a nonmetal (such as NaCl, $CaCl_2$, or Al_2O_3), the ions are simple monatomic ions and their oxidation numbers are equal to their charges. The formulas of the ions formed by the representative elements were given in Table 2.6 on page 65. If necessary, review them and learn how to use the periodic table to obtain their correct charges. (Read the discussion on page 65.)

If you recognize a polyatomic ion in a formula, for example, SO_4^{2-} in $Cr_2(SO_4)_3$, you can use its charge as the *net* oxidation number of the ion. For the "SO_4" in this example, we can assign it a net oxidation number of −2, which means that the chromium must have an oxidation number of +3 (applying Rule 3, the summation rule: $[2 \times (+3)] + [3 \times (-2)] = 0$).

Redox is redefined on page 521 in terms of changes in oxidation number.

Oxidation: an increase in oxidation number (oxidation state)
Reduction: a decrease in oxidation number (oxidation state)

Study Example 12.7 to review how we use oxidation numbers to identify oxidation and reduction.

Self-Test

2. Assign oxidation numbers to each atom in the following:
   (a) NO_3^- ________________
   (b) $SbCl_5$ ________________
   (c) $CaHAsO_4$ ________________
   (d) ClF_3 ________________
   (e) I_3^- ________________
   (f) S_8 ________________
3. Determine whether the following changes are oxidation, reduction, or neither oxidation nor reduction.
   (a) SO_3^{2-} to SO_4^{2-} ________________
   (b) Cl_2 to ClO_3^- ________________
   (c) N_2O_4 to NH_3 ________________
   (d) PbO to $PbCl_4^{2-}$ ________________
   (e) Ag to Ag_2S ________________
4. Consider the balanced equation,

$$3H_2O + 3Cl_2 + NaI \rightarrow 6HCl + NaIO_3$$

   (a) Which substance is oxidized? ________________
   (b) Which substance is reduced? ________________
   (c) Which substance is the oxidizing agent? ________________
   (d) Which substance is the reducing agent? ________________

New Terms

Write the definitions of the following terms, which were introduced in this section. If necessary, refer to the Glossary at the end of the text.

oxidation number oxidation state

12.3 Balancing Equations for Redox Reactions: The Ion-Electron Method

Review

In applying the ion-electron method, we divide an equation for a redox reaction into two half-reactions, balance the half-reactions separately, and then combine the balanced half-reactions to give the balanced overall net ionic equation. The method is not difficult to apply if you proceed in a stepwise fashion, without skipping any of the steps. If you can't obtain a balanced equation, it is probably because you haven't remembered to do things in sequence. Be sure you learn the seven steps given on page 525 for reactions that occur in acidic solutions.

In applying the ion-electron method, one of the most frequent causes for error is not remembering to write the correct charges on the formulas. You need these charges to get the correct numbers of electrons, so if you forget to write the charges on the ions, you will surely get wrong answers.

Another common problem students have is computing the net charge on each side of a half-reaction. Suppose we have reached the following stage in balancing a half-reaction

$$6H_2O + N_2H_5^+ \rightarrow 2NO_3^- + 17H^+$$

To obtain the charge contributed by each substance, multiply its coefficient by its charge. Thus, on the right side of this half-reaction the net charge is

$$2 \times (1-) + 17 \times (1+) = 15+$$

The 2 is the coefficient of the NO_3^- and the 1– is the charge on the NO_3^-. Similarly, 17 is the coefficient of H^+ and 1+ is the charge on H^+.

In determining the number of electrons that must be added to balance a half-reaction, be especially careful when charges on opposite sides of the half-reaction have *opposite* algebraic signs. For example, consider the following half-reaction just before we add electrons to it.

$$7H^+ + SO_3^{2-} \rightarrow HS^- + 3H_2O$$

The net charge on the left is 5+, the net charge on the right is 1–. The number of electrons we must add equals the algebraic difference between them,

$$\text{number of electrons to be added} = (5+) - (1-) = 6$$

The electrons are added to the more positive side, so the half-reaction properly balanced is

$$6e^- + 7H^+ + SO_3^{2-} \rightarrow HS^- + 3H_2O$$

Notice that the net charge is the same on both sides.

Reactions in Basic Solutions

To balance an equation for basic solution, first balance it as if the solution were acidic. Then follow the three-step conversion to basic solution described on page 526.

Self-Test

5. Balance the following half-reactions for acid solution.
   (a) $NO \rightarrow NO_3^-$ ____________________
   (b) $Br_2 \rightarrow BrO_3^{2-}$ ____________________
   (c) $P_4 \rightarrow HPO_3^{2-}$ ____________________
6. Balance the following half-reactions for basic solution.
   (a) $Cl_2 \rightarrow OCl^-$ ____________________
   (b) $AsO_4^{3-} \rightarrow AsH_3$ ____________________
   (c) $S_2O_4^{2-} \rightarrow SO_4^{2-}$ ____________________
7. Balance the following reaction that occurs in acidic solution.

$$H_2SeO_3 + I^- \rightarrow I_2 + Se$$

____

8. Balance this reaction for basic solution.

$$SeO_3^{2-} + I^- \rightarrow I_2 + Se$$

____

New Terms

Write the definitions of the following terms, which were introduced in this section. If necessary, refer to the Glossary at the end of the text.

half-reaction
ion-electron method

12.4 Reactions of Metals with Acids

Review

Every acid releases hydrogen ions, H^+ (which, of course, actually exist in solution as H_3O^+). Hydrogen ion is a mild oxidizing agent and can oxidize many metals. The products are the metal ion and H_2 gas. For example, tin reacts with H^+ to give H_2 and Sn^{2+}.

$$Sn(s) + 2H^+(aq) \rightarrow Sn^{2+}(aq) + H_2(g)$$

Many metals dissolve in acids such as HCl. Some examples are iron, zinc, tin, aluminum, and magnesium. These metals are said to be *more active* than H^+. (In the next section, you will learn more about how you can tell whether a given metal will dissolve in acids.)

There are some metals that are not attacked by H^+, and they will not dissolve in acids that have H^+ as the strongest oxidizing agent. Such of acids are called *nonoxidizing acids*. An example is HCl, which gives H^+ and Cl^- in solution. The Cl^- ion cannot gain any more electrons, so it cannot act as an oxidizing agent. Solutions of HCl therefore have H^+ as the only oxidizing agent.

Metals that won't dissolve in nonoxidizing acids often will dissolve in *oxidizing acids* such as HNO_3. Copper is the example that's given in the text. Notice that when the nitrate ion serves as the oxidizing agent, hydrogen gas is *not* among the products. Instead, the reduction product comes from the NO_3^- ion. Also, you should remember that when concentrated nitric acid is used as an oxidizing agent, the reduction product is usually NO_2, and when dilute nitric acid is used, the reduction product is usually NO.

Self-Test

9. Hydrobromic acid, HBr, is a nonoxidizing acid. Why must this be true?

   ____

10. Cadmium reacts with dilute solutions of sulfuric acid with the evolution of hydrogen. Write a chemical equation for the reaction.

    ____

11. In the reaction described in Question 10, which is the oxidizing agent and which is the reducing agent?

    Oxidizing agent ____________________

    Reducing agent ____________________

12. Mercury dissolves in concentrated nitric acid, with the evolution of a reddish-brown gas, just as in the reaction shown in Figure 12.2 for copper. The oxidation state of the mercury after reaction is +2. Write a balanced net ionic equation for the reaction

    ____

13. Construct the balanced molecular equation for the reaction described in

Question 12.

New Terms

Write the definitions of the following terms, which were introduced in this section. If necessary, refer to the Glossary at the end of the text.

nonoxidizing acid

oxidizing acid

12.5 Displacement of One Metal by Another from Compounds

Review

If one metal is more active than another, then the more active metal will be able to reduce the ion of the less active metal. In the reaction, the more active metal is oxidized. This is what happens when metallic zinc is placed into a solution that contains a soluble copper salt. The zinc (which is the more active metal) is oxidized to zinc ion and the copper ion is reduced to metallic copper.

The activity series on page 532 lists metals in order of increasing activity (that is, in order of increasing ability to be oxidized and to serve as a reducing agent). Remember that any metal in this table is able to displace the ion of any metal above it from compounds. For instance, if you turn to Table 12.1 (page 532), you will see that aluminum can reduce the ions of manganese, zinc, chromium, iron, and ions of all the other metals above it in the table.

Table 12.1 can also be used to determine if a given metal will dissolve in a nonoxidizing acid. If the metal is *below* hydrogen in the table, a nonoxidizing acid will react with it to give the metal ion and hydrogen gas. If a metal is above hydrogen in this table, an oxidizing acid must be used to dissolve it.

Self-Test

14. Write molecular equations for any reaction that will occur when the following reactants are combined:

    (a) Manganese metal added to a solution of $CoCl_2$.

    ____

    (b) Iron metal added to a solution of $AuCl_3$.

    ____

(c) Cadmium metal added to a solution of $MgCl_2$.

____

15. Which metal ions will be reduced if an excess amount of iron powder is added to a solution that contains a mixture of $Ca(NO_3)_2$, $Cu(NO_3)_2$, $AgNO_3$, KNO_3, $Pb(NO_3)_2$, and $Hg(NO_3)_2$.

____

16. Which of the following metals will dissolve in a solution of HBr: aluminum, tin, cobalt, gold, mercury, manganese?

____

New Terms

Write the definition of the following term, which was introduced in this section. If necessary, refer to the Glossary at the end of the text.

activity series

12.6 Periodic Trends in the Reactivity of Metals

Review

The ease with which a metal is oxidized is roughly inversely proportional to the metal's ionization energy. If the ionization energy is large, then it is difficult to remove the metal's electrons and the metal is difficult to oxidize. Study Figure 12.6 and note the following:

1 The most easily oxidized metals are those in Group IA; the next most easily oxidized are in Group IIA.

2 The alkali metals, and the alkaline earth metals from calcium through barium react with water to release hydrogen and form the metal hydroxide.

3 The least reactive metals — those most difficult to oxidize — are located in the center of the periodic table in Period 6 (the third row of transition metals).

Self-Test

17. Based on their locations in the periodic table, choose the metal that is the more easily oxidized in each of the following pairs.

(a) Zr or Os ________ (d) Sr or Mo ________

(b) Hg or Sn ________ (e) Sc or W ________

(c) Fe or Os ________ (f) Ir or Hf ________

18. Write a molecular equation and a net ionic equation for the reaction of barium with water.

____

____

New Terms

none

12.7 Periodic Trends in the Reactivity of Nonmetals

Review

The ability of nonmetals to serve as oxidizing agents parallels their electronegativities. The strongest oxidizing agent is fluorine; the second strongest is oxygen. Oxygen can displace other nonmetals from their compounds as illustrated in the reaction of O_2 with CuS described on page 535. Study the reactions of the halogens on page 535.

Self-Test

19. Write a balanced equation for the reaction in which oxygen displaces sulfur from the compound H_2S. (Assume the sulfur is oxidized to SO_2.)

____

New Terms

none

12.8 Molecular Oxygen as an Oxidizing Agent

Review

Molecular oxygen, O_2, is a very good oxidizing agent. In this section you learn about several kinds of reactions that are brought about by O_2. The purpose is to enable you to make reasonable predictions of the outcome of combustion reactions.

1 **Reactions of O_2 with organic compounds.** Organic compounds normally contain carbon, hydrogen, oxygen, and perhaps several other elements. When these compounds are burned in a plentiful supply of O_2, the carbon forms CO_2, the hydrogen forms H_2O, and if any sulfur is present in the compound, it is changed to SO_2.

If the combustion takes place in a limited supply of O_2, hydrogen still is changed to H_2O, but the carbon is only oxidized to CO. In extremely limited supplies of O_2, elemental carbon (soot) is formed.

2 **Reactions of O_2 with metals.** Many metals combine with oxygen directly. Aluminum, iron, and magnesium are given as examples in the text. The products are metal oxides.

3 **Reactions of O_2 with nonmetals.** These reactions give nonmetal oxides. An important nonmetal that does not react with O_2 is nitrogen.

Self-Test

20. Complete and balance the following equations:

(a) $C_2H_6 + \quad O_2 \rightarrow$ ____________________

(b) $C_6H_{12}O_6 + \quad O_2 \rightarrow$ ____________________

(c) $C_3H_6O + \quad O_2 \rightarrow$ ____________________

(d) $C_2H_5SH + \quad O_2 \rightarrow$ ____________________

(e) $Al + \quad O_2 \rightarrow$ ____________________

(f) $S + \quad O_2 \rightarrow$ ____________________

(g) $Ca + \quad O_2 \rightarrow$ ____________________

21. When burned in an excess supply of oxygen, phosphorus forms the compound P_4O_{10}. Write a chemical equation for the reaction.

____

22. Write a balanced chemical equation for the reaction of benzene, C_6H_6, with a severely limited amount of O_2.

____

New Terms

Write the definitions of the following terms, which were introduced in this section. If necessary, refer to the Glossary at the end of the text.

organic compound hydrocarbon combustion

12.9 Redox Reactions in the Laboratory

Review

Problems that deal with the stoichiometry of redox reactions are handled in the same way as other stoichiometry problems. We begin with a balanced chemical equation because the coefficients give us the ratios by moles in which the various substances react. Study Examples 12.11 and 12.12.

In redox titrations, a convenient oxidizing agent is MnO_4^- because it serves as its own indicator in acidic solutions. Study Example 12.13 in detail, concentrating especially on the Analysis sections.

Thinking It Through

For the following, identify the information needed to solve the problem and show (or explain) what must be done with it.

1 A sample of tin ore with a mass of 0.225 g was dissolved in acid and all the tin converted to Sn^{2+}. The solution required 24.33 mL of 0.0200 *M* $KMnO_4$ to reach an end point in a titration in which the Sn^{2+} was converted to Sn^{4+}. What was the percentage of SnO_2 in the original ore sample?

Self-Test

23. (a) Write a balanced net ionic equation for the reaction of sulfurous acid with permanganate ion in an acidic solution. The products of the reaction are sulfate ion and manganese(II) ion.

____

(b) How many milliliters of 0.150 *M* $KMnO_4$ are required to react with a solution that contains 0.440 g of H_2SO_3?

____

24. The reaction of MnO_4^- with Sn^{2+} in acidic solution follows the equation:

$$2MnO_4^- + 5Sn^{2+} + 16H^+ \rightarrow 2Mn^{2+} + 5Sn^{4+} + 8H_2O$$

(a) How many grams of $KMnO_4$ must be used to react with 35.0 g of $SnCl_2$?

____

(b) How many milliliters of 0.0500 *M* $KMnO_4$ solution would be needed to react with all the tin in 25.0 mL of 0.250 *M* $SnCl_2$ solution?

____

(c) A 0.500 g sample of solder, which is an alloy containing lead and tin, was dissolved in acid and all the tin was converted to Sn^{2+}. The solution was then titrated with 0.0200 *M* $KMnO_4$ solution. The titration required 27.73 mL of the $KMnO_4$ solution. What is the percentage by weight of tin in the solder?

____

25. Sodium chlorite, $NaClO_2$, is used as a bleaching agent. A 2.00 g sample of a bleach containing this compound was dissolved in an acidic solution and treated with excess NaI. The net ionic equation for the reaction that took place is

$$4H^+ + ClO_2^- + 6I^- \rightarrow Cl^- + 2I_3^- + 2H_2O$$

The solution containing the I_3^- was then titrated with 28.80 mL of 0.200 *M* $Na_2S_2O_3$ solution. The reaction that took place during the titration was

$$I_3^- + 2S_2O_3^{2-} \rightarrow 3I^- + S_4O_6^{2-}$$

What is the percentage by mass of $NaClO_2$ in the bleach?

____

New Terms

none

Answers to Thinking It Through

1 First, we need to write a balanced chemical equation for the reaction of Sn^{2+} with MnO_4^- to give Sn^{4+} and Mn^{2+}. The product of molarity and volume for the permanganate solution gives us the moles of MnO_4^- used. The coefficients of the equation are then used to calculate the number of moles of Sn^{2+} that reacted. Since 1 mol of Sn comes from 1 mol of SnO_2, the number of moles of Sn^{2+} obtained is equal to the number of moles of SnO_2 that were in the ore sample. Multiply this value by the formula mass of SnO_2 to obtain the mass of SnO_2 in the ore sample. The mass of SnO_2 multiplied by

100% and then divided by 0.225 g gives the percentage of SnO_2 in the ore sample. [The answer is 81.5 % SnO_2]

Answers to Self-Test Questions

1. (a) $Mg \rightarrow Mg^{2+} + 2e^-$; $F_2 + 2e^- \rightarrow 2F^-$ (b) Mg , (c) F_2, (d) F_2, (e) Mg
2. (a) N, +5; O, –2 (b) Sb, +5; Cl, –1 (c) Ca, +2; H, +1; As, +5; O, –2
   (d) Cl, +3; F, –1 (e) I, $-\frac{1}{3}$ (f) S, zero
3. (a) oxidation, (b) oxidation, (c) reduction, (d) neither, (e) oxidation
4. (a) NaI, (b) Cl_2, (c) Cl_2, (d) NaI
5. (a) $2H_2O + NO \rightarrow NO_3^- + 4H^+ + 3e^-$
   (b) $6H_2O + Br_2 \rightarrow 2BrO_3^- + 12H^+ + 10e^-$
   (c) $12H_2O + P_4 \rightarrow 4H_3PO_3 + 12H^+ + 12e^-$
6. (a) $4OH^- + Cl_2 \rightarrow 2OCl^- + 2H_2O + 2e^-$
   (b) $8e^- + 7H_2O + AsO_4^{3-} \rightarrow AsH_3 + 11OH^-$
   (c) $8OH^- + S_2O_4^{2-} \rightarrow 2SO_4^{2-} + 4H_2O + 6e^-$
7. $4H^+ + 4I^- + H_2SeO_3 \rightarrow Se + 2I_2 + 3H_2O$
8. $3H_2O + 4I^- + SeO_3^{2-} \rightarrow Se + 2I_2 + 6OH^-$
9. Br^-cannot accept another electron to become Br^{2-}, so Br^- cannot be an oxidizing agent.
10. $Cd + H_2SO_4 \rightarrow CdSO_4 + H_2$
11. Cd is the reducing agent, H^+ is the oxidizing agent.
12. $Hg + 2NO_3^- + 4H^+ \rightarrow Hg^{2+} + 2NO_2 + 2H_2O$
13. $Hg + 4HNO_3 \rightarrow Hg(NO_3)_2 + 2NO_2 + 2H_2O$
14. (a) $Mn + CoCl_2 \rightarrow MnCl_2 + Co$, (b) $2AuCl_3 + 3Fe \rightarrow 3FeCl_2 + 2Au$
    (c) $Cd + MgCl_2 \rightarrow$ no reaction
15. Cu^{2+}, Ag^+, Pb^{2+}, Hg^{2+}
16. Al, Sn, Co, Mn
17. (a) Zr, (b) Sn, (c) Fe, (d) Sr, (e) Sc, (f) Hf
18. $Ba + 2H_2O \rightarrow Ba(OH)_2 + H_2$; $Ba(s) + 2H_2O \rightarrow Ba^{2+} + 2OH^- + H_2(g)$
19. $2H_2S + 3O_2 \rightarrow 2H_2O + 2SO_2$
20. (a) $2C_2H_6 + 15O_2 \rightarrow 12CO_2 + 6H_2O$ (b) $C_6H_{12}O_6 + 6O_2 \rightarrow 6CO_2 + 6H_2O$
    (c) $C_3H_6O + 4O_2 \rightarrow 3CO_2 + 3H_2O$
    (d) $2C_2H_5SH + 9O_2 \rightarrow 4CO_2 + 6H_2O + 2SO_2$
    (e) $2Al + 3O_2 \rightarrow Al_2O_3$ (f) $S + O_2 \rightarrow SO_2$
    (g) $2Ca + O_2 \rightarrow 2CaO$
21. $4P + 5O_2 \rightarrow P_4O_{10}$
22. $2C_6H_6 + 3O_2 \rightarrow 12C + 6H_2O$
23. (a) $5H_2SO_3 + 2MnO_4^- \rightarrow 5SO_4^{2-} + 2Mn^{2+} + 4H^+ + 3H_2O$ (b) 14.3 mL
24. (a) 23.3 g $KMnO_4$, (b) 50.0 mL, (c) 32.9% Sn
25. 6.51% $NaClO_2$

Tools you have learned

Remove this chart from the Study Guide and keep it handy when tackling homework problems.

Tool	*Function*
Rules for assigning oxidation numbers	To assign oxidation numbers to the atoms in a chemical formula.
Definition of redox according to changes in oxidation numbers	To identify oxidation and reduction and the oxidizing and reducing agent in a reaction.
Ion-electron method Follow the steps outlined on p 525 for acidic solutions, and on p 526 for basic solutions.	Enables you to balance net ionic equations for redox reactions.
Activity series	Enables you to determine whether one metal will displace another from its compounds in solution.
Periodic trends in reactivities of metals	Enables you to use the periodic table to roughly identify the locations of the metals that are easily oxidized and those that are difficult to oxidize.
Periodic trends in reactivity of nonmetals	Enables you to use the periodic table to compare the oxidizing abilities of the nonmetallic elements.
Products of reactions with oxygen These reactions are discussed on pages 536 to 539.	Enables you to anticipate the products for reactions in which oxygen is one of the reactants.

Summary of Useful Information

Rules for Assigning Oxidation Numbers

1. The oxidation number of any free element is zero, regardless of how complex its molecules might be.
2. The oxidation number of any simple, monatomic ion is equal to the charge on the ion.
3. The sum of all the oxidation numbers of the atoms in a molecule or ion must equal the charge on the particle.
4. In its compounds, fluorine has an oxidation number of −1.
5. In its compounds, hydrogen has an oxidation number of +1.
6. In its compounds, oxygen has an oxidation number of −2.

Ion-Electron Method — Acidic Solution

Step 1. Divide the equation into two half-reactions.

Step 2. Balance atoms other than H and O.

Step 3. Balance O by adding H_2O.

Step 4. Balance H by adding H^+.

Step 5. Balance net charge by adding e^-.

Step 6. Make e^- gain equal e^- loss; then add half-reactions.

Step 7. Cancel anything that's the same on both sides.

Additional Steps for Basic Solution

Step 8. Add to *both* sides of the equation the same number of OH^- as there are H^+.

Step 9 Combine H^+ and OH^- to form H_2O

Step 10. Cancel any H_2O that you can.

Chapter 13

THERMODYNAMICS

In this chapter we turn our attention to what it is that causes events of any kind, whether they be physical or chemical, to occur spontaneously—that is, by themselves without outside assistance. Spontaneous events are crucial for our existence, because without them nothing would ever happen and we wouldn't exist. Understanding the factors that contribute to spontaneity is important because such events provide the driving force for all natural changes.

Learning Objectives

As you study of this chapter, keep in mind the following objectives:

1. To learn what thermodynamics means and to discover the kinds of questions it seeks to answer.
2. To learn how thermodynamics deals with the exchange of energy between a system and its surroundings.
3. To learn why the heat and work that a system exchanges with its surroundings are not state functions, even though the net energy change is a state function.
4. To learn why the heat of reaction at constant volume is not necessarily equal to the heat of reaction at constant pressure.
5. To learn what a spontaneous change is and how everything that happens can be traced to some spontaneous change somewhere.
6. To learn what influence energy changes have on the tendency for an event to occur spontaneously.
7. To learn the meaning of the term entropy, S, and to see how an entropy increase favors a spontaneous change.
8. To learn how entropy changes can be calculated from absolute entropies.
9. To learn how the Gibbs free energy, G, interrelates the energy and entropy factors in determining the spontaneity of a chemical or physical change.
10. To learn how to calculate standard free energy changes from ΔH° and ΔS° and from standard free energies of formation.

11 To learn what relationship exists between the free energy change and the work that is available from a chemical reaction.

12 To see how free energy is related to equilibrium.

13 To learn how ΔG° can be used to predict the outcome of a chemical reaction.

13.1 Introduction

Review

Thermodynamics is the study of energy changes, with a focus on the way energy flows between system and surroundings. Studying thermodynamics allows us to understand why spontaneous events occur and gives us insight into answers to many questions that plague modern society.

New Terms

Write the definition of the following term, which was introduced in this section. If necessary, refer to the Glossary at the end of the text.

thermodynamics

13.2 Energy Changes in Chemical Reactions—A Second Look

Review

The internal energy, E, is the sum of all the system's kinetic and potential energies. The change in the internal energy is given by the *first law of thermodynamics*: $\Delta E = q + w$, where q is the heat added to the system and w is the work done on the system.

It is important to keep the algebraic signs of q and w straight. Both are positive when the energy is being added to the system (i.e., when heat is added or the system has work done on it). This is summarized on page 555.

The values of q and w depend on how a change is carried out, as explained in the text, so q and w are not state functions. ΔE is a state function, however.

A system can do work by expanding against an opposing pressure, and the amount of work can be calculated as $w = -P\Delta V$, where ΔV is the change in volume. (Here w is expressed as a negative quantity because the system loses energy when it does work.) If a system cannot change volume during a reaction, then the heat of reaction is equal to ΔE for the change.

$\Delta E = q_V$ (where q_V is the heat of reaction at constant volume)

The enthalpy, H, is defined to deal with changes that occur at constant pressure. If the only kind of work a system can do is expansion work against the opposing atmospheric pressure, then ΔH is equal to the heat of reaction, and it is called the heat of reaction at constant pressure.

$\Delta H = q_P$ (where q_P is the heat of reaction at constant pressure)

For most reactions, the difference between ΔE and ΔH is very small and can be neglected. To convert between ΔE and ΔH, we can use the equation

$$\Delta H = \Delta E + \Delta n\, RT$$

where Δn is the change in the number of moles of *gas* on going from the reactants to products. (If the energy is to be in joules or kilojoules, remember to use $R = 8.314$ J mol^{-1} K^{-1}.)

Self-Test

1. Give the definition of enthalpy in terms of the internal energy, the pressure, and the volume.

   ____

2. The product of pressure in pascals (Pa, which has the units newtons per square meter, N/m^2) times volume in cubic meters (m^3) gives work in units of joules. In Chapter 8, the standard atmosphere was defined in terms of the pascal as 1 atm = 101,325 Pa. With this information, calculate the amount of work, in joules, done by a gas when it expands from a volume of 500 mL to 1500 mL against a constant opposing pressure of 5.00 atm.

   (Remember that 1 mL = 1 cm^3.) ________________________

3. Suppose that during an exothermic reaction at constant volume, a gas is consumed, so that the pressure decreases. How would ΔE for this reaction compare to its ΔH? Why?

   ____

   ____

4. How would the values of ΔE and ΔH compare for the reaction:

   $$CO_2(g) + H_2(g) \rightarrow CO(g) + H_2O(g)$$

   Explain your answer. ____________________________

   ____

5. At 25 °C and a pressure of 1 atm, the reaction

$$2NO_2(g) \rightarrow N_2O_4(g)$$

has $\Delta H = -57.9$ kJ. What is the value of ΔE for this reaction at the same temperature?

____

New Terms

Write the definitions of the following terms, which were introduced in this section. If necessary, refer to the Glossary at the end of the text.

first law of thermodynamics

internal energy

pressure-volume work

13.3 Spontaneous Change

Review

A spontaneous change is one that takes place all by itself without continual assistance. Once conditions are right, it proceeds on its own. A nonspontaneous change needs continual help to make it happen, and some spontaneous change must occur first to drive the nonspontaneous one. Everything we see happen is ultimately caused by a spontaneous event of some kind.

Self-Test

6. Which of the following do you recognize as spontaneous events?

   (a) Ice melts on a hot day. ____________

   (b) Water boils in a home in Alaska. ____________

   (c) A scratched fender on an automobile rusts. ____________

   (d) Dirty clothes become clean. ____________

New Terms

Write the definition of the following term, which was introduced in this section. If necessary, refer to the Glossary at the end of the text.

spontaneous change

13.4 Enthalpy Changes and Spontaneity

Review

When a change is accompanied by a decrease in the potential energy of the system (i.e., when the change is exothermic), it tends to occur spontaneously. Some examples are shown in the photographs on page 561. A potential energy decrease is a factor in *favor* of spontaneity. It is not the sole factor, however, because there are changes that are endothermic and nevertheless spontaneous.

In chemical reactions at constant pressure and temperature, the energy change is given by ΔH. For a pressure of 1 atm and a temperature of 25 °C, it is ΔH°, the standard enthalpy change. When ΔH° is negative, the process occurs by a decrease in its potential energy and it tends to be spontaneous.

Self-Test

7. Which of the following chemical changes *tend* to be spontaneous at 25 °C and 1 atm, based on their enthalpy changes. (Refer to Table 4.2 on page 158 of the text.)

   (a) $2PbO(s) \rightarrow 2Pb(s) + O_2(g)$

   (b) $NO(g) + SO_3(g) \rightarrow NO_2(g) + SO_2(g)$

   (c) $Fe_2O_3(s) + 3CO(g) \rightarrow 3CO_2(g) + 2Fe(s)$

   Answer: ____________

8. Give an example based on observations you have made in your daily life of a change that is endothermic and spontaneous?

   ____

New Terms

none

13.5 Entropy and Spontaneous Change

Review

In this section we see that statistical probability and randomness are important factors to consider in analyzing spontaneous events. In general, an increase in randomness or disorder favors a spontaneous change. Systems tend to proceed spontaneously from conditions of low statistical probability (an ordered state) to conditions of high probability (a disordered state).

The thermodynamic quantity related to probability and randomness is the entropy, S. An increase in entropy corresponds to an increase in randomness and therefore favors spontaneity. Another way of saying this is that a change tends to occur spontaneously if $\Delta S > 0$ (ΔS is positive).

The entropy change can often be predicted for a change. In general, the entropy of a system increases when

- the temperature of the system increases
- there is an increase in the volume of the system
- there is a change from solid → liquid, liquid → gas, or solid → gas

For a chemical reaction, ΔS will be positive when

- there is an increases in the number of moles of gas in the system
- there is a decrease in the complexity of the molecules (which is accompanied by an increase in the number of particles in the system.)

The *second law of thermodynamics* tells us that any spontaneous event is accompanied by an overall increase in the entropy of the universe. Earlier we saw that everything that happens is the net result of a spontaneous change of some sort, so every time something happens in the world the entropy, or disorder of the universe, increases. If we do anything to decrease disorder somewhere, it is met by an even greater increase in disorder somewhere else. That is why human activities pollute and why energy is always becoming less and less available for use.

Thinking It Through

1 The reaction $N_2O_4(g) \rightarrow N_2(g) + 2O_2(g)$ is exothermic. Is this reaction expected to be spontaneous? (Base your answer on the enthalpy and entropy changes involved.)

Self-Test

9. If two containers holding different gases are connected together, the gases will gradually diffuse until the composition of the mixture is the same in both containers. Explain this in probability terms.

____

____

____

____

10. What is the sign of ΔS for each of the following changes?
    (a) Solid iodine vaporizes. ____________
    (b) Waste oil is spilled into the ground. ____________
    (c) Ocean water is desalinated for irrigation. ____________
    (d) A brick wall is built. ____________
    (e) Crude oil is separated into fuel oil, gasoline, and its other components. ____________

11. What is the sign of ΔS for each of the following reactions?
    (a) $2HgO(s) \rightarrow 2Hg(g) + O_2(g)$ ____________
    (b) $CaO(s) + CO_2(g) \rightarrow CaCO_3(s)$ ____________
    (c) $H_2(g) + I_2(s) \rightarrow 2HI(g)$ ____________
    (d) $2C_2H_6(g) + 7O_2(g) \rightarrow 4CO_2(g) + 6H_2O(g)$ ____________

New Terms

Write the definitions of the following terms, which were introduced in this section. If necessary, refer to the Glossary at the end of the text.

entropy
second law of thermodynamics

13.6 The Third Law of Thermodynamics

Review

At 0 K the entropy of any pure crystalline substance is zero. This is a statement of the *third law of thermodynamics*. Because the zero point on the entropy scale is known, the absolute amount of entropy that a substance possesses can be determined. (This can be compared with energy or enthalpy, where it is impossible to figure out exactly how much energy a substance has because it is impossible to know when a substance has zero energy.)

Standard entropies—entropies at 25 °C and 1 atm—can be used to calculate standard entropy changes using Equation 13.6 on page 569.

Self-Test

12. Calculate the standard entropy change, in J/K, for the following reactions.

    (a) $2NaCl(s) \rightarrow 2Na(s) + Cl_2(g)$ __________

    (b) $CO(g) + 2H_2(g) \rightarrow CH_3OH(l)$ __________

    (c) $CH_4(g) + 2O_2(g) \rightarrow CO_2(g) + 2H_2O(g)$ __________

    (d) $2KCl(s) + H_2SO_4(l) \rightarrow K_2SO_4(s) + 2HCl(g)$ __________

13. If entropy changes were the *only* factors involved, which of the reactions in the preceding question would occur spontaneously at 25 °C and a pressure of 1 atm?

    ____

New Terms

Write the definitions of the following terms, which were introduced in this section. If necessary, refer to the Glossary at the end of the text.

third law of thermodynamics
standard entropy
standard entropy change
standard entropy of formation

13.7 The Gibbs Free Energy

Review

The Gibbs free energy is a composite function of the enthalpy and entropy. It allows us to gauge quantitatively how important these individual factors are in determining the spontaneity of an event.

At constant temperature and pressure,

$$\Delta G = \Delta H - T\Delta S$$

A change will be spontaneous when the enthalpy term (ΔH) and the entropy term ($T\Delta S$) combine to give a negative value for ΔG; that is, when the free energy decreases.

You should be able to analyze how (or if) temperature will affect the spontaneity of an event when you know the signs of ΔH and ΔS. Study the summary in Figure 13.11.

Thinking It Through

2 In the first Thinking It Through exercise, you learned that the reaction

$$N_2O_4(g) \rightarrow N_2(g) + 2O_2(g)$$

is exothermic. What is the expected sign of ΔG for this reaction and how do we expect the algebraic sign of ΔG to be affected by the temperature?

Self-Test

14. For which of the following is ΔG *always* positive, regardless of the temperature?

    (a) ΔH positive, ΔS negative

    (b) ΔH negative, ΔS negative ____________________

15. Based on thermodynamics and your own intuition, arrange the following in order of their increasing likelihood of occurring spontaneously. Give the signs of ΔH and ΔS for each.

    (a) A pile of bricks slides downhill and ends up as a brick wall.

    ____

    (b) A brick wall slides downhill and produces a jumbled pile at the bottom.

    ____

    (c) A brick wall slides uphill and ends up as a jumbled pile of bricks.

    ____

    (d) A pile of bricks slides uphill and forms a brick wall.

    ____

    Order: ____

16. Each of the events in Question 15 can be accomplished (although not all are spontaneous themselves). Arrange them in order of their increasing ease of being accomplished.

    ____

New Terms

Write the definitions of the following terms, which were introduced in this section. If necessary, refer to the Glossary at the end of the text.

Gibbs free energy
exergonic
endergonic

13.8 Standard Free Energies

Review

As with other thermodynamic quantities, the standard free energy change is the free energy change at 25 °C and 1 atm. We can compute it in two ways. One is from ΔH° and ΔS°.

$$\Delta G^\circ = \Delta H^\circ - (298\ \text{K})\Delta S^\circ$$

The other is from tabulated values of standard free energies of formation.

$$\Delta G^\circ = (\text{sum } \Delta G_f^\circ \text{ products}) - (\text{sum } \Delta G_f^\circ \text{ reactants})$$

These are not difficult calculations. Study Examples 13.4 and 13.5 and work Practice Exercises 6 and 7; then try the Self-Test below.

Self-Test

17. Using Table 4.2 on page 158 and Table 13.1 on page 568, calculate ΔG° (in kilojoules) for the reaction, $CaO(s) + H_2O(l) \rightarrow Ca(OH)_2(s)$

____

18. Use standard free energies of formation in Table 13.2 on page 574 to calculate ΔG° (in kilojoules) for the following reactions:

    (a) $2C_2H_5OH(l) + 6O_2(g) \rightarrow 4CO_2(g) + 6H_2O(l)$

    ____

    (b) $2HNO_3(l) + 2HCl(g) \rightarrow Cl_2(g) + 2NO_2(g) + 2H_2O(l)$

    ____

New Terms

Write the definitions of the following terms, which were introduced in this section. If necessary, refer to the Glossary at the end of the text.

standard free energy change

standard free energy of formation

13.9 Free Energy and Maximum Work

Review

The free energy change for a reaction is equal to the maximum amount of energy that, theoretically, can be recovered as useful work. In all real situations, however, somewhat less than this maximum is actually be obtained. This is because the maximum work can only be gotten if the process occurs reversibly. A reversible process takes an infinite length of time and consists of an infinite number of small changes in which the driving "force" is very nearly balanced by an opposing "force." No real process from which work is extracted occurs reversibly, so we always obtain less than the maximum. Nevertheless, ΔG gives us a goal to aim at, and allows us to gauge the efficiency of the way we are using a reaction to obtain work.

Example 13.1 Maximum Energy and Maximum Work

Problem

At 25 °C and 1 atm, what is the maximum amount of heat that could be extracted from the reaction of 1 mol $CaO(s)$ with $H_2O(l)$?

$$CaO(s) + H_2O(l) \rightarrow Ca(OH)_2(s)$$

What is the maximum amount of work that could be obtained from this reaction?

Solution

If you worked Question 17 in the Study Guide, you already have all of the data that you need. For that question we found the following:

$$\Delta H^\circ = -65.2 \text{ kJ}$$

$$\Delta S^\circ = -34 \text{ J/K} \quad (T\Delta S = 10 \text{ kJ})$$

$$\Delta G^\circ = -55 \text{ kJ (rounded)}$$

If the reaction is carried out so that no work is done, all the energy escapes as heat and the heat evolved is equal to ΔH°. In other words, 65.2 kJ of heat is given off. If the reaction is carried out reversibly so that the maximum work is obtained, this work is equal to ΔG°, or 55 kJ of work.

Notice that we get less energy as work than as heat. What happens to the rest of the energy? The answer is that it goes to decreasing the entropy. ΔS° is negative, and when the reaction is used to produce work, some of the available energy (10 kJ) is used to increase the order of the system. We get what's left over.

Self-Test

19. Carbon monoxide is sometimes used as an industrial fuel.

    (a) What is the maximum heat obtained at 25 °C and 1 atm by burning 1 mol of CO? The reaction is: $2CO(g) + O_2(g) \rightarrow 2CO_2(g)$

    ____

    (b) What is the maximum work available at 25 °C and 1 atm from the combustion of 1 mol of $CO(g)$?

    ____

    (c) What happens to the difference between the heat and work that is available?

    ____

New Terms

Write the definition of the following term, which was introduced in this section. If necessary, refer to the Glossary at the end of the text.

reversible process

13.10 Free Energy and Equilibrium

Review

At equilibrium, the total free energy of the products equals the total free energy of the reactants and ΔG for the system is equal to zero.

$$\Delta G = 0 \quad \text{at equilibrium}$$

Since $\Delta G = 0$ at equilibrium, and since ΔG equals the maximum amount of work obtainable from a change, at equilibrium we can obtain no work from a system.

Equilibria in Phase Changes

For a phase change, equilibrium can occur at only one temperature. At this temperature, T,

$$T = \frac{\Delta H}{\Delta S}$$

Without much error, T can be calculated using ΔH° and ΔS° (which apply, strictly speaking, at 25 °C) because ΔH and ΔS do not change very much with temperature.

Equilibrium involving a phase change at 1 atm can only occur at one temperature. For example, pure water at a pressure of 1 atm can be in equilibrium with ice *only* if its temperature is 0 °C. It can be in equilibrium with water vapor that has a pressure of 1 atm *only* when the temperature is 100 °C. At 25 °C and a pressure of 1 atm, (and with the absence of other gases such as air) water will exist entirely as a liquid. No equilibrium exists. (See Figure 13.1 below.)

A typical free energy diagram for a phase change is illustrated in Figure 13.12 on page 579. Notice that $G_{\text{products}} = G_{\text{reactants}}$ only at one temperature and that is the only temperature at which equilibrium can exist.

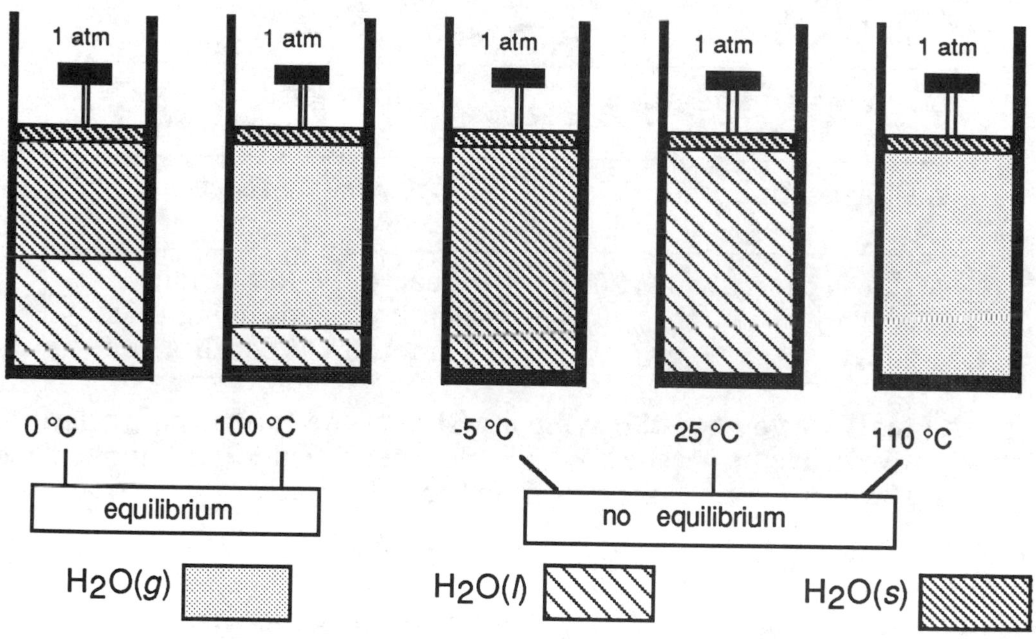

Figure 13.1 *At a pressure of 1 atm, equilibrium between any two phases of water can only occur at a single temperature. At temperatures of –5, 25, and 110 °C, only a single phase can exist.*

Equilibria in Homogeneous Chemical Systems

Nearly all homogeneous chemical reactions are able to exist in a state of equilibrium at 25 °C, and the position of equilibrium is determined by the value of ΔG°. Study the free energy diagrams in Figures 13.13 and 13.14 on page 580. Notice that when ΔG° is positive, the position of equilibrium lies near the reactants. On the other hand, when ΔG° is negative, the position of equilibrium lies close to the products.

When ΔG° has a reasonably large negative value (–20 kJ, or so), the position of equilibrium lies far in the direction of the products, and when the reaction occurs it will appear to go essentially to completion. On the other hand, if ΔG° has a reasonably large positive value, hardly any products will be present at equilibrium. In other words, when the reactants are mixed, they will not *appear* to react at all. For all practical purposes, no reaction occurs when ΔG° is positive and has a value in excess of about 20 kJ. Therefore, we can use the sign and magnitude of ΔG° as an indicator of whether or not we expect to observe the formation of products in a reaction.

Summary

ΔG° large and negative	reaction goes to very nearly to completion.
ΔG° large and positive	virtually no reaction occurs.
ΔG° less than about ±20 kJ	reactants and products are both present in significant amounts at equilibrium.

In the text, we've used the symbol ΔG_T° to stand for the equivalent of ΔG°, but at a temperature other than 25 °C. Because ΔH and ΔS change little with temperature, we can approximate ΔG_T° by the equation

$$\Delta G_T^\circ = \Delta H^\circ - T\Delta S^\circ$$

The sign and magnitude of ΔG_T° can be used in the same way as for ΔG° in predicting the outcome of a reaction.

Keep in mind that even though ΔG° or ΔG_T° may predict that a reaction should be "spontaneous," it tells us nothing about how rapidly the reaction will be. The reaction of H_2 with O_2 has a very negative value of ΔG°, so it is very "spontaneous." However, at room temperature the reaction is so slow that no reaction is observed. Thus, a change must not only be spontaneous, it must occur relatively fast for us to actually observe the change.

Thinking It Through

For the following, identify the information needed to solve the problem and show (or explain) what must be done with it.

3 For the reaction

$$2PCl_3(g) + O_2(g) \rightleftharpoons 2POCl_3(g),$$

How will the amount of PCl_3 present at equilibrium at 25 °C compare with the amount present at 45 °C?

Self-Test

20. At the boiling point of water, liquid and vapor are in equilibrium at a pressure of 1 atm. Use ΔH° and ΔS° for the reaction,

$$H_2O(l) \rightleftharpoons H_2O(g)$$

to calculate the boiling point of water. How does the answer compare to the actual boiling point? Does this support the statement that ΔH and ΔS are nearly temperature independent?

____

____

21. What would you expect to observe at 25 °C if 2 mol CO and 1 mol O_2 were mixed and allowed to react according to the equation:

$$2CO(g) + O_2(g) \rightarrow 2CO_2(g)?$$

____

22. What would you expect to observe at 25 °C if 2 mol of $N_2O(g)$ were mixed with 1 mol $O_2(g)$ in order to form $NO(g)$ by the reaction:

$$2N_2O(g) + O_2(g) \rightarrow 4NO(g)?$$

____

23. What should we expect to observe at 25 °C if we were to check a mixture of N_2O and O_2 for the reaction:

$$2N_2O(g) + 3O_2(g) \rightarrow 4NO_2(g)?$$

____

24. Assuming that ΔH and ΔS are approximately independent of temperature, calculate the "standard" free energy change, ΔG°_T, for the reaction in Question 17 at 1 atm and 50 °C

$$\Delta G^\circ_{323} = \underline{\hspace{6cm}}$$

Does this reaction proceed farther toward completion at this higher temperature?

____

New Terms

none

Answers to Thinking It Through

1 Two factors determine spontaneity, the energy change and the entropy change. The reaction is exothermic, so the energy change is in favor of the reaction being spontaneous. Since four molecules are being formed from two, the entropy increases, so this factor is also in favor of spontaneity. Because both the energy and entropy changes favor spontaneity, we can anticipate that the reaction should be spontaneous.

2. The sign of ΔG is determined by the algebraic signs of ΔH and $T\Delta S$. The reaction is exothermic, so ΔH is negative. Three molecules are formed from one, so ΔS is positive, which means that $T\Delta S$ is positive. The sign of ΔG is determined by $\Delta G = (-) - (+)$. Therefore, we conclude that ΔG must be negative and the reaction must be spontaneous.

   Since T is a positive quantity, the $T\Delta S$ term will be positive regardless of the temperature, so we conclude that ΔG will be negative regardless of the temperature.

3 The position of equilibrium is determined by the the value of ΔG°_T. For the reaction at 25 °C, we can use tabulated values of ΔG°_f for the reactants and products to calculate ΔG°_{298}. To obtain ΔG°_T for the reaction at 45 °C, we use values of ΔH° and ΔS° for the reaction at 25°C and then apply the equation $\Delta G^\circ_T = \Delta H^\circ - T\Delta S^\circ$. To obtain ΔH°, we can use tabulated values of ΔH°_f and apply Hess's law. To obtain ΔS° we use tabulated value of S° and perform a Hess's-law type of calculation.

   Once we know the values of ΔG°_T at the two temperatures, we can determine which way the reaction will shift with temperature. If ΔG°_T becomes more negative as T increases, then at the higher temperature there will be more product and less reactants.

[If you wish to work through to an answer on this question, the necessary data is to be found in Appendix E. An alternative solution is presented here, too. We calculate ΔH°_{298} and ΔS°_{298} just as we would in the other solution. We use these values to calculate ΔG° at both temperatures and then compare. (This eliminates any inconsistencies in the data.)

$$\Delta H^\circ_{298} = [2(-1109.7 \text{ kJ})] - [2(-282.0 \text{ kJ})]$$
$$= -1644 \text{ kJ}$$
$$\Delta S^\circ_{298} = [2(324)] - [2(311.8) + 1(205)] \text{ (all J mol}^{-1}\text{ K}^{-1})$$
$$= -181 \text{ J mol}^{-1}\text{ K}^{-1}$$
$$= -0.181 \text{ kJ mol}^{-1}\text{ K}^{-1}$$

Now we calculate ΔG°.

$$\Delta G^\circ_{298} = -1644 \text{ kJ} - (298 \text{ K})(-0.181 \text{ kJ mol}^{-1}\text{ K}^{-1})$$
$$= -1590 \text{ kJ}$$
$$\Delta G^\circ_{318} = -1644 \text{ kJ} - (318 \text{ K})(-0.181 \text{ kJ mol}^{-1}\text{ K}^{-1})$$
$$= -1586 \text{ kJ}$$

Since ΔG° is less negative at the higher temperature, the reaction does not proceed as far toward completion.]

Answers to Self-Test Questions

1. $H = E + PV$
2. 507 J
3. $\Delta H > \Delta E$; at constant pressure there would be a volume decrease. Therefore, at constant pressure, the system absorbs some energy as work which is being done on it during the volume decrease. This extra energy can appear as extra heat given off during the reaction at constant pressure.
4. No volume change would occur at constant pressure, so ΔH and ΔE are the same.
5. $\Delta E = -55.5$ kJ
6. (a) spontaneous, (b) nonspontaneous, (c) spontaneous, (d) nonspontaneous
7. (a) nonspontaneous, $\Delta H^\circ = +438.4$ kJ (b) nonspontaneous, $\Delta H^\circ = +41.8$ kJ (c) spontaneous, $\Delta H^\circ = -26.4$ kJ
8. The melting of ice on a warm day, or the evaporation of a puddle of water.
9. When the gas molecules in one container have the entire volume of the other available to it, a state of low probability exists until the gas expands into the other container as well.
10. (a) positive, (b) positive, (c) negative, (d) negative, (e) negative
11. (a), (c)
12. (a) +180.2 J/K, (b) −332.3 J/K, (c) −5.2 J/K, (d) +227.2 J/K

13. (a) spontaneous, (b) nonspontaneous, (c) nonspontaneous (d) spontaneous
14. (a) ΔG is positive at all temperatures.
15. (a) $\Delta H = (-)$, $\Delta S = (-)$ (b) $\Delta H = (-)$, $\Delta S = (+)$ (c) $\Delta H = (+)$, $\Delta S = (+)$ (d) $\Delta H = (+)$, $\Delta S = (-)$ Order: (d) < (c) < (a) < (b)
16 (d) < (c) < (a) < (b) [(b) occurs spontaneously.]
17. $\Delta G^\circ = -55$ kJ
18. (a) $\Delta G^\circ = -2651$ kJ, (b) $\Delta G^\circ = -20.4$ kJ,
19. (a) $\Delta H^\circ = -283$ kJ/mol CO; heat evolved = 283 kJ, (b) $\Delta G^\circ = -257.1$ kJ/mol of CO; max. work done = 257.1 kJ, (c) $T\Delta S = -25.9$ kJ. This is the work needed to increase the order in the system (decrease the entropy).
20. $\Delta H^\circ = 44.1$ kJ, $\Delta S^\circ = 118.7$ J/K, $T_b = 372$ K = 99 °C. The actual boiling point is 100 °C = 373 K. The answer from $\Delta H^\circ/\Delta S^\circ$ is quite close, which supports the statement.
21. $\Delta G^\circ = -514.2$ kJ; the reaction should go very nearly to completion.
22. $\Delta G^\circ = +139.6$ kJ. No reaction should be observed.
23. $\Delta G^\circ = +0.17$ kJ. Substantial amounts of both N_2O and NO_2 should be present at equilibrium.
24. $\Delta G^\circ_T = -54$ kJ at 50 °C (323 K). ΔG°_T is less negative than ΔG°, so the reaction should not proceed as far toward completion at the higher temperature.

Tools you have learned

Remove this chart from the Study Guide and keep it handy when tackling homework problems.

Tool	***Function***
Predicting the sign of ΔS	Enables you to determine whether the entropy change favors spontaneity.
Standard entropies	Used to calculate the value of ΔS° for a reaction.
Sign of ΔG	Determines whether or not a change is spontaneous.
$\Delta G^\circ = \Delta H^\circ - (298\ K)\Delta S$	Calculate ΔG° from ΔH° and ΔS° values.
Standard free energies of formation	Calculate ΔG° for a reaction from tabulated values of ΔG_f°.
Equilibrium condition of $\Delta G = 0$	When this condition is fulfilled, a system is at equilibrium, so this condition is a criterion for equilibrium.
Value of ΔG° for a reaction Only when ΔG° is negative do we expect to observe significant amounts of products.	The sign of ΔG° tells us whether or not a reaction can be observed.
$\Delta G_T^\circ = \Delta H_{298}^\circ - T\Delta S_{298}^\circ$	To calculate the value of ΔG° at temperatures other than 25 °C.

Summary of Important Equations

First law of thermodynamics

$$\Delta E = q - w$$

Definition of enthalpy

$$H = E + PV$$

Calculating work done *by* a system.

$$w = -P\Delta V$$

Converting between ΔE and ΔH

$$\Delta H = \Delta E + \Delta n\, RT$$

Remember, Δn is the change in the number of moles of *gas*.

Calculating ΔS°

$$\Delta S^\circ = (\text{ sum of } S^\circ \text{ of products}) - (\text{ sum of } S^\circ \text{ of reactants})$$

Gibbs free energy change

$$\Delta G = \Delta H - T\Delta S$$

Calculating ΔG° for a reaction from ΔG°_f

$$\Delta G^\circ = (\text{sum } \Delta G^\circ_f \text{ of products}) - (\text{sum } \Delta G^\circ_f \text{ of reactants})$$

Calculating ΔG° at a temperature other than 25°C

$$\Delta G^\circ_T = \Delta H^\circ_{298} - T\,\Delta S^\circ_{298}$$

Chapter **14**

KINETICS: THE STUDY OF RATES OF REACTION

In Chapter 13 you learned that we can predict whether a reaction is possible. However, many reactions that thermodynamics tells us should be spontaneous do not appear to occur. An example is the decomposition of nitrogen oxides into N_2 and O_2. For these substances, the rates of their decompositions are just too slow. In this chapter we study why some reactions are fast and others are slow.

Learning Objectives

As you study of this chapter, keep in mind the following goals:

1. To learn how the speeds of reactions vary over wide ranges and to learn about the information we gain by the study of speeds of reactions.
2. To learn about the kinds of things that influence how fast a reaction occurs.
3. To learn how rates of reaction are expressed and how they are measured.
4. To learn how the rate of reaction is related quantitatively to the concentrations of the reactants.
5. To learn how to calculate the concentration of a reactant at any time after the start of a reaction for first- and second-order reactions.
6. To learn how the speed of a reaction is related to the time it takes for half of a reactant to disappear.
7. To learn why the rate of a reaction increases with increasing temperature.
8. To learn how to deal quantitatively with the effect of temperature on the rate of reaction.
9. To learn how the rate law for a reaction can give clues to the sequence of chemical steps that ultimately give the products in a reaction.
10. To learn how substances called catalysts affect the rate of a re-

action, and to learn how homogeneous and heterogeneous catalysts work.

14.1 Speeds at Which Reactions Occur

Review

The term *kinetics* implies action or motion. For a chemical reaction, this motion is the speed at which the reaction takes place, and by that we mean the speed at which the reactants are consumed and the products formed.

The speeds of reactions range from very rapid to extremely slow. We call the speed of reaction the *rate of reaction* or the *reaction rate*. One of the benefits of studying the rate of a reaction and the factors that control it is some insight into the *mechanism* of the reaction—the individual chemical steps that produce the net overall change described by the balanced equation.

Self-Test

1. Why is it unlikely that the combustion of octane, C_8H_{18}, a component of gasoline, occurs by a single step as given by the equation

$$2C_8H_{18}(g) + 25O_2(g) \rightarrow 16CO_2(g) + 18H_2O(g)$$

____

2. Why would chemical manufacturers be interested in studying the factors that affect the rate of a reaction?

____

New Terms

Write the definitions of the following terms, which were introduced in this section. If necessary, refer to the Glossary at the end of the text.

rate of reaction   mechanism

14.2 Factors that Affect Reaction Rates

Review

There are five factors that control the rate of a reaction.

1 *The nature of the reactants*: Some substances, because of their chemical bonds, just naturally react faster than others under the same conditions.

2 *Ability of reactants to meet*: Many reactions are carried out in solution where the reactants can mingle on a molecular level. For heterogeneous reactions, particle size is the controlling factor. For a given mass of reactant, the smaller the particle size, the larger the area of contact with the other reactants in other phases, and the faster the reaction.

3 *Concentrations of the reactants*: The rates of most reactions increase with increasing reactant concentration.

4 *Temperature*: With very few exceptions, reactions proceed faster as their temperature is raised.

5 *Catalysts*: The rates of many reactions are increased by the presence of substances called catalysts. A catalyst is a substance that is intimately involved in the reaction, but which is not used up during the reaction.

Thinking It Through

For the following, describe the information needed to answer the question.

1 If you wish to greatly increase the rate at which heat can be obtained from the combustion of coal, what might you do? Answer the question in terms of the factors that affect the rate of a reaction.

Self-Test

3. Elemental potassium reacts more rapidly with moisture than does sodium under the same conditions. Which of the factors discussed in this section is responsible for this?

____

4. Insects move more slowly in autumn than in the summer. Why?

____

5. What is one reason why coal miners are concerned about open flames during the mining operation?

____

6. Fire fighters are taught about the "fire triangle."

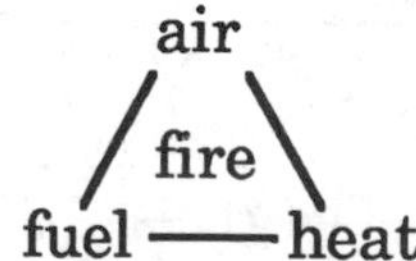

Eliminate any one of the three corners, and the fire is extinguished. Which factors discussed in this section affect the fire triangle?

---

7. Why is pure oxygen more dangerous to work with than air?

New Terms

Write the definitions of the following terms, which were introduced in this section. If necessary, refer to the Glossary at the end of the text.

heterogeneous reaction — homogeneous reaction

14.3 Measuring the Rate of Reaction

Review

A *rate* is always a ratio in which units of time appear in the denominator. A reaction rate has units of molar concentration in the numerator, so the units of reaction rate are mol L^{-1} s^{-1} (mole per liter per second). The rate of a reaction generally changes with time as the reactants are used up. The rate can be measured at any particular instant by determining the slope of the concentration versus time curve. Study Figure 14.2 on page 598.

The rate is usually measured by monitoring the substance whose concentration is most easily measured. Once we know the rate at which one substance is changing, we can calculate the rate for any other substance. This is because the relative rates of formation of the products and rates of disappearance of the reactants are related by the coefficients of the balanced equation. Review Example 14.1 on page 597 to see how such calculations are made.

Self-Test

8. In Figure 14.2, what is the rate at which HI is reacting at t = 200 seconds?

---

9. What would be the rate at which H_2 is forming at t = 200 s?

---

10. In the combustion of octane

$$2C_8H_{18}(g) + 25O_2(g) \rightarrow 16CO_2(g) + 18H_2O(g)$$

what would be the rate of formation of CO_2 if the concentration of octane was changing at a rate of -0.25 mol L^{-1} s^{-1}?

____

11. If octane is burning at a constant rate of 0.25 mol s^{-1}, how many moles of it will burn in 15 minutes?

____

New Terms

Write the definition of the following term, which was introduced in this section. If necessary, refer to the Glossary at the end of the text.

rate

14.4 Concentration and Rate

Review

Concentration and rate are related by the *rate law* for a reaction. For example, for the reaction

$$xA + yB \rightarrow \text{products}$$

in which x and y are coefficients of reactants A and B, respectively, the rate law will be of the form

$$\text{rate} = k[A]^n[B]^m$$

Remember that square brackets, [ ], around a chemical formula stands for the molar concentration (units, mol L^{-1}) of the substance. The proportionality constant, k, is the *rate constant*. The exponents give the *order* with respect to each reactant, and their sum is the *overall order* of the reaction.

It is very important to remember that the exponents (n and m in this example) are not necessarily equal to the coefficients x and y, and can only be known for sure if they are determined from experimentally measured data. These data have to show how the rate changes when the concentrations change. In analyzing data like that in Table 14.2, observe how the rate changes when the concentration of one of the reactants changes while the concentrations of the other reactants are held constant.

Study Table 14.3 and note how it applies to Examples 14.3 to 14.5. After working Practice Exercises 6 and 7, try the following Self-Test.

Self-Test

12. What is the order of the reaction for the dimerization of isoprene in Example 14.4?

    ____

13. In Practice Exercise 7, what is the order with respect to A and B, and what is the overall order of the reaction?

    ____

14. What is the value of the rate constant for the dimerization of isoprene in Example 14.4

    ____

15. When the concentration of a particular reactant was increased by a factor of 10, the rate of the reaction was increased by a factor of 1000. What is the order of the reaction with respect to that reactant?

    ____

16. For a reaction, $2A + D \rightarrow$ products, the following data were obtained:

Initial Concentration (M)		Initial Rate
A	B	(mol L^{-1} s^{-1})
0.20	0.10	1.0×10^{-2}
0.40	0.10	2.0×10^{-2}
0.20	0.40	1.6×10^{-1}

(a) What is the rate law for the reaction? ________________

(b) What is the value of the rate constant? (Give the correct units, too) ________________

New Terms

Write the definitions of the following terms, which were introduced in this section. If necessary, refer to the Glossary at the end of the text.

rate law
rate constant
order of a reaction

14.5 Concentration and Time

Review

For a first-order reaction, the rate constant can be obtained from a graph of the natural logarithm of the reactant concentration, ln [A], versus time. The slope of the line equals the rate constant. Alternatively, Equation 14.5 relates concentration and time to the rate constant.

Finding the time required for the concentration to drop to some particular value is simple, as shown in Example 14.6 in the text. Just substitute the initial and final concentrations into the concentration ratio, take the logarithm of the ratio, substitute the value of the rate constant, and then solve for t. Notice that the units of t are of the same kind as the units of the rate constant;if k has units of s^{-1}, then t will be in seconds; if k has units of hr^{-1}, then t will be in hours.

Finding the concentration after a specified time is illustrated in Example 14.7. Notice that we calculate the logarithm of the concentration ratio from k and t. Taking the antilogarithm gives the numerical value for the ratio. Then we substitute the known concentration and solve for the concentration we want to find.

For second-order reactions, a graph of the reciprocal of the concentration versus time gives a straight line. The slope of this line equals k. Calculations for second order reactions are done with Equation 14.6 on page 609. This equation doesn't involve logarithms, so it is easier to manipulate algebraically. Study Example 14.8 page 609 to learn how to use it.

Half-Lives

The *half-life* of a reaction, $t_{1/2}$, is the time required for half of a given reactant to disappear. For a first-order reaction, $t_{1/2}$ is independent of the initial reactant concentration. The value of $t_{1/2}$ is inversely proportional to the initial reactant concentration for a second-order reaction.

You should be able to use Equations 14.7 and 14.8, and you should be able to use the way $t_{1/2}$ varies with initial concentration to determine whether a reaction is first or second order.

Thinking It Through

For the following, identify the information needed to solve the problem and show (or explain) what must be done with it.

2 The decomposition of SO_2Cl_2 is a first order reaction. In an experiment, it was found that the SO_2Cl_2 concentration was 0.022 M after 300 min. The value of k at the temperature at which the experiment was

performed is 2.2×10^{-5} s^{-1}. How many *moles* of the reactant were originally present in the apparatus, which has a volume of 350 mL?

Self-Test

17. A certain reactant disappears by a first-order reaction that has a rate constant $k = 3.5 \times 10^{-3}$ s^{-1}. If the initial concentration of the reactant is 0.500 mol/L, how long will it take for the concentration to drop to 0.200 mol/L?

____

18. For the reactant described in Question 17, what will the concentration of the reactant be after 20.0 min if its initial concentration is 2.50 mol/L?

____

19. A certain reaction follows the stoichiometry

$$2D \rightarrow \text{products}$$

It has the rate law Rate = $k[D]^2$. The rate constant for the reaction equals 5.0×10^{-3} L mol^{-1} s^{-1}.

(a) How many seconds will it take for the concentration of D to drop from 1.0 mol/L to 0.60 mol/L?

____

(b) If the initial concentration of D is 1.0 mol/L, what will its concentration be after 30 min?

____

20. In a certain reaction, the half-life of a particular reactant is 30 minutes. If the initial concentration of the reactant is 2.0 mol/L, and, if the reaction is first order,

(a) What will be the concentration of the reactant after 2 hours?

____

(b) How long will it take for the concentration to be reduced to 0.0303 mol/L?

____

21. A certain first-order reaction has $t_{1/2}$ = 30 minutes. What is the rate constant for the reaction?

____

22. The decomposition of SO_2Cl_2 has $k = 2.2 \times 10^{-5}\ s^{-1}$ (Example 14.3). What is $t_{1/2}$ for this reaction in seconds and in minutes?

____

23. At a certain temperature the half-life for the decomposition of N_2O_5 was 350 seconds when the N_2O_5 concentration was 0.200 mol/L. When the concentration was 0.400 mol/L, the half-life was 5.83 minutes. What is the order of the decomposition reaction?

____

New Terms

Write the definitions of the following terms, which were introduced in this section. If necessary, refer to the Glossary at the end of the text.

half-life

$t_{1/2}$

14.6 Theories About Reaction Rates

Review

Collision theory postulates that the rate of a reaction is proportional to the number of effective collisions per second between the reactant molecules or ions. The number of effective collisions per second is less than the total number of collisions per second for two principal reasons:

1 For some reactions, it is important that the reactant molecules be in the correct orientation when they collide.

2 A minimum kinetic energy, called the *activation energy* (E_a), must be possessed by the reactant molecules in a collision to overcome the repulsions between their electron clouds and thereby permit the electronic rearrangements necessary for the formation of new product molecules.

The activation energy requirement explains why the rate of a reaction increases with increasing temperature. Study Figure 14.7 and the discussion on pages 613 and 614.

In *transition state theory*, we follow the energy of the reactants as they are transformed to the products. Study Figures 14.8, 14.10 and 14.11. You should be able to identify the activation energy for both the forward and reverse reactions, the potential energies of the reactants and products, and the heat of reaction. You should also be able to locate the *transition state* on the diagram.

Self-Test

24. Without peeking at the text, sketch and label the potential energy diagram for
    (a) an exothermic reaction.
    (b) an endothermic reaction.

25. Where is the transition state located on the energy diagram for a reaction?

    ____

26. One step in the reaction $NO_2(g) + CO(g) \rightarrow NO(g) + CO_2(g)$ is believed to involve the collision of two NO_2 molecules to give NO and NO_3 molecules

$$NO_2 + NO_2 \rightarrow NO + NO_3$$

    What might be a reasonable structure for the activated complex in this collision?

27. On the basis of the transition state theory, explain why the temperature of a collection of molecules rise as an exothermic reaction occurs within it?

    ____

    ____

28. Why do reactions having a low activation energy usually occur faster than ones having a high activation energy?

    ____

    ____

New Terms

Write the definitions of the following terms, which were introduced in this section. If necessary, refer to the Glossary at the end of the text.

collision theory
activation energy
transition state theory
reaction coordinate
transition state
activated complex

14.7 Measuring the Activation Energy

Review

The activation energy is related to the rate constant by the *Arrhenius equation,*

$$k = A\,e^{-E_a/RT}$$

As described in Example 14.11, plotting the natural logarithm of the rate constant versus the reciprocal of the absolute temperature yields a straight line whose slope is equal to $-E_a/R$. After studying Example 14.11, work Question 31 of the Self-Test at the end of this section. To do that, set up a table with headings of "ln k" and "$1/T$" and compute these values from the data given. Then choose a piece of graph paper, plot the data, measure the slope, and calculate E_a.

To calculate E_a from rate constants at two different temperatures (which is actually less accurate than the graphical procedure) you will need to know Equation 14.11. Examples 14.12 and 14.13 illustrate how this equation is used, and you should study them carefully before beginning the Practice Exercises.

In working with Equation 14.11, it is helpful to note that if a negative value is obtained for E_a when you calculate the activation energy, you probably interchanged the 1 and 2 subscripts on either the rate constants or the temperatures. The only effect that such an error will have on the computed E_a is to change its sign. Since the activation energy must be positive, just change its sign to positive.

When you use the activation energy to calculate a rate constant at some temperature, given the rate constant at some other temperature, keep in mind that the value of k is always larger at the higher temperature. After finishing the calculation, make sure your values of k fit this rule. If they don't, then you switched the subscripts 1 and 2 on the k's in the ratio of rate constants.

A final point to be especially careful about in these calculations is to use kelvin temperatures in Equation 14.11, not Celsius temperatures.

Self-Test

29. A certain first-order reaction has a rate constant $k = 1.0 \times 10^{-2}\ s^{-1}$ at 30 °C. At 40 °C its rate constant is $k = 2.5 \times 10^{-2}\ s^{-1}$. Calculate the activation energy for this reaction in kJ/mol.

____

30. A certain second-order reaction has an activation energy of 105 kJ/mol. At 25 °C the rate constant for the reaction has a value of 2.3×10^{-3} L $mol^{-1}\ s^{-1}$. What would its rate constant be at a temperature of 45 °C?

____

31. In the table below are tabulated values of the rate constant for a reaction at 5 °C intervals from 25 °C to 100 °C. Use these data to determine the activation by the graphical method.

Temp (°C)	$k(\text{s}^{-1})$	Temp (°C)	$k(\text{s}^{-1})$
25	0.0240	65	0.206
30	0.0324	70	0.259
35	0.0433	75	0.325
40	0.0573	80	0.406
45	0.0751	85	0.502
50	0.0978	90	0.619
55	0.126	95	0.757
60	0.162	100	0.922

New Terms

Write the definitions of the following terms, which were introduced in this section. If necessary, refer to the Glossary at the end of the text.

Arrhenius equation

frequency factor

14.8 Collision Theory and Reaction Mechanisms

Review

Usually, a net overall reaction occurs as a sequence of simple elementary processes, the slowest of which determines how fast the products are able to form. This is called the rate-determining step, and the rate law of the overall reaction is the same as the rate law for the rate-determining step.

If we know the stoichiometry for an elementary process, we can predict its rate law; the coefficients of the reactants are equal to their exponents in the rate law. Remember, however, that this works *only* if the elementary process is known. When we first begin to study a reaction we don't know what its mechanism is, so we can't predict with any hope of confidence what the exponents in the rate law will be.

Determining a mechanism involves guessing what the elementary processes are and then comparing the predicted rate law, based on the mechanism, with the rate law determined from experimental data. If the two rate laws match, the mechanism may be correct. If they don't match, the search for a mechanism must continue.

Self-Test

32. Suppose an elementary process in a mechanism is: $2A + M \rightarrow P + Q$. What is the rate law for this step?

____

33. The following mechanism has been proposed for the reaction

$$(CH_3)_3CBr + OH^- \rightarrow (CH_3)_3COH + Br^-$$

Step 1 $(CH_3)_3CBr \rightarrow (CH_3)_3C^+ + Br^-$ (slow)

Step 2 $(CH_3)_3C^+ + OH^- \rightarrow (CH_3)_3COH$ (fast)

If this mechanism is correct, what is the expected rate law for the overall reaction?

____

34. What would be the rate law for the overall reaction in Question 33 if it occurred in a single step (i.e., if the overall reaction were actually an elementary process)?

____

New Terms

Write the definitions of the following terms, which were introduced in this section. If necessary, refer to the Glossary at the end of the text.

elementary process

rate-determining step

rate-limiting step

14.9 Catalysts

Review

Catalysts open alternative pathways (mechanisms) for reactions. These paths have lower activation energies than the uncatalyzed mechanisms, so catalyzed reactions occur faster. A homogeneous catalyst is in the same phase as the reactants; a heterogeneous catalyst is in a different phase than the reactants. A homogeneous catalyst, like the NO_2 described in the lead chamber process, is used up in one step of the mechanism and then regenerated in a later step. A heterogeneous catalyst functions by adsorption of the reactants on its surface where the reactants are able to react with a relatively low activation energy.

Self-Test

35. How do catalyst "poisons" work? ______________________________

____

New Terms

Write the definitions of the following terms, which were introduced in this section. If necessary, refer to the Glossary at the end of the text.

homogeneous catalyst heterogeneous catalyst

Answers to Thinking It Through

1 We could make use of Factors 2 and 3 above. Combustion of coal is a heterogeneous reaction, so we can increase the rate by providing a greater area of contact between the reactants. This could be done by pulverizing the coal before burning it. To further increase the ability of oxygen to reach the coal, we could blow air into the reaction mixture. We could also increase the concentration of the oxygen in the air.

2 Since the reaction is first order, we use Equation 14.5. The concentration we wish to find corresponds to $[A]_o$ in the equation (i.e., $[SO_2Cl_2]_o$). First we multiply k by t. Then we take the antilogarithm to obtain the ratio of concentrations, Solving for $[SO_2Cl_2]_o$

$$\frac{[SO_2Cl_2]_o}{[SO_2Cl_2]_t} = e^{kt}$$

Solving for $[SO_2Cl_2]_o$ and substituting 0.022 M for $[SO_2Cl_2]_t$ gives us the the initial SO_2Cl_2 concentration. Multiplying this by the volume, 0.350 L, gives the number of moles. [The answer is 0.011 mol.]

Answers to Self-Test Questions

1. It would require the simultaneous collision of 27 molecules, which is an unlikely event.
2. By adjusting conditions, they can make their reactions go faster and more efficiently.
3. Nature of the reactants

4. Their biochemical reactions are slower in the cooler weather.
5. The possibility of a coal dust explosion.
6. Ability of reactants to meet; the effect of temperature.
7. The O_2 is less concentrated in air than in pure O_2.
8. Approximately 1.2×10^{-4} mol L^{-1} s^{-1}
9. 0.6×10^{-4} mol L^{-1} s^{-1}
10. 2.0 mol L^{-1} s^{-1}
11. 225 mol C_8H_{18}
12. second order
13. second order with respect to both A and B, fourth order overall.
14. 7.92 L mol^{-1} s^{-1}
15. third order
16. (a) rate $=k[A]^1 [B]^2$ (b) $k = 5.0$ L^2 mol^{-2} s^{-1}
17. 2.6×10^2 seconds
18. 0.037 mol/L
19. (a) 1.3×10^2 seconds, (b) 0.10 mol/L
20. (a) 0.125 M, (b) 6 half-lives = 3.0 hours
21. 3.8×10^{-4} s^{-1}
22. 3.2×10^4 s = 525 min
23. first order
24. (a) see Figure 14.8 (b) see Figure 14.10
25. At the high-point on the energy curve.
26.
```
 O           O
  \         /
   N---O---N
  /
 O
```
27. As the potential energy decreases, the average kinetic energy rises, which means the temperature rises.
28. When E_a is small, a large fraction of molecules have at least this minimum energy that they need to react.
29 $E_a = 72.2$ kJ mol^{-1}
30. 3.3×10^{-2} L mol^{-1} s^{-1}
31. Data to be graphed:

ln k	$1/T$	ln k	$1/T$
−3.73	0.00335	−1.58	0.00296
−3.43	0.00329	−1.35	0.00291
−3.14	0.00324	−1.12	0.00287
−2.86	0.00319	−0.902	0.00283
−2.59	0.00314	−0.689	0.00279
−2.33	0.00309	−0.480	0.00275
−2.07	0.00305	−0.278	0.00272
−1.82	0.00300	−0.0809	0.00268

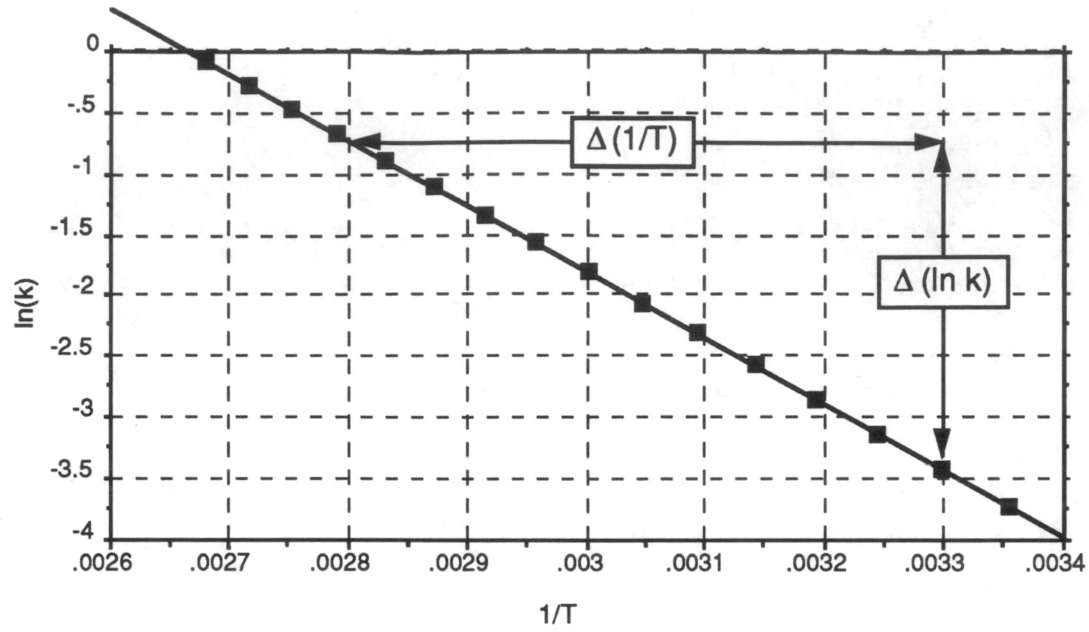

$\Delta(1/T) = 0.00330 - 0.00280 = 0.00050$

$\Delta(\ln k) = -3.40 - (-0.75) = -2.65$

$$\text{slope} = \frac{\Delta(\ln k)}{\Delta(1/T)} = \frac{-2.65}{5.0 \times 10^{-4}\ \text{K}^{-1}} = -5.3 \times 10^{3}\ \text{K} = \frac{-E_a}{R} \quad \text{(Note units for slope.)}$$

Using $R = 8.314\ \text{J mol}^{-1}\ \text{K}^{-1}$, $E_a = -8.314(-5.3 \times 10^{3}) = 44 \times 10^{3}$ J/mol

$= 44$ kJ/mol

The activation energy is approximately 44 kJ/mol.

32. rate = $k[A]^2\,[M]$
33. rate = $k[(CH_3)_3CBr]$
34. rate = $k[(CH_3)_3CBr]\ OH^-]$
35. The poison becomes attached to the catalyst's surface and prevents adsorption of the reactants.

Tools you have learned

Remove this chart from the Study Guide and keep it handy when tackling homework problems.

Tool	*Function*
Concentration versus time data	Determine the rate law for a reaction
Rate constant with equations below $\ln\frac{[A]_0}{[A]_t} = kt$ $\frac{1}{[B]_t} - \frac{1}{[B]_0} = kt$	Rate constant is used in conjunction with concentration versus time equations for first and second order reactions to determine the time it takes for concentration to drop to some value, or the concentration after a specified time.
Half-lives and rate constant	Determine concentration after specified number of half-life periods; determine the half-life of a reactant.
Arrhenius equation	Determine E_a for a reaction; determine rate constant at some temperature given rate constant at some other temperature.

Summary of Important Equations

Concentration versus time, first order reaction

$$\ln \frac{[A]_o}{[A]_t} = kt$$

Concentration versus time, second order reaction

$$\frac{1}{[B]_t} - \frac{1}{[B]_o} = kt$$

Half-life, first order reaction

$$t_{1/2} = \frac{\ln 2}{k}$$

Half-life, second order reaction

$$t_{1/2} = \frac{1}{k\,[B]_o}$$

Arrhenius equation

$$k = A\,e^{-E_a/RT}$$

$$\ln\left(\frac{k_2}{k_1}\right) = \frac{-E_a}{R}\left(\frac{1}{T_2} - \frac{1}{T_1}\right)$$

Remember, if you mix up the 1's and 2's, the only effect will be to change the sign of E_a. However, E_a must be positive. Also, k is always larger at the higher temperature.

Chapter **15**

CHEMICAL EQUILIBRIUM—GENERAL CONCEPTS

In Chapter 13 you learned how we can predict whether a reaction is possible, and in Chapter 14 you learned the factors that determine how fast spontaneous reactions are able to proceed. In this chapter we shall explore the fate of most chemical reactions, namely, dynamic equilibrium. For most reactions, the concentrations eventually level off at constant values, which are the equilibrium concentrations in the reaction mixture.

Learning Objectives

As you study of this chapter, keep in mind the following goals:

1 To become totally familiar with the concept of dynamic equilibrium.

2 To learn how the composition of an equilibrium mixture is independent of where we begin along the path from reactants to products, provided that the overall composition of the system is the same each time.

3 To learn how to construct an equation called the equilibrium law that relates the equilibrium concentrations of the reactants and products in a chemical system to a constant called the equilibrium constant, K_c.

4 To learn how the equilibrium law for gaseous reactions can be written using partial pressures and related to an equilibrium constant K_p.

5 To learn how to relate equilibrium constants quantitatively to thermodynamic data.

6 To learn how to convert between K_p and K_c for gaseous reactions.

7 To be able to use the magnitude of the equilibrium constant to make a qualitative estimate of the extent of reaction.

8 To be able to write the equilibrium law for a heterogeneous reaction.

9 To learn how to apply Le Châtelier's principle to chemical equilibria. You should be able to predict the effects of adding or removing a reactant or product, changes in volume, changes in temperature, addition of a catalyst, and addition of an inert gas at constant volume.

10 To learn how to calculate K_c from data relating to equilibrium concentrations and to learn how to use the value of K_c to calculate equilibrium concentrations.

15.1 Dynamic Equilibrium in Chemical Systems

Review

Equilibrium is established in a chemical system when the rate at which the reactants combine to form the the products is equal to the rate at which the products react to form the reactants. We call it a dynamic equilibrium because the reaction hasn't ceased; instead there are two opposing reactions occurring at equal rates.

When equilibrium is reached in a chemical system, the concentrations of the reactants and products attain steady, constant values that do not change with time.

Self-Test

1. If we were able to follow a particular carbon atom in a solution of $HC_2H_3O_2$, part of the time it would exist in an acetate ion, $C_2H_3O_2^-$ and part of the time it would exist in a molecule of acetic acid, $HC_2H_3O_2$. Why is this so?

   ____

   ____

   ____

New Terms

none

15.2 Reaction Reversibility

Review

In the example used in this section, 0.0350 mol N_2O_4 and 0.0700 mol NO_2 each contain the same total amount of nitrogen and oxygen. The two "initial" systems differ in how the nitrogen and oxygen atoms are distributed among the molecules. When these gases are allowed to come to equilibrium from either direction, the same equilibrium composition of NO_2 and N_2O_4 is reached. This means that for a given *overall* composition, the same equilibrium composition will always be reached regardless of whether we begin with reactants, products, or a mixture of them. The reaction can go in either direction, and the direction in which it proceeds is determined by how the concentrations must change in order to become equal to the appropriate equilibrium concentrations.

New Terms

none

15.3 The Equilibrium Law for a Reaction

Review

As you learned in Chapter 14, the molar concentration of a substance is represented symbolically by placing the formula for the substance in brackets. Thus, $[CO_2]$ stands for the molar concentration (moles per liter) of CO_2.

The mass action expression for a reaction is a fraction. Its numerator is constructed by multiplying together the molar concentrations of the *products*, each raised to power that is equal to its coefficient in the balanced chemical equation for the equilibrium. The denominator is obtained by multiplying together the molar concentrations of the *reactants*, each raised to a power equal to its coefficient. For example, for the reaction

$$2CO(g) + O_2(g) \rightleftharpoons 2CO_2(g)$$

the mass action expression is

$$\frac{[CO_2]^2}{[CO]^2[O_2]}$$

The numerical value of the mass action expression is called the *reaction quotient*, *Q*, and at equilibrium, the reaction quotient has a value that we call the *equilibrium constant Kc*. Remember that the "c" in K_c means that the mass action expression is written with molar concentrations. Equating the mass action expression to the equilibrium constant gives the equilibrium law for the reaction. Thus, for the reaction above, the equilibrium law is

$$\frac{[CO_2]^2}{[CO]^2\,[O_2]} = K_c$$

For any given reaction, the numerical value of K_c varies with temperature. But at a given temperature, the reaction quotient will always be the same when the system is at equilibrium, regardless of the individual concentrations. In other words, the concentrations can have *any* values; the only requirement is that they must satisfy the equilibrium law when the system reaches equilibrium.

Manipulating Equations for Chemical Equilibria

The following rules apply when we manipulate chemical equilibria. Notice that the rules are different than the ones we used when we applied Hess's law. (Study the examples on pages 644 and 645.)

- Change the direction of a reaction:
  Take the reciprocal of K
- Multiply by a factor
  Raise K to a power equal to the factor
- Add two equations
  Multiply their Ks

Self-Test

2. Write the equilibrium law for the following reactions:

   (a) $C_2H_4(g) + H_2(g) \rightleftharpoons C_2H_6(g)$

   ____

   (b) $2N_2O(g) + 3O_2(g) \rightleftharpoons 4NO_2(g)$

   ____

   (c) $2HCrO_4^-(aq) \rightleftharpoons Cr_2O_7^{2-}(aq) + H_2O(l)$

   ____

   (d) $CH_3OH(l) + CH_3CO_2H(l) \rightleftharpoons CH_3CO_2CH_3(l) + H_2O(l)$

   ____

3. We saw that at 440 °C, $K_c = 49.5$ for the reaction,

$$H_2(g) + I_2(g) \rightleftharpoons 2HI(g).$$

Which of the following mixtures are at equilibrium at 440 °C?

(a) $[H_2] = 0.0122\ M$, $[I_2] = 0.0432\ M$, $[HI] = 0.154\ M$

(b) $[H_2] = 0.708\ M$, $[I_2] = 0.0115\ M$, $[HI] = 0.635\ M$

(c) $[H_2] = 0.0243\ M$, $[I_2] = 0.0226\ M$, $[HI] = 0.165\ M$

____

4. We saw that at 440 °C, $K_c = 49.5$ for the reaction,

$$H_2(g) + I_2(g) \rightleftharpoons 2HI(g).$$

(a) What is K_c for the reaction $2HI(g) \rightleftharpoons H_2(g) + I_2(g)$?

____

(b) What is K_c for the reaction $\frac{1}{2}H_2(g) + \frac{1}{2}I_2(g) \rightleftharpoons HI(g)$?

____

(c) Write the equation for the reaction we obtain if we added the equations in parts a and b. What is the value of the equilibrium constant for this final equation?

____

New Terms

Write the definitions of the following terms, which were introduced in this section. If necessary, refer to the Glossary at the end of the text.

mass action expression
reaction quotient
equilibrium constant
equilibrium law
K_c

15.4 Equilibrium Laws for Gaseous Reactions

Review

The partial pressure of a gas in a mixture is proportional to its concentration. Therefore, the mass action expression for reactions involving gases can be written using partial pressures in place of concentrations. When partial pressures are used in the mass action expression, the equilibrium constant is designated as K_p.

Thinking It Through

For the following, identify the information needed to solve the problem and show (or explain) what must be done with it.

1 What is the molar concentration of nitrogen in air that has a total pressure of 760 torr at 25 °C? Air is composed of 79% N_2 by volume

Self-Test

5. Write the K_p expression for the reaction

$$2N_2O(g) + 3O_2(g) \rightleftharpoons 4NO_2(g)$$

____

6. A chemical reaction has the following equilibrium law.

$$K_p = \frac{P^2_{BrF_3}}{P^3_{F_2}\,P_{Br_2}}$$

(a) What is the chemical equation for the reaction?

____

(b) What is the expression for K_c?

____

7. The partial pressure of O_2 in air is approximately 160 torr. What is the concentration of O_2 in air at 25 °C expressed in mol/L?

____

New Terms

Write the definition of the following term, which was introduced in this section. If necessary, refer to the Glossary at the end of the text.

K_p

15.5 Calculating Equilibrium Constants from Thermodynamic Data

Review

For a given composition in a reaction mixture, the value of ΔG for the reaction is related to the reaction quotient by Equation 15.4.

$$\Delta G = \Delta G^\circ + RT \ln Q \qquad (15.4)$$

If you must apply this equation, remember that to calculate Q we use partial pressures for gaseous reactions and molar concentrations if the reaction is in solution. Equation 15.4 can be used to determine where a reaction stands relative to equilibrium:

ΔG is negative	reaction spontaneous in forward direction
ΔG is zero	the reaction is at equilibrium
ΔG is positive	reaction spontaneous in reverse direction

Example 15.3 illustrates how Equation 15.4 is applied.

The principal emphasis in this section is that the equilibrium constant is related to ΔG° for the reaction. You should learn Equation 15.5.

$$\Delta G^\circ = -RT \ln K \qquad (15.5)$$

In these equations, K is K_p for reactions involving gases. It is K_c for reactions in liquid solution.

In using Equation 15.5, be sure to choose the value of R that matches the energy units of ΔG°. If ΔG° is in kilojoules, use $R = 8.314$ J mol^{-1} K^{-1} . If ΔG° is in kilocalories, use R = 1.987 cal mol^{-1} K^{-1}. Also, remember to change kilojoules to joules (or kilocalories to calories).

In this section we also see how we can estimate the value of the thermodynamic equilibrium constant at temperatures other than 25 °C. This is done by calculating the value of ΔG_T° from ΔH° and ΔS°. The appropriate equation is

$$\Delta G_T^\circ = \Delta H^\circ - T\Delta S^\circ$$

Once you've obtained ΔG_T° in this way, then you use Equation 15.5 to calculate the value of K.

Thinking It Through

For the following, identify the information needed to solve the problem and show (or explain) what must be done with it.

2 The reaction, $2C_4H_{10}(g) + 13O_2(g) \rightleftharpoons 8CO_2(g) + 10H_2O(g)$, has $\Delta G^\circ = -5406$ kJ at 25 °C. A certain mixture of these gases has the following partial pressures: for C_4H_{10}, 3×10^{-6} torr; for O_2, 12.0 torr; for CO_2, 359 torr; for H_2O, 375 torr. Is this reaction mixture at equilibrium? If not, which way must the reaction proceed to reach equilibrium?

Self-Test

8. The reaction, $2N_2O(g) + 3O_2(g) \rightleftharpoons 4NO_2(g)$, has $\Delta G^\circ = +0.17$ kJ. In a reaction mixture the gases involved in the reaction have the following partial pressures: N_2O, 220 torr; O_2, 120 torr; NO_2, 458 torr. Is this reaction mixture at equilibrium? If not, which direction must the reaction proceed to reach equilibrium?

____

____

9. The reaction, $NO(g) + NO_2(g) + H_2O(g) \rightleftharpoons 2HNO_2(g)$, has $K_p = 1.56$ atm^{-1} at 25 °C. What is ΔG° for this reaction expressed in kilojoules?

____

10. At 25 °C, $K_p = 4.8 \times 10^{-31}$ for the reaction, $N_2 + O_2 \rightleftharpoons 2NO$. What is ΔG° for this reaction expressed in kJ?

____

11. The reaction, $2N_2O(g) + 3O_2(g) \rightleftharpoons 4NO_2(g)$, has $\Delta G^\circ = +0.17$ kJ. What is K_p for this reaction?

____

12. Use the data in Table 16.2 on page 675 of the text to compute the value of K_p for the reaction: $C_2H_2(g) + 2H_2(g) \rightleftharpoons C_2H_6(g)$

____

13. The reaction, $2C_4H_{10}(g) + 13O_2(g) \rightleftharpoons 8CO_2(g) + 10H_2O(g)$, has $\Delta G^\circ = -1292$ kcal at 25 °C. What is the value of K_p for this reaction?

____

14. The oxidation of sulfur dioxide to sulfur trioxide by molecular oxygen, $2SO_2(g) + O_2(g) \rightleftharpoons 2SO_3(g)$, has a standard heat of reaction, $\Delta H^\circ = -196.6$ kJ and a standard entropy of reaction $\Delta S^\circ = -189.6$ J/K. What is the value of K_p for this reaction

   (a) at 25 °C ______________________________

(b) at 500 °C ________________

New Terms

Write the definition of the following term, which was introduced in this section. If necessary, refer to the Glossary at the end of the text.

thermodynamic equilibrium constant

15.6 The Relationship Between K_p and K_c

Review

For reactions in which there is a change in the number of moles of gas on going from the reactants to the products, the numerical values of K_p and K_c are not the same, and in this section you learn how to convert between them. The equation that you need to remember in order to do this is Equation 15.6. In applying this equation, be careful to compute Δn_g correctly; it is the difference in the total number of moles of *gas* between the product side and the reactant side of the equation. It is also important to remember to use R = 0.0821 L atm mol^{-1} K^{-1}. Any other value of R will give incorrect answers.

Thinking It Through

For the following, identify the information needed to solve the problem and show (or explain) what must be done with it.

3 The reaction $2H_2(g) + O_2(g) \rightleftharpoons 2H_2O(g)$ has $K_c = 9.1 \times 10^{80}$ at 25 °C. What is the value of ΔG°_{298} for this reaction?

Self-Test

15. For which of the following reactions will K_p be numerically equal to K_c?

(a) $PCl_5(g) \rightleftharpoons PCl_3(g) + Cl_2(g)$

(b) $H_2(g) + Cl_2(g) \rightleftharpoons 2HCl(g)$

(c) $2CO(g) + O_2(g) \rightleftharpoons 2CO_2(g)$ ______

16. For the reactions in Question 15, which will have

(a) $K_p > K_c$ at 25 °C? ______

(b) $K_p < K_c$ at 25 °C? ______

17. The reaction $2H_2(g) + O_2(g) \rightleftharpoons 2H_2O(g)$ has $K_c = 9.1 \times 10^{80}$ at 25 °C. What is the value of K_p for this reaction?

____

18. At 300 °C, the reaction $2SO_2(g) + O_2(g) \rightleftharpoons 2SO_3(g)$ has $K_p = 1.0 \times 10^8$. What is the value of K_c for this reaction at this temperature?

____

New Terms

none

15.7 The Significance of the Magnitude of *K*

Review

When the equilibrium constant (either K_p or K_c) is large, the reaction goes far toward completion by the time equilibrium is reached. We express this by saying that the position of equilibrium lies far to the right. When K is small, very little products are formed when equilibrium is reached, so we say the position of equilibrium lies far to the left. See the summary on page 654 of the text.

By comparing *K*'s for reactions of similar stoichiometry, we are able to compare the extent to which the reactions proceed toward completion. Even if the stoichiometries differ, comparisons are still valid if the *K*'s are vastly different.

Self-Test

19. For the reaction

$$CH_4(g) + H_2O(g) \rightleftharpoons CO(g) + 3H_2(g)$$

$K_c = 1.78 \times 10^{-3}$ at 800 °C, $K_c = 4.68 \times 10^{-2}$ at 1000 °C,and $K_c = 5.67$ at 1500 °C. From these data, should more or less CO and H_2 form as the temperature of this equilibrium is increased? Explain.

____

____

20. Hypochlorite ion, OCl^-, is the active ingredient in liquid laundry bleach. It has a tendency to react with itself as follows (although the reaction is slow at room temperature).

$$3OCl^- \rightleftharpoons 2Cl^- + ClO_3^- \qquad K_c = 10^{27}$$

If a bottle of bleach is allowed to come to equilibrium, how should the relative concentrations of the reactants and products compare?

New Terms

none

14.9 Heterogeneous Equilibria

Review

A heterogeneous equilibrium is one in which not all of the reactants and products are in the same phase. The mass action expression for a heterogeneous reaction is normally written without terms for the concentrations of pure liquids or solids. This is because the number of moles per liter for such substances does not depend on the amount of the substance in the reaction mixture. Here are two examples. For the reaction

$$Cl_2(g) + 2NaBr(s) \rightleftharpoons Br_2(l) + 2NaCl(s)$$

$$K_c = \frac{1}{[Cl_2]} \quad \text{and} \quad K_p = \frac{1}{P_{Cl_2}}$$

For the reaction

$$NH_4Cl(s) \rightleftharpoons NH_3(g) + HCl(g)$$

$$K_c = [NH_3]\,[HCl]$$

and

$$K_p = P_{NH_3}P_{HCl}$$

Self-Test

21. Write the equilibrium law for the following reactions in terms of K_c.

(a) $PCl_3(l) + Cl_2(g) \rightleftharpoons PCl_5(s)$

(b) $CaCO_3(s) + SO_2 \rightleftharpoons CaSO_3(s) + CO_2(g)$

22. Write the equilibrium law for K_c the reaction

$$PbCl_2(s) \rightleftharpoons Pb^{2+}(aq) + 2Cl^-(aq)$$

23. Write the equilibrium law for K_c the reaction

$$3Zn(s) + 2Fe^{3+}(aq) \rightleftharpoons 2Fe(s) + 3Zn^{2+}(aq)$$

New Terms

Write the definitions of the following terms, which were introduced in this section. If necessary, refer to the Glossary at the end of the text.

homogeneous reaction
heterogeneous reaction

15.9 Le Châtelier's Principle and Chemical Equilibria

Review

Le Châtelier's principle states that if a system in equilibrium is upset by an outside disturbance, the system changes in a direction that counteracts the disturbance and, if possible, returns the system to equilibrium.

1. A reaction shifts in a direction away from the side of the reaction to which a substance is added. It shifts toward the side from which a substance is removed.
2. A decrease in the volume of a gaseous reaction increases the pressure and shifts the equilibrium toward the side with the fewer number of molecules of *gas*. If both sides have the same number of molecules of gas, a volume change will not affect the equilibrium. Solids and liquids are virtually incompressible, so pressure changes have no effect on them.

3. An increase in temperature shifts an equilibrium in a direction that absorbs heat. The value of K increases with increasing temperature for a reaction that is endothermic in the forward direction. If you remember this, it's easy to figure out what happens for an exothermic reaction as well. Remember that temperature is the *only* thing that affects the value of K for a reaction.
4. Catalysts have absolutely no effect on an equilibrium. They only affect how fast a system gets to equilibrium.
5. Adding an inert gas, without simultaneously changing the volume, will have no effect on the position of equilibrium.

Self-Test

24. Self-Test Question 19 gives values of K_c at three different temperatures for the reaction

$$CH_4(g) + H_2O(g) \rightleftharpoons CO(g) + 3H_2(g)$$

Is this reaction, as read from left to right, exothermic or endothermic? Explain.

____

____

25. For the reaction in Question 24, state how the amount of CO at equilibrium will be affected by
    (a) adding CH_4 ______________
    (b) adding H_2 ______________
    (c) removing H_2O ______________
    (d) decreasing the volume ______________
    (e) adding helium at constant volume

    ____

26. For the reaction, $F_2 + Cl_2 \rightleftharpoons 2ClF$ + heat, describe how the amount of Cl_2 will be affected by
    (a) adding F_2 ______________
    (b) adding ClF ______________
    (c) decreasing the volume ______________

(d) raising the temperature ____________________

(e) adding a catalyst ____________________

27. How will the value of K_p change for the reaction in Question 26 if the temperature is lowered?

____

New Terms

none

15.10 Equilibrium Calculations

Review

As described in the text, the calculations that you learn how to perform in this section can be divided into two categories—calculating K_c from equilibrium concentrations, and vice versa. These are not really very difficult *if you approach them systematically*. It is important that you not attempt to take shortcuts; that's where many students make mistakes or become lost and can't finish the problem.

In working equilibrium problems there are some important rules to follow, and the concentration table that we construct under the chemical equation helps you follow them.

1 The concentrations that you substitute into the equilibrium law *must* correspond to equilibrium concentrations. These are the only quantities that satisfy the equilibrium law.

2 Since we are working with K_c, quantities in the concentration table should have the units of molar concentration (mol L^{-1}). This means that if you are not given the molar concentration, but instead a certain number of moles in a certain volume, you should immediately change the data to mol L^{-1}. For example, if you are told that 5.0 mol of a certain reactant is in a volume of 2.0 L, change the data to 5.0 mol/2.0 L = 2.5 mol L^{-1} before entering it into the appropriate place in the concentration table.

3 An important difference between the two kinds of calculations that you are learning to perform is that when you are asked to calculate K_c, the equilibrium constant is the unknown and you must therefore find *numerical values* for *all* the entries in the "Equilibrium concentration" row of the table. On the other hand, when K_c is known and you are be-

ing asked to find an equilibrium concentration, then an x (or other symbol) appears in some way in the "Equilibrium concentration" row.

4 The initial concentrations in a reaction mixture are determined by the person doing the experiment. When you are given the composition of a mixture in a problem, these are initial concentrations, unless stated otherwise. In filling in this line of the table, we imagine that the reaction mixture can be prepared before we allow the reaction to take place. Then we let the reaction proceed to equilibrium and see how the concentrations change.

5 The changes in concentration are controlled by the stoichiometry of the reaction. When we want to compute K_c and have to calculate what the equilibrium concentrations are, the entries in this column are *numbers* whose ratios are the same as the ratios of the coefficients (see page 662). When you are given K_c and initial concentrations, this is where the x's first appear. The following points are useful to remember:

- The coefficients of x can be the same as the coefficients in the balanced equation; this ensures that they are in the right ratio.
- The "changes" for the reactants all must have the same algebraic sign, and these signs must be the opposite of the algebraic signs of the "changes" for the products. (In other words, if the changes for the reactants are positive, the changes for the products are negative.)
- If the initial concentration of some reactant or product is zero, its change *must* be positive.

6 Equilibrium concentrations are obtained by algebraically adding the "Change in concentration" to the "Initial concentration." In cases where K_c is *very* small, the extent of reaction from left to right will also be *very* small. This allows equilibrium concentration expressions such as $(0.200 + x)$ or $(0.100 - x)$ to be simplified. The initial concentrations of the reactants hardly change at all as the reaction proceeds, so x can be expected to have a very small value. In this case, a quantity such as $(0.200 + x)$ will be very nearly equal to 0.200 after the very small value of x is added to 0.200.

$$(0.200 + x) \approx 0.200$$

This kind of simplifying assumption often makes a difficult problem very easy. Review the discussion of this on page 670.

7 Look for ways of simplifying the algebra in solving for x. Sometimes this isn't possible, as we see in Example 15.14, where the quadratic equation is used.

Thinking It Through

For the following, identify the information needed to solve the problem and show (or explain) what must be done with it.

4 **The reaction $2HI(g) \rightleftharpoons H_2(g) + I_2(g)$ has $K_c = 1.6 \times 10^{-2}$. A mixture is prepared containing 0.100 M HI, 0.010 M H_2, and 0.020 M I_2. When this mixture reaches equilibrium, what will be the molar concentration of HI?**

Self-Test

28. At 25 °C a mixture of Br_2 and Cl_2 in carbon tetrachloride reacted according to the equation,

$$Br_2 + Cl_2 \rightleftharpoons 2BrCl.$$

The following equilibrium concentrations were found: $[Br_2] = 0.124\ M$, $[Cl_2] = 0.237\ M$, $[BrCl] = 0.450\ M$. What is the value of K_c for this reaction?

____

29. Referring to Question 28, what is the value of K_c for the reaction

$$2BrCl \rightleftharpoons Br_2 + Cl_2 \text{ at } 25\ °C?$$

____

30. A mixture of $N_2O(g)$, $O_2(g)$, and $NO_2(g)$ was prepared in a 4.00-liter container and allowed to come to equilibrium according to the equation

$$2N_2O(g) + 3O_2(g) \rightleftharpoons 4NO_2(g)$$

The mixture originally contained 0.512 mol N_2O.

(a) As the reaction came to equilibrium, the concentration of N_2O decreased by 0.0450 mol/L. What was the equilibrium concentration of N_2O?

____

(b) By how much did the O_2 concentration change?

____

(c) By how much did the NO_2 concentration change?

____

31. At a certain temperature, a mixture of $N_2O(g)$ and $O_2(g)$ was prepared having the following initial concentrations: $[N_2O] = 0.0972\ M$, $[O_2] = 0.156\ M$. The mixture came to equilibrium following the equation,

$$2N_2O(g) + 3O_2(g) \rightleftharpoons 4NO_2$$

At equilibrium, the NO_2 concentration was found to be 0.0283 *M*. What is K_c for this reaction?

32. The reaction, $2BrCl \rightleftharpoons Br_2 + Cl_2$, has $K_c = 0.145$ in CCl_4 at 25 °C. If 0.220 mol of BrCl is dissolved in 250 mL of CCl_4, what will be the concentrations of BrCl, Cl_2 and Br_2 when the reaction reaches equilibrium?

33. At 25 °C, $K_c = 1.6 \times 10^{-17}$ for the reaction,

$$N_2(g) + 2O_2(g) \rightleftharpoons 2NO_2(g)$$

In air, the concentrations of N_2 and O_2 are: $[N_2] = 0.0320\ M$, $[O_2] = 0.00860\ M$. Taking these as initial concentrations, what should be the equilibrium concentration of NO_2 in air?

New Terms

none

Answers to Thinking It Through

1 The molar concentration of a gas is given by the equation

$$M = \frac{n}{V} = \frac{P}{RT}$$

where *P* is the pressure, in this case, the partial pressure of N_2. Air is 79% N_2 by volume. This means that if we had 1 liter of air, 79% of it would be N_2; when expanded to the full 1 liter volume, it would exert 79% of the total pressure. Therefore, the partial pressure of the N_2 is 79% of 760 torr. We can change the calculated partial pressure of N_2 to atmospheres by multiplying by the factor (1 atm/760 torr). We use $R = 0.0821$ L atm mol^{-1} K^{-1} to make the units cancel correctly, and we must express the temperature in kelvins. Substituting values into the equation above gives the molarity of the N_2. [The answer is 0.0323 *M*]

2 We can use the equation $\Delta G = \Delta G^\circ + RT \ln Q$ to calculate ΔG. If ΔG is zero, the reaction mixture is at equilibrium; if ΔG is negative the reaction must proceed in the forward direction to reach equilibrium; and if ΔG is positive, then the reaction must proceed in the reverse direction.

To calculate ΔG we need the value of the reaction quotient Q. This is the value of the mass action expression for the system, written using partial pressures.

$$\text{mass action expression} = \frac{P^2_{CO_2} P^{10}_{H_2O}}{P^2_{C_4H_{10}} P^{13}_{O_2}}$$

Substituting partial pressures expressed in atmospheres (obtained from partial pressures in torr by dividing by 760 torr/atm) gives the value of Q. We also need to use $T = 298$ K and $R = 8.314$ J mol^{-1} K^{-1}, and $\Delta G^\circ = -5406 \times 10^3$ J. These quantities are substituted into the equation above to calculate ΔG. [$\Delta G = -5209$ kJ, so the reaction is spontaneous in the forward direction.]

3 To calculate ΔG°, we need to have the value of K_p, not the value of K_c, because this is a reaction between gases. Therefore, we have to convert K_c to K_p. Rearranging equation 15.6 to solve for K_p gives

$$K_p = K_c\,(RT)^{-\Delta n_g}$$

We substitute $R = 0.0821$ L atm mol^{-1} K^{-1}, $T = 298$ K, $K_c = 9.1 \times 10^{80}$, $\Delta n_g = -1$, and then solve for K_p. Then we substitute K_p into the equation

$$\Delta G^\circ = -RT \ln K_p$$

using $R = 8.314$ J mol^{-1} K^{-1} and $T = 298$ K. The answer will be in joules. [The calulated value of ΔG° is -4.70×10^3J, or -470 kJ.]

4 First we write the equilibrium law for the reaction

$$K_c = \frac{[H_2][I_2]}{[HI]^2} = 1.6 \times 10^{-2}$$

We also must set up the concentration table, which is shown at the top of the next page.. Since we don't know which way the reaction will proceed to equilibrium, let's assume the reaction will go to the left, to form more HI.

	$2HI(g)$ $\rightleftharpoons$	$H_2(g)$ +	$I_2(g)$
Initial concentration (M)	0.100	0.010	0.020
Change in concentration (M)	$+2x$	$-x$	$-x$
Equilibrium concentrations (M)	$0.100 + 2x$	$0.010 - x$	$0.020 - x$

Next, we substitute equilibrium concentrations into the mass action expression.

$$K_c = \frac{(0.10 + 2x)^2}{(0.010 - x)(0.020 - x)}$$

The equation is not a perfect square, but the highest order in x term is is x^2. Therefore, we can expand the equation and substitute values into the quadratic formula. Two values will be obtained for x, but only one of them will make sense physically. This is the one we use to substitute into the expression for [HI], namely, [HI] = (0.100 + $2x$).

Answers to Self-Test Questions

1. The $C_2H_3O_2^-$ ion that it is in can pick up an H^+ and become a $HC_2H_3O_2$ molecule. Later this molecule can lose H^+ to become a $C_2H_3O_2^-$ ion again. This can be repeated over and over.

2. (a) $\frac{[C_2H_6]}{[C_2H_4]\,[H_2]} = K_c$ (b) $\frac{[NO_2]^4}{[N_2O]^2\,[O_2]^3} = K_c$

   (c) $\frac{[Cr_2O_7^{2-}]\,[H_2O]}{[HCrO_4^-]^2} = K_c$ (d) $\frac{[CH_3CO_2CH_3]\,[H_2O]}{[CH_3OH]\,[CH_3CO_2H]} = K_c$

3. Only (b) and (c) are at equilibrium.
4. (a) $K_c = 0.0202$,
   (b) $K_c = 7.04$,
   (c) $HI(g) \rightleftharpoons \frac{1}{2} H_2(g) + \frac{1}{2} I_2(g)$, $K_c = 0.142$

5 $K_p = \frac{P_{NO_2}^4}{P_{N_2O}^2 P_{O_2}^3}$

6. (a) $3F_2(g) + Br_2(g) \rightleftharpoons 2BrF_3(g)$
   (b) $K_c = \frac{[BrF_3]^2}{[F_2]^3\,[Br_2]}$
7. 8.60×10^{-3} mol L^{-1}

8. Calculated $\Delta G = +15{,}022$ J. Mixture is not at equilibrium. Reverse reaction is spontaneous, so reaction goes to the left to reach equilibrium.
9. $\Delta G^\circ = +1.10$ kJ
10. $\Delta G^\circ = +173$ kJ
11. $K_p = 0.93$
12. $K_p = 2.9 \times 10^{42}$
13. $K_p = 10^{948}$
14. (a) $K_p = 5.2 \times 10^2$ (b) $K_p = 3.5 \times 10^3$
15. Reaction b
16. (a) Reaction a, (b) Reaction c
17. $K_p = 3.7 \times 10^{79}$
18. $K_c = 4.7 \times 10^9$
19. More CO_2 and H_2 should form, because K_c increases with increasing temperature.
20. At equilibrium the concentrations of Cl^- and ClO_3^- should be large compared to the concentration of OCl^-.
21. (a) $K_c = \dfrac{1}{[Cl_2]}$ (b) $K_c = \dfrac{[CO_2]}{[SO_2]}$
22. $K_c = [Pb^{2+}][Cl^-]^2$
23. $K_c = \dfrac{[Zn^{2+}]^3}{[Fe^{3+}]^2}$
24. Endothermic, because K_c increases with increasing temperature.
25. (a) increase , (b) decrease, (c) decrease, (d) decrease, (e) no change
26. (a) decrease, (b) increase, (c) no change, (d) increase, (e) no change
27. K_p increases as the temperature is lowered.
28. $K_c = 6.89$
29. $K_c = 0.145$
30. (a) 0.083 *M* (The initial N_2O concentration was 0.512 mol/4.00 L = 0.128 *M*.) (b) decreased by 0.0675 mol/L (c) increased by 0.0900 mol/L
31. $K_c = 0.0378$
32. [BrCl] = 0.500, $[Cl_2] = 0.190$ *M*, $[Br_2] = 0.190$ *M*
33. $[NO_2] = 3.1 \times 10^{-12}$ *M*

Tools you have learned

Remove this chart from the Study Guide and keep it handy when tackling homework problems.

Tool	*Function*
Reaction quotient	When equal to K, a system is at equilibrium.
Chemical equation	Enables you to construct the equilibrium law for a reaction. For gaseous reactions, can be expressed as K_c or K_p.
Manipulating equilibrium equations	Determine K for a reaction by manipulating a chemical equilibrium or by combining other chemical equilibria.
$\Delta G = \Delta G^\circ + RT \ln Q$	Determine whether a reaction is at equilibrium
$\Delta G^\circ = -RT \ln K$	Relate ΔG° to the equilibrium constant; K_p for gaseous reactions, K_c for reactions in solution.
$K_p = K_c\,(RT)^{\Delta n_g}$	Convert between K_p and K_c.
Magnitude of the Equilibrium constant	Magnitude of K is related to extent to which reaction proceeds toward completion when equilibrium is reached.
Equation for heterogeneous reaction	Construct equilibrium law of the reaction, omitting pure solids and liquids from the mass action expression.
Le Châtelier's Principle	Analyze the effects of disturbances on the position of equilibrium.

Tool	Function
Concentration table	Used in solving problems in chemical equilibria. Incorporates initial concentrations, changes in concentration, and equilibrium concentration expressions.
Simplifying assumptions	Used to simplify the algebra when working equilibrium problems for reactions for which K is very small.

Summary of Important Equations

The equilibrium law from the coefficients of a balanced equation

For the general chemical equation

$$dD + eE \rightleftharpoons fF + gG$$

$$\frac{[F]^f\,[G]^g}{[D]^d\,[E]^e} = K_c$$

Free energy change and the reaction quotient

$$\Delta G = \Delta G^\circ + RT \ln Q$$

Standard free energy change and the equilibrium constant

$$\Delta G^\circ = -RT \ln K$$

Relationship between K_p and K_c

$$K_p = K_c\,(RT)^{\Delta n_g}$$

Chapter 16

ACID-BASE EQUILIBRIA

A great deal of the chemical properties of all matter depend on where two small particles are—the electron and the proton—and on what they can do and where they can move. All of electrochemistry, for example, involves electron transfers. In this chapter we concentrate our attention on the other tiny particle, the proton. All of acid-base chemistry as viewed either by Arrhenius or by Brønsted revolves around what substances can supply protons, what can accept them, and how their transfers can be arranged or prevented.

This is a chapter where learning how to do various calculations is of paramount importance. The large number of worked examples have been very carefully prepared, and if you master one kind of calculation before going on to the next, the way to success will be smooth. The trickiest parts involve making simplifying assumptions about what we can safely ignore in certain calculations. It is particularly important that you understand these assumptions, because they really do simplify the calculations. Good luck in your study of this chapter. There are very few places in either the chemical or biological sciences where acid-base equilibria in water aren't vitally involved.

Learning Objectives

1. To learn the extent of the self-ionization of water in terms of its ion-product constant (K_w); how to calculate [H_3O^+] given [OH^-], or [OH^-] given [H_3O^+]; and how acidic, basic, and neutral solutions are defined.
2. To learn the equations that define pH and pOH; to calculate either pH or pOH given the other (at 25 °C) or given a value of [H^+] or [OH^-]; to calculate [H^+] from pH or pOH; and to learn some ways by which pH values are determined experimentally.
3. To learn to calculate the pH (or pOH) given the molar concentration of a dilute solution of a strong acid or strong base.
4. To learn how to write the equation for K_a for an acid; to use K_a data to classify an acid as weak, moderate or strong; to calculate the K_a of an acid from a pH value and $[HA]_{initial}$; to calculate percentage ionization of a weak acid; and to calculate val-

ues of $[H^+]$ (or pH) from data on K_a and initial concentrations of a weak acid.

5. To learn how to calculate pK_a from K_a and how to use pK_a values to judge the relative strengths of acids.
6. To learn how to write equations for K_b for a base; to calculate K_b of a base from a pH value of its solution at some given initial concentration; and to calculate the pH of a solution of a base from its initial concentration and its K_b.
7. To learn how to calculate values of pK_b from K_b and how to use pK_b values to judge the relative strengths of bases.
8. To learn the meaning of expressions such as "the solution is buffered at pH 7.35," or "these species in water make up a buffer system"; to use molarity data for buffer systems to calculate the pH of a buffered solution; to learn how to select a buffer pair for holding a particular pH relatively constant at some predetermined value; and to do the calculations required for preparing such a solution.
9. To learn how to calculate the pK (pK_a or pK_b) of one member of a conjugate pair from the value of pK for the other.
10. To be able to judge from the formula of a salt if one (or both) of its ions can affect the pH of an aqueous solution; and to be able to calculate the pH of the solution.
11. To see how the pH's of solutions of various kinds of acids or bases change as a titrant is added and acid-base neutralization occurs; and to learn the principles involved in the choice of a good acid-base indicator.

16.1 The Ion Product of Water

Review

Water ionizes very slightly: $2H_2O \rightleftharpoons H_3O^+ + OH^-$ and the *ion product,* $[H_3O^+][OH^-]$ or K_w, is a constant at any given temperature. For example, at 25 °C, $K_w = 1.00 \times 10^{-14}$. Regardless of the temperature or the presence of solutes in an aqueous solution, when $[H_3O^+] = [OH^-]$, the solution is *neutral*; when $[H_3O^+]$ is greater than $[OH^-]$, the solution is *acidic*; and when $[H_3O^+]$ is less than $[OH^-]$, the solution is *basic*. Be sure to learn these facts.

Self-Test

1. When $[H_3O^+] = 3.80 \times 10^{-6}\ M$ at 25 °C,
   (a) What is $[OH^-]$? ________
   (b) Is the solution acidic, basic, or neutral? ________
2. Which is the likeliest value of K_w at 60 °C?
   (a) 9.6×10^{-16} (c) 1.00×10^{-14}
   (b) 9.6×10^{-15} (d) 9.6×10^{-14} ________
3. At 5 °C, $K_w = 1.85 \times 10^{-15}$. If an aqueous solution has $[H_3O^+] = 4.30 \times 10^{-8}\ M$, is the solution acidic, basic, or neutral?

   ____

New Terms

Write the definitions of the following terms, which were introduced in this section. If necessary, refer to the Glossary at the end of the text.

acidic solution
basic solution
ion-product constant of water
K_w
neutral solution

16.2 The pH Concept

Review

Learning how to do the calculations described in the Objectives is the major goal, and for these the central equations that must be learned are:

$$\text{pH} = -\log\,[H^+]$$

$$\text{pOH} = -\log\,[OH^-]$$

$$\text{pH} + \text{pOH} = 14.00 \text{ (at 25 °C)}$$

The simplicity of the third equation lets us give less emphasis to the question of pOH because any time that we might need it we can obtain it simply by subtracting the pH from 14.00 (assuming we're working at 25 °C). Now is the time to get it firmly fixed in your mind that as the pH decreases, the acidity increases; as the pH increases, the acidity decreases. When the pH is less than 7, the solution is acidic; when it's greater than 7, the solution is basic.

Thinking It Through

1 An aqueous solution with a pH of 3 has how many times the molar concentration of H^+ as an aqueous solution with a pH of 6? ________________

Self-Test

4. An aqueous solution at 25 °C with a pOH of 10 has a molar concentration of hydrogen ions, $[H^+]$, equal to

   (a) 1.00×10^{-10} *M* (c) 10.00 *M*

   (b) 1.00×10^{-4} *M* (d) 4.00 *M* __________

5. When $[H^+] = 4.8 \times 10^{-6}$, the pH is

   (a) 5.20 (b) 6.48 (c) 5.32 (d) 4.80 __________

6. When the pH = 9.65 at 25 °C, the value of $[H^+]$ is

   (a) 2.24×10^{-10} *M* (c) 4.35×10^{-10} *M*

   (b) 9.65×10^{-14} *M* (d) 3.50×10^{-9} *M* ______

7. When pOH = 7.35 at 25 °C, what is $[H^+]$? ______

8. What color does each of the following indicators have in strongly acidic and in strongly basic solutions? (Use Table 16.3 in the text.)

	Color in	
Indicator	Acid	Base
phenolphthalein	________	________
bromothymol blue	________	________
thymol blue	________	________

New Terms

Write the definitions of the following terms, which were introduced in this section. If necessary, refer to the Glossary at the end of the text.

acidic solution (in pH terms at 25 °C) pH

basic solution (in pH terms at 25 °C) pOH
neutral solution (in pH terms at 25 °C)

16.3 Solutions of Strong Acids and Bases

Review

When we have dilute solutions of strong acids or bases, the calculation of $[H^+]$ or pH is particularly simple because we can assume 100% ionization (or dissociation). For monoprotic acids, for example, the value of $[H^+]$ is identical with the molar concentration of the acid. Be sure to memorize the names and formulas of the five, strong, monoprotic acids given on page 688. Also be sure to learn that the hydroxides of all group IA metals are strong, water-soluble bases and that the hydroxides of all group IIA metals are strong, sparingly soluble bases.

Self Test

9. Calculate the pH of each solution.
   (a) 0.25 *M* HCl ____________
   (b) 0.25 *M* NaOH ____________
   (c) 0.0010 *M* $Ca(OH)_2$
   (Note: $Ca(OH)_2$ is a strong base.) ____________

New Terms

none

16.4 Acid Ionization Constants

Review

For the equilibrium involving a weak acid in water, we use the equation:

$$HA \rightleftharpoons H^+ + A^- \qquad (1)$$

We write the *acid ionization constant* from this equation:

$$K_a = \frac{[H^+][A^-]}{[HA]} \qquad (2)$$

And the pK_a for the weak acid is given by the equation:

$$pK_a = -\log K_a \qquad (3)$$

Being able to construct Equations (1) and (2) for any acid is an essential skill in chemistry.

When we compute the strengths of acids, the higher the K_a the stronger is the acid. When K_a is smaller than about 1×10^{-3}, the acid is weak; if it's greater than 1, it's strong. In between values of 10^{-3} and 1, the acid is classified as moderate.

In terms of values of pK_a, when pK_a is greater than 3, the acid is weak; when it is between 0 and 3 it is a moderate acid. (Strong acids have negative pK_a values.)

The values of molar concentrations of the various species in the equations for K_a are their values *after* the equilibrium has been established, not the values calculated from the quantities used to prepare the solutions. However, when we work with any weak acid, and its concentration is at least 1×10^{-3} *M* or higher, we make one enormously useful simplification. Whenever we have to use values of K_a and $[HA]_{initial}$ to calculate either $[H^+]$ or pH, we can let $[HA]_{initial}$ equal $[HA]_{equilibrium}$. We thus avoid having to solve a quadratic equation. We can summarize the steps in the major calculations explained in Section 16.4 in the text as follows.

Required to calculate: K_a and pK_a

The given data are: pH and $[HA]_{initial}$

Steps:

1. Calculate $[H^+]$ from the pH. This gives $[H^+]_{eq}$.
2. Recognize that $[H^+]_{eq}$ also equals $[A^-]_{eq}$. (For every H^+ obtained from HA we can't help but get one A^-, also.)
3. Let $[HA]_{eq} = [HA]_{initial} - [H^+]_{eq}$
4. Now we have values for all parts of the equation for K_a, so calculate K_a. From this we calculate pK_a

Required to calculate: $[H^+]$ and pH

The given data are: K_a and $[HA]_{initial}$

Steps:

1. The given value of $[H^+]$ is actually the value of $[H^+]_{eq}$, so we can let $[H^+] = [H^+]_{eq} = [A^-]_{eq} = x$. If the value of x is less than 5% of the value

of $[HA]_{initial}$, make the assumption given in step 2, next. If not, use the quadratic solution or the method of successive approximations.

2. (a) When x is less than 5% of $[HA]_{initial}$, set $[HA]_{initial}$ equal to $[HA]_{eq}$. (This simplification lets us avoid solving a quadratic equation.)

   (b) Use $K_a = \frac{x^2}{[HA]_{initial}}$ to find x, which is $[H^+]$.

3. Use the calculated value of $[H^+]$ to calculate the pH.

Required to calculate: Percentage ionization

The given data are: molar concentration, M, of a weak acid and K_a

Steps:

1. Calculate from M and K_a the value of $[H^+]$.
2. Note that $[H^+]$ = amount of HA that ionized.
3. Find percentage ionization as

$$\frac{\text{(amount of HA ionized)}}{\text{(amount of HA initially available)}} \times 100\%$$

In all cases, the systematic way to apply these steps is through the construction of concentration tables as were used in the worked examples.

Thinking It Through

2 Consider a solution of acetic acid that has a concentration of 0.620 M. Calculate $[H^+]$ and the pH for the solution. (Use $K_a = 1.8 \times 10^{-5}$.)

Self-Test

10. A new organic acid was discovered and at 25 °C a 0.115 M solution of this acid in water had a pH of 3.42. Calculate the value of K_a for this acid.

    ____

11. What is the value of pK_a for the bicarbonate ion if its K_a is 4.7×10^{-11}?

    ____

12. Which is the stronger acid, nitrous acid ($K_a = 7.1 \times 10^{-4}$) or formic acid ($K_a = 1.8 \times 10^{-4}$)?

____

13. Which is the stronger acid, hydrocyanic acid, HCN ($pK_a = 9.20$) or the ammonium ion, NH_4^+ ($pK_a = 9.24$)?

____

14. Calculate the percentage ionization of HCN in a solution that is 0.10 *M* HCN

____

New Terms

Write the definitions of the following terms, which were introduced in this section. If necessary, refer to the Glossary at the end of the text.

acid ionization constant, K_a $\qquad$ pK_a
percentage ionization

16.5 Base Ionization Constants

Review

The reaction of a Brønsted base with water sets up the following equilibrium, where *B* is a general base.

$$B + H_2O \rightleftharpoons HB^+ + OH^- \qquad (4)$$

The *base ionization constant,* K_b, of the base is associated with this equilibrium:

$$K_b = \frac{[HB^+][OH^-]}{[B]} \qquad (5)$$

The value of pK_b is found by the equation:

$$pK_b = -\log K_b \qquad (6)$$

If the base is negatively charged, like the acetate ion ($C_2H_3O_2^-$), then the charges in Equation (4) would have to be adjusted. (Now *HB* would be $HC_2H_3O_2$, for example, but OH^- would still be the other product.)

Be sure you are able to construct Equations (4) and (5) for any given base. It's an essential skill in chemistry. The steps laid down in the worked examples in Section 16.5 can be summarized as follows.

Required to calculate: K_b and pK_b

The Given data are: pH and $[B]_{initial}$

Steps:

1. Use the pH to find pOH, and from pOH calculate $[OH^-]$.
2. Recognize that $[OH^-]$ is the same as $[OH^-]_{eq}$, so $[OH^-] = [BH^+]_{eq}$.
3. Find $[B]_{eq}$ from: $[B]_{eq} = [B]_{initial} - [OH^-]$
4. Now calculate K_b and then pK_b.

Required to calculate: $[H^+]$ or $[OH^-]$ and pH

The given data are: K_b (or pK_b) and $[B]_{initial}$

Steps:

1. Recognize that $[OH^-]_{eq} = [BH^+]_{eq}$, and set equal to x.
2. Assume that $[B]_{eq} = [B]_{initial}$ thereby avoiding a quadratic equation. (We shall not encounter moderate bases requiring a quadratic solution.)
3. Use the given value of K_b to find x by: $K_b = \dfrac{x^2}{[B]_{initial}}$
4. Because $x = [OH^-]_{eq}$, use x to find pOH and from pOH find pH.

Self-Test

15. Calculate the pH of a 0.425 *M* solution of NH_3 in water. (Use $K_b = 1.8 \times 10^{-5}$.)

    ____

16. A new drug obtained from the seeds of a Uraguyan bush was found to be a weak organic base. A 0.100 *M* solution in water of this drug had a pH of 10.80. What is the K_b of this drug?

    ____

17. The K_b for the fluoride ion is 1.5×10^{-11}. What is its pK_b?

____

New Terms

Write the definitions of the following terms, which were introduced in this section. If necessary, refer to the Glossary at the end of the text.

base ionization constant, K_b $\qquad$ pK_b

16.6 Buffers: The Control of pH

Review

Be sure to notice that a buffer doesn't necessarily hold a solution at a *neutral* pH (7.00); it simply helps to maintain a reasonably steady pH at whatever the value selected—which could be on the acidic or on the basic side of 7.00. Also, a buffer doesn't maintain an absolutely constant pH; it prevents large changes even if strong acids or bases enter the system.

A buffer pair is often a solution of a salt such as NaA and the conjugate acid of A^-, in which A^- is a true Brønsted base and its conjugate, HA, is a weak acid. A typical example is $NaC_2H_3O_2$ and $HC_2H_3O_2$. The base, A^-, can tie up extra protons that enter, and the weak acid, HA, can neutralize extra OH^- ions that get into the solution.

Another kind of buffer pair is an ammonium salt of a weak acid plus ammonia; e.g., NH_4Cl and NH_3. The ion, NH_4^+, can donate a proton to any OH^- ion that enters; and NH_3 can tie up any proton that gets in.

Two factors determine the pH that a buffer pair will hold:

1. the pK_a of the Brønsted acid in the pair (or the pK_b of the Brønsted base if a system such as NH_4/NH_4Cl is used); and

2. the log of the *ratio* of the *initial* values of the molar concentrations of the substances used to establish the buffer.

If the buffer system is made up of a weak acid, HA, and its salt MA (where M is some metal ion and A is the anion), then

$$\text{pH} = \text{p}K_a + \log\frac{[\text{anion}]}{[\text{acid}]}$$

If the buffer is made up of a weak base, B, and $B\text{H}^+$ where B is a base (e.g., as in NH_3 and NH_4Cl), then

$$pOH = pK_b = \log\frac{[\text{cation}]}{[\text{base}]}$$

The terms in brackets, [ ], are *initial* molar concentrations.

To prepare a buffer for an acidic pH, we first have to decide the pH we want that is to be held reasonably constant. Then we search for a weak acid that has a value of pK_a quite close to this pH. We do this because pH = pK_a when the log term equals zero—which it does when the value in the numerator equals that of the denominator. Then we use the pH value we want and the pK_a that we have to accept when we pick the acid, and we calculate the ratio: [anion]/[acid], that will have a logarithm that will fit the pK_a to the pH. For maximum effectiveness, this log should be somewhere between 1.0 and –1.0, which means that the ratio, [anion]/[acid], has to be somewhere between 10 and 1/10th. Once this ratio is known, then we have to pick the actual quantities of the solutes. For a large capacity buffer we would want relatively high concentrations of each, but this may make the system toxic to some living organism being studied (if that's the final use of the buffered system). Just remember that it's the *ratio* of [anion] to [acid] that counts in determining the pH being buffered, but it's the actual quantities of these species that determines the buffer's capacity. Similar steps are used when we want a buffer with an alkaline pH.

Thinking It Through

3 You are a member of a team of scientists that decides that the lab should have on hand 1 L of an aqueous buffer for a pH of 4.5. What questions or problems have to be answered before the chemicals can be assembled, measured, and mixed?

Self-Test

18. Calculate the pH of a buffered solution made up as 0.125 *M* $KC_2H_3O_2$ and 0.100 *M* $HC_2H_3O_2$. The pK_a for acetic acid is 4.74.

19. The pK_a for formic acid, $HCHO_2$, is 3.74. What mole ratio of sodium formate, $NaCHO_2$, to formic acid is needed to prepare a solution buffered at a pH of 3.00?

20. How many grams of NH_4Cl have to be added to 200 mL of 0.25 *M* NH_3 to make a solution that is buffered at pH 9.10?

New Term

Write the definition of the following term, which was introduced in this section. If necessary, refer to the Glossary at the end of the text.

buffer

16.7 Conjugate Acid-Base Pairs and Their Values of K_a and K_b

Review

There is a very simple relationship between pK_a and pK_b for a *conjugate* acid-base pair:

$$pK_a + pK_b = 14.00 \text{ (at 25 °C)} \qquad (7)$$

The value of 14.00 comes from the value of pK_w, the negative of the log of the ion product constant of water. At 25 °C, $K_w = 1.0 \times 10^{-14}$, so

$$pK_w = -\log K_w \qquad (8)$$

$$= 14.00$$

Equation 7 shows us why the generalizations about conjugate pairs are valid: the stronger the acid, the weaker is its conjugate base and the stronger the base, the weaker is its conjugate acid.

Self-Test

21. What is the pK_a for HF, the conjugate acid of the fluoride ion for which $pK_b = 10.82$?

____

New Term

Write the definition of the following term, which was introduced in this section. If necessary, refer to the Glossary at the end of the text.

pK_w

16.8 Solutions of Salts: Ions as Weak Acids and Bases

Review

When an anion of a salt is a moderately strong Brønsted base, like the acetate ion or the cyanide ion, it reacts to some extent with water to generate some hydroxide ion and so raise the pH of the solution. We can tell if a given anion can do this simply by figuring out the conjugate acid of the anion and seeing if it's on the list of strong acids. If the conjugate acid is a weak acid, then the anion in question is at least a moderately strong base.

None of the anions of the strong acids reacts with water. All of the anions of weak acids do to some extent. To find how such anions affect the pH of the solution we just do the kind of pH calculations appropriate for any Brønsted base.

As for the cations of salts, they are either metal ions or are proton-donating cations, like NH_4^+. Metal ions exist in water as hydrated ions, and hydrated ions are often proton-donors. Those that are not proton-donors are the hydrated ions with 1+ or 2+ charges from Groups IA and IIA (except Be^{2+}, whose hydrate is acidic). Thus we expect all metal ions except those from Group IA and below Be in Group IIA to supply some hydrogen ions in water and so lower the pH. We have learned how to calculate the pH of a solution of a weak Brønsted acid like NH_4^+, but we have not studied a pH calculation for a hydrated metal ion (nor will we).

Self-Test

22. Decide if each ion is a Brønsted base toward water (and therefore will affect the pH of a solution).

(a) HCO_3^- ____________________

(b) SO_3^{2-} ____________________

(c) NO_3^- ____________________

23. Predict the behavior of each salt in water. State if it will cause an aqueous solution to be acidic, basic, or neutral.

(a) KNO_3 __________ (d) $NaNO_3$ __________

(b) $NaHCO_3$ __________ (e) K_3PO_4 __________

(c) K_2SO_3 __________ (f) $NaNO_2$ __________

24. Will any ion from potassium acetate, $KC_2H_3O_2$, react with water?

If so, calculate the pH of a 0.120 *M* solution.________________

25. Will any ion from ammonium bromide, NH_4Br, react with water?

____

If so, calculate the pH of a 0.240 *M* solution.________________

26. Will any ion from potassium iodide, KI, react with water?

____

If so, calculate the pH of a 0.100 *M* solution.________________

New Terms

none

16.9 Acid-Base Titrations Revisited

Review

The nature and appearance of an acid-base titration curve is a function of the relative strengths of the acid and base, because these determine how the ions being produced react with water (if at all) as acids or bases.

> When the acid and the base being titrated are both strong, the titration curve is symmetrical about pH 7, the equivalence point.
>
> When the acid is weak (e.g., acetic acid) and the base is strong (e.g., NaOH or KOH), the equivalence point is greater than 7 because the anion that forms reacts with waters to give a slightly alkaline pH.
>
> When the base is weak (e.g., ammonia) and the acid is strong (e.g., hydrochloric acid), the equivalence point is less than 7 because the cation that forms gives a slightly acidic pH.

The calculations required to obtain the points for a titration curve are identical in kind to those studied earlier in the chapter. Before any titrant is added, the situation is that of a dilute solution of some given acid or base having some previously determined ionization constant.

As soon as titrant is added, you have to take into account three factors—the total volume changes, so concentrations change for this reason alone; an initial concentration of what is being titrated decreases because of the neutralization; and, *if the salt being formed is one with an ion that is basic or acidic*, there is a buffer effect requiring a buffer-type calculation.

The indicator to be picked ideally is one whose color change is at its own midpoint when the titration reaches the equivalence point. In the ideal situation, the pH value at the equivalence point equals the K_{In} value for the acid-base indicator (its own acid ionization constant).

Self-Test

27. Calculate the pH of the solution that exists in a titration after 22.00 mL of 0.10 *M* NaOH have been added to 25.00 mL of 0.10 *M* $HC_2H_3O_2$.

    ____

28. Calculate the pH of the solution that forms in a titration by the addition of 24.00 mL of 0.20 *M* KOH to 25.00 mL of 0.20 *M* HCl.

    ____

29. Calculate the pH of the solution that has been produced during a titration when 26.00 mL of 0.10 *M* NaOH have been added to 25.00 mL of 0.10 *M* $HC_2H_3O_2$.

    ____

30. Calculate the pH of the solution that has formed during a titration when 28.00 mL of 0.10 *M* HCl has been added to 25.00 mL of 0.10 *M* NH_3.

    ____

31. If you knew of two indicators, X and Y, that underwent their color changes in the same range, and that X changed from colorless to blue and Y from orange to red, which indicator would be the better choice and why?

    ____

32. If you were asked to carry out a titration of dilute hydrobromic acid with potassium hydroxide, which indicator would be the better choice (and why)—thymol blue or bromothymol blue?

    ____

New Terms

Write the definitions of the following terms, which were introduced in this section. If necessary, refer to the Glossary at the end of the text.

end point

equivalence point

Answers to Thinking It Through

1 The definition of pH needs to be recalled and translated into $[H^+]$. When pH = 3, $[H^+] = 1 \times 10^{-3}$ mol/L; when pH = 6, $[H^+] = 1 \times 10^{-6}$ mol/L. The ratio of these two molar concentrations of H^+,

$$\frac{1 \times 10^{-3} \text{ mol/L}}{1 \times 10^{-6} \text{ mol/L}}$$

is 1000 : 1, so the solution with a pH of 3 has 1000 times the H^+ concentration as the solution with a pH of 6.

2 To solve this we need the equation for the chemical equilibrium:

$$HC_2H_3O_2(aq) + H_2O(l) \rightleftharpoons H_3O^+(aq) + C_2H_3O_2^-(aq)$$

and the equation for the acid ionization constant:

$$K_a = \frac{[H^+]\,[C_2H_3O_2^-]}{[HC_2H_3O_2]} = 1.8 \times 10^{-5}$$

Then we construct a concentration table to work our way toward values for the molarities in the K_a equation. Letting H^+ represent H_3O^+:

	$HC_2H_3O_2(aq)$ $\rightleftharpoons$	$H^+(aq)$ +	$C_2H_3O_2^-(aq)$
Initial concentrations	0.620	0	0
Changes in concentration caused by the ionization	$-x$	$+x$	$+x$
Final concentrations at equilibrium	$(0.620 - x)$ $= 0.620$	x	x

Therefore,

$$K_a = \frac{(x)(x)}{(0.620)} = 1.8 \times 10^{-5}$$

$$x = 3.3 \times 10^{-3}$$

Hence, $[H^+] = 3.3 \times 10^{-3}$ mol/L. The pH is the negative of the log of this value or 2.48

3 The desired pH is on the acidic side of 7, so a combination of a weak *acid* and its (perhaps sodium) salt should be used. The first question,

therefore, is "which acid?" This is decided by finding an acid with a pK_a value in the range of 4.5 ± 1.0. As studied in the text, the best pK_a is one that is related to the desired pH by the equation

$$pH = pK_a \pm 1$$

The next question is "what ratio of acid molarity to anion molarity (salt molarity) would bring the pH of the buffer to 4.5."
We know that for a buffer, from equation 16.17, page 711 of the text,

$$pH = pK_a - \log\frac{[HA]_{init}}{[A^-]_{init}}$$

The values of pH and pK_a would be inserted into this equation and the ratio solved for.

The next question is "what actual number of moles of the acid and the salt should be used to provide the desired ratio?" This depends on how much capacity is needed, which depends on an estimate by the research group of how much incursion of strong acid or base into the buffer solution can be tolerated. As stated on page 714 of the text, you need about 10 times as many moles of a buffer component as the estimated moles of acid (or base) that are expected to invade the system.

Another question—actually one of the first if the buffer is meant for a biological system—is "how toxic would the proposed buffer components be to the biological system being studied?"

Answers to Self-Test Questions

1. (a) 2.63×10^{-9} *M* (b) acidic (because $[H_3O^+]$ is greater than $[OH^-]$
2. d. K_w at 60 °C has to be larger than K_w at 25 °C.
3. The solution is neutral, since $[H_3O^+] = [OH^-] = 4.30 \times 10^{-8}$ *M*
4. b. (When pOH = 10, pH = 4 and $[H^+] = 1.00 \times 10^{-4}$)
5. c. (When $[H^+] = 4.68 \times 10^{-6}$ *M*, pH = –log (4.68×10^{-6})
6. a. (When pH = 9.65, $[H^+] = 1 \times 10^{-9.65} = 2.24$
7. 2.2×10^{-7} *M*
8.

Indicator	Color in Acid	Color in Base
phenolphthalein	none	red
bromothymol blue	yellow	blue
thymol blue	yellow	blue

9. (a) 0.60, (b) 13.40, (c) 11.30
10. $K_a = 1.26 \times 10^{-6}$
11. 10.33

12. Nitrous acid
13. HCN
14. 0.0079%
15. pH = 11.44
16. $K_b = 4.01 \times 10^{-6}$
17. 10.82
21. 3.18
19. pH = 4.84
20. $[CHO_2^-]/[HCHO_2] = 0.18$
21. 3.9 g of NH_4Cl
22. (a) HCO_3^- is a Brønsted base, (b) SO_3^{2-} is a Brønsted base, (c) NO_3^- is not a Brønsted base
23. (a) neutral, (b) basic, (c) basic, (d) neutral, (e) basic, (f) basic
24. Yes, $C_2H_3O_2^-$, pH = 8.92
25. Yes, NH_4^+, pH = 4.93
26. No
27. pH = 5.61
28. pH = 2.39
29. pH = 11.30 (caused solely by the OH^- ion provided by the excess NaOH)
30. pH = 2.25 (caused solely by the H^+ ion provided by the excess HCl)
31. Indicator X, because a very dramatic color change is easier to notice.
32. Bromothymol blue, because the equivalence point for the titration of a strong acid and a strong base is at pH 7, so we need an indicator whose color change occurs at or near pH 7.

Tools you have learned

Remove this chart from the Study Guide and keep it handy when tackling homework problems.

Tool	*Function*
Ion Product Constant of Water $K_w = [H^+][OH]$	Calculate $[H^+]$ or $[OH^-]$ from the value of the other and the value of K_w (14.00 at 25 °C)
Acidic solution requirements $[H^+] > [OH^-]$, or pH < 7 (at 25 °C)	Use a value of $[H^+]$ or pH to describe a solution as acidic.
Basic solution requirements $[H^+] < [OH^-]$, or pH > 7 (at 25 °C)	Use a value of $[H^+]$ or pH to describe a solution as basic.
Neutral solution requirements $[H^+] = [OH^-]$, or pH = 7 (at 25 °C)	Use a value of $[H^+]$ or pH to describe a solution as neutral.
pH or pOH $pH = -\log [H^+]$ $pOH = -\log [OH^-]$ pH + pOH = 14.00 (at 25 °C)	Calculate pH from $[H^+]$ or $[H^+]$ from pH. Calculate pOH from $[OH^-]$ or $[OH^-]$ from pH. Calculate pOH from pH or pH from pOH (at 25 °C)

Acid ionization constant: K_a and pK_a For the equilibrium: $HA \rightleftharpoons H^+ + A^-$ $K_a = \frac{[H^+][A^-]}{[HA]}$ $pK_a = -\log K_a$	Write the equilibrium expressions for the ionization of a weak acid. Calculate K_a and pK_a from $[H^+]$ and the initial concentration of HA. Calculate $[H^+]$ and pH from K_a (or pK_a) and the initial concentration of HA.
Relative Acid Strengths $K_a < 10^{-3}$ (or $pK_a > 3$): weak acid $K_a = 1 - 10^{-3}$ (or $pK_a = 0\text{-}3$): moderate acid $K_a > 1$: strong acid	Classify acids. Decide what approach to use in calculations involving the values of $[H^+]$ or the pH of solutions of acids.
Percent ionization of an acid, HA percent ionization = $\frac{\text{amt of HA ionized}}{\text{amt of HA initially available}} \times 100\%$	Calculate the percent ionization of a weak acid.
Base ionization constant: K_b and pK_b For the equilibrium: $B{:}^- + H_2O \rightleftharpoons BH + OH^-$ $K_b = \frac{[BH][OH^-]}{[B{:}^-]}$ $pK_b = -\log K_b$	Write the equilibrium expressions for the ionization of a weak base. Calculate K_b and pK_b from $[H^+]$ and the initial concentration of the base. Calculate $[OH^-]$ or $[H^+]$ and pH from K_b (or pK_b) and the initial concentration of the base.
Buffer pairs	From the identities of the components of a buffer, write the equations that show how the buffer maintains the pH.

Buffer calculation $$pH = pK_a - \log\frac{[HA]_{init}}{[A^-]_{init}}$$ For the best buffer action: $pH = pK_a \pm 1$	Given the identities of the components of a buffer (and so the pK_a of HA) and their concentrations, calculate the pH of the buffer solution. Select the best weak acid to use in preparing a buffer.
pK_a and pK_b for conjugate acid-base pairs $pK_a + pK_b = 14.00$ (at 25 °C)	Calculate pK_b and then K_b for a weak base from the value of pK_a (or K_a) of its conjugate acid.
Effect of a salt on pH Acidic cations lower the pH. Basic anions increase the pH.	To judge from the formula of a salt if it can affect the pH of a solution and in what way.

Summary of Important Equations

Ion Product Constant of Water

$$K_w = [H^+][OH]$$

pH and pOH

$$pH = -\log [H^+]$$

$$pOH = -\log [OH^-]$$

$$pH + pOH = 14.00 \text{ (at 25 °C)}$$

Acid ionization constant; K_a and pK_a

For the equilibrium:

$$HA \rightleftharpoons H^+ + A^-$$

$$K_a = \frac{[H^+][A^-]}{[HA]}$$

$$pK_a = -\log K_a$$

Percent ionization of a weak acid

$$\text{percent ionization} = \frac{\text{amt of H}A\text{ ionized}}{\text{amt of H}A\text{ initially available}} \times 100\%$$

Base ionization constant; K_b and pK_b

For the equilibrium:

$$B{:}^{-} + H_2O \rightleftharpoons BH + OH^{-}$$

$$K_b = \frac{[BH][OH^{-}]}{[B{:}^{-}]}$$

$$pK_b = -\log K_b$$

Buffers

$$pH = pK_a - \log\frac{[HA]_{init}}{[A^{-}]_{init}}$$

For the best buffer action:

$$pH = pK_a \pm 1$$

Conjugate acid-base pairs.

$$pK_a + pK_b = 14.00 \text{ (at 25 °C)}$$

Chapter **17**

SOLUBILITY AND SIMULTANEOUS EQUILIBRIA

At the heart of this chapter are the methods used to deal with equilibria more complex than those of simple monoprotic acids (Chapter 16). Because reasonable and workable simplifications are usually warranted, however, the methods turn out to be quite simple. Be sure to spot the simplifications, understand their rationale, and learn when to use them.

Learning Objectives

Throughout your study of this chapter, keep in mind the following objectives:

1 To learn how to calculate $[H^+]$ (or the pH) and $[A^{2-}]$ at equilibrium in a solution of any weak, diprotic acid, H_2A, given its values of K_{a_1} and K_{a_2}, and $[H_2A]_{initial}$.

2 To learn how to estimate the pH of a solution of a salt of a polyprotic acid given its molar concentration and the relevant ionization constant.

3 To learn how to deal quantitatively with solubility equilibria; to be able to write the equilibrium law for solubility; and to be able to perform the following kinds of calculations: (1) calculate the equilibrium constant from solubility data; (2) calculate the solubility, given the equilibrium constant; and (3) use the equilibrium constant to determine whether or not a precipitate will form in a solution having a certain composition.

4 To learn how solubility products are written and used when working with sparingly soluble oxides and sulfides of metals.

5 To learn how selective precipitation can work to remove one metal ion from a solution while leaving another in solution.

6 To review what complex ions are and to learn how to represent the equilibria involved in their formation and decomposition. You should also learn how the formation of complex ions can affect the solubility of salts.

17.1 Polyprotic Acids

Review

When H_2A is a weak acid (for example, H_2CO_3 or H_2S), the calculation of $[H^+]$ can be done—this is the key simplification—using only K_{a_1} and $[H_2A]_{initial}$ just as if it were a monoprotic acid. This works only if H_2A is *weak*, because then the contribution to $[H^+]$ from the second ionization (that is, the ionization of HA^-) is too small to survive the rounding of results to the correct number of significant figures.

In calculating $[A^{2-}]$ for solutions containing only H_2A, all we have to do—here's another simplification—is to write the symbol for the units, M (mol/L), after the numerical value of K_{a_2}. It works out this way because of a fortuitous cancellation of contributing terms.

Self-Test

1. Write the equilibrium equations and the expressions for K_{a_1} and K_{a_2} for the step-wise ionization of H_2CO_3.

2. What are the values of $[H^+]$, pH, and $[Asc^{2-}]$ in a solution of ascorbic acid (which we may represent as H_2Asc) with a concentration of 0.100 M? For ascorbic acid, $K_{a_1} = 7.9 \times 10^{-5}$ and $K_{a_2} = 1.6 \times 10^{-12}$.

New Terms

none

17.2 The pH of Solutions of Salts of Polyprotic Acids

Review

The text used the case of the carbonate ion to illustrate how to handle the reaction with water of an ion that produces yet another ion that can also react with water. Thus, the reaction of CO_3^{2-} with water gives HCO_3^- and OH^-, but HCO_3^- can also react as a Brønsted base with water, in a second equilibrium, to give more OH^- plus H_2CO_3. Similar situations occur with the HPO_4^{2-} ion and the S^{2-} ion.

On the surface, these appear to present complicated problems in calculating $[H^+]$ and pH, but there are useful simplifications when the acid is weak. One is that the second equilibrium contributes too little to the change in pH to matter. Thus *there is only one relevant equilibrium*, the first equilibrium involving the principal ion of the salt and not the second equilibrium involving the conjugate acid of the principal ion. There is thus only one relevant ionization constant—that of the salt's principal ion. The calculation is therefore no more than the kind we have already learned how to do—finding the pH of a solution of a weak base, for example, a solution of CO_3^{2-}, or HPO_4^{2-}, or S^{2-}.

Self-Test

3. Calculate the pH of a 0.10 *M* solution of Na_2X at 25 °C. The first ionization constant of H_2X is 9.5×10^{-8} and the second ionization constant is 1.3×10^{-14}. (A quadratic equation will have to be solved.)

____

New Terms

none

17.3 Solubility Equilibria for Salts

Review

The equilibrium law for the solubility equilibrium of a salt involves the *ion product,* the product of ion molarities raised to powers obtained from the subscripts in the formula. The "insoluble" salt, $Ca_3(PO_4)_2$, for example, gives the following equilibrium in a saturated solution.

$$Ca_3(PO_4)_2(s) \rightleftharpoons 3Ca^{2+}(aq) + 2PO_4^{3-}(aq)$$

The ion product for $Ca_3(PO_4)_2(s)$ is $[Ca^{2+}]^3[PO_4^{3-}]^2$. When the solution is *saturated*, the value of the ion product is a constant (at a given temperature) and is called the *solubility product constant*, K_{sp}.

$$K_{sp} = [Ca^{2+}]^3[PO_4^{3-}]^2 \quad \text{(saturated solution)}$$

(1) Calculating K_{sp} from solubility.

If you're given the *molar solubility* of a salt, you are told how many moles of it dissolve in one liter to give a saturated solution. This information is used to construct the "change" row in a concentration table. The concentrations of the ions change according to how many are produced when the salt dissociates, as shown in Examples 17.3 and 17.4. Notice that no entry appears under the formula for the solid. Also note that the initial concentrations are zero in these examples. This is because the salt is dissolving in pure water. In Example 17.5, the $PbCl_2$ is dissolving in a solution that already contains a dissolved salt that gives one of the ions involved in the equilibrium. The concentration of the *common ion,*provided by the salt and already in solution is given in the "initial concentration" row. As usual, the equilibrium concentration is obtained by algebraically adding the initial concentration to the change in concentration.

(2) Calculating solubility from K_{sp}

In these problems we take the molar solubility to be x. That's our unknown. Then the change in the concentration of each ion is just x multiplied by the number of those ions produced when one formula unit of the salt dissociates (Examples 17.6 and 17.7).

A salt is less soluble in a solution that contains one of its ions than it is in pure water. That's the *common ion effect.* In doing these calculations, we enter the concentration of the ion produced by the already-dissolved solute in the "initial concentration" row. The unknown is still the molar solubility, x. We make entries in the "change" row as before.

When you set up the equilibrium concentrations, you find expressions such as $(0.10 + 2x)$ in Example 17.8. In these cases, we will assume that x or $2x$ is very small, because we are dealing with "insoluble" salts. This allows us to make simplifying assumptions that give very easy arithmetic.

(3) Predicting when a precipitate will form.

A precipitate can only form if a solution is supersaturated, and this condition is fulfilled only when the ion product of the salt in question exceeds the value of K_{sp} for the salt. Remember the summary on page 752 of the text just prior to Example 17.9.

When considering what happens when two solutions are mixed, you must first determine the effect of dilution on the concentrations of the ions. Then compute the value of the ion product and compare it to K_{sp} to determine whether a precipitate will form.

Thinking It Through

1 Will a precipitate of $CaSO_4$ form if 40.0 mL of 2.0×10^{-3} *M* $CaCl_2$ is mixed with 60.0 mL of 3.0×10^{-2} *M* Na_2SO_4?

Self-Test

4. The molar solubility of $MnCO_3$ in pure water is 2.24×10^{-5} mol/L. What is K_{sp} for $MnCO_3$?

____

5. The molar solubility of lead iodate, $Pb(IO_3)_2$, is 4.0×10^{-5} *M*. What is the value of K_{sp} for $Pb(IO_3)_2$?

____

6. The molar solubility of $Cu(OH)_2$ in 0.10 *M* NaOH is 4.8×10^{-18} mol/L. What is K_{sp} for $Cu(OH)_2$?

____

7. For $PbCO_3$, $K_{sp} = 7.4 \times 10^{-14}$. What is the molar solubility of $PbCO_3$ in pure water?

____

8. What is the molar solubility of $PbCO_3$ in 0.020 *M* Na_2CO_3? (For $PbCO_3$, $K_{sp} = 7.4 \times 10^{-14}$)

____

9. What is the molar solubility of $PbCl_2$ in 0.30 *M* $CaCl_2$ solution? (For $PbCl_2$, $K_{sp} = 1.7 \times 10^{-5}$)

____

10. Referring to Question 9, what is the molar solubility of $PbCl_2$ in 0.30 *M* NaCl solution?

____

11. Will a precipitate of $PbCl_2$ form in a solution that contains 0.20 *M* Pb^{2+} and 0.030 *M* Cl^-? For $PbCl_2$, $K_{sp} = 1.7 \times 10^{-5}$.

____

12. Will a precipitate of $CaSO_4$ form in a solution containing 0.0030 *M* Ca^{2+} and 0.0010 *M* SO_4^{2-}? See Table 17.2 on page 745 for K_{sp} data.

____

New Terms

Write the definitions of the following terms, which were introduced in this section. If necessary, refer to the Glossary at the end of the text.

common ion
common ion effect
ion product
molar solubility
solubility product constant

17.4 Solubility Equilibria for Metal Oxides and Sulfides

Review

We cannot treat the sparingly soluble oxides and sulfides of metal ions in the same way as other sparingly soluble ionic compounds because once the anion—oxide ion or sulfide ion—is released from the crystal it reacts virtually quantitatively with water. The OH^- ion appears in the place of O^{2-} and the HS^- ion forms instead of S^{2-}. This section carries the discussion further only with the equilibria involving metal sulfides.

For a metal sulfide of the general form *M*S, the solubility equilibrium is

$$MS(s) + H_2O \rightleftharpoons M^{2+}(aq) + HS^-(aq) + OH^-(aq)$$

and the solubility product constant is

$$K_{sp} = [M^{2+}][HS^-][OH^-]$$

If the solution is made acidic, then we have to rewrite the equilibrium, because both HS^- and OH^- are neutralized by acids. In dilute acid, the equilibrium involving our general sulfide, *M*S, is

$$MS(s) + 2H^+(aq) \rightleftharpoons M^{2+}(aq) + H_2S(aq)$$

The solubility product expression is rewritten to reflect this equilibrium, and the *acid solubility product constant* or K_{spa}, is

$$K_{spa} = \frac{[M^{2+}][H_2S]}{[H^+]^2}$$

(When the metal sulfide is not of the form *M*S, the equilibrium equation and the equation for K_{spa} must, of course, be altered accordingly.)

One group of metal sulfides—the *acid-insoluble sulfides*—have values of K_{spa} so small that no acid exists concentrated enough to bring them into solution simply by converting the sulfide ion to H_2S. Their K_{spa} values range from about 10^{-32} to 10^{-5}. When K_{spa} is about 10^{-4} or greater, then the sulfides will dissolve in acid, and they form the group of the *acid-soluble sulfides.* Because of the huge differences in K_{spa} values among the metal sulfides, the adjustment of the pH of a medium containing two metal sulfides can bring one into solution but leave the other insoluble.

Thinking It Through

2. A solution contains both $AgNO_3$ and $Co(NO_3)_2$ at concentrations of 0.010 *M* each. It is to be saturated with hydrogen sulfide. Is it possible to adjust the pH of the solution so as to prevent one of the cations from precipitating when the H_2S is added? If so, which cation remains dissolved and what pH should be used? (Consult Table 17.3, page 756) for values of K_{spa} as needed.

Self Test Questions

13. Consider lead(II) sulfide.

    (a) Write its solubility equilibrium and K_{sp} equation for a saturated solution in water.

    ____

    (b) Write its solubility equilibrium and K_{spa} equation for a saturated solution in aqueous acid.

    ____

    ____

14. What value of pH permits the selective precipitation of the sulfide of just one of the two metal ions in a solution that has a concentration of 0.020 *M* Pb^{2+} and 0.020 *M* Fe^{2+}?

    ____

New Term

Write the definition of the following term, which was introduced in this section. If necessary, refer to the Glossary at the end of the text.

acid solubility product constant, K_{spa}

17.5 Complex Ion Equilibria

Review

Recall that *complex ions,* which are often simply referred to as *complexes*, are formed when a metal ion becomes surrounded by and bonded to one or more neutral molecules or negative ions. The general term *ligand* refers to the molecules or ions that become attached to the metal ion. Complex ions are quite common and are formed by many metals, especially the transition metals. In this chapter, we examine the equilibria involved in their formation and the effect that complex ion formation has on the solubility of salts that contain metal ions that form complexes.

Chemists who study complexes have not decided on any one method of describing how stable they are. Some prefer to write chemical equations for the formation of a complex, and then use the size of the equilibrium constant as a measure of the complex's stability. The equilibrium constant for a reaction in which the complex is *formed* from the metal ion and its ligands is called a *formation constant* or *stability constant*. In other words, if you come across an equilibrium constant and it is called a formation constant, the chemical equation that goes with it is one for the *formation* of the complex. It also means that in the mass action expression, the concentration of the complex appears in the numerator, and the concentrations of the metal ion and the ligands appear in the denominator. If a formation constant is large, it means that a relatively large amount of complex is formed from the metal ion and the ligands and that the complex is quite stable.

Other chemists prefer to think of the stability of a complex in terms of how easily it decomposes. The reactions that they write are for the decomposition of complexes into their metal ions and ligands. The equilibrium constants for these reactions are called *instability constants* because the larger the value of K, the more fully decomposed the complex is at equilibrium, and therefore the more unstable the complex is. Formation constants and instability constants are related to each other in a very simple way—one is just the reciprocal of the other.

The effect of the formation of a complex on the solubility of a salt can be predicted qualitatively by Le Châtelier's principle, as described on page 760. Let's look at another example, though, to be sure you understand the thinking involved. Suppose we wish to anticipate the effect of adding sodium cyanide, NaCN, on the solubility of cadmium hydroxide, $Cd(OH)_2$. Also, suppose that we know that Cd^{2+} forms a very stable complex with cyanide ion, CN^-, which has the formula $Cd(CN)_4{}^{2-}$. We begin by writing chemical equations for the solubility equilibrium and for the formation of the complex.

$$Cd(OH)_2(s) \rightleftharpoons Cd^{2+}(aq) + 2OH^-(aq)$$

$$Cd^{2+}(aq) + 4CN^-(aq) \rightleftharpoons Cd(CN)_4{}^{2-}(aq)$$

We now realize that as we add CN^- to the reaction mixture, it will combine with Cd^{2+} and form the complex. This reduces the amount of cadmium ion in the solution, and therefore upsets the solubility equilibrium. In an effort to replace the lost Cd^{2+} and to restore equilibrium, some of the $Cd(OH)_2$ will dissolve. Therefore, adding CN^- to a solution in which $Cd(OH)_2$ is in equilibrium with Cd^{2+} and OH^- will increase the amount of $Cd(OH)_2$ that is dissolved. This is another way of saying that forming a complex with the cadmium ion increases the solubility of the $Cd(OH)_2$.

Determining the effect of complex ion formation on solubility is a bit more complicated than other equilibrium problems that you have encountered up till now. Example 17.12 illustrates the strategy. Note that we obtain the overall K_c for the system by multiplying K_{sp} by K_{form}. Furthermore, in any of the problems that you will encounter in this book, you can assume that the complex ion is so stable that the amount of uncomplexed metal ion in the solution is negligible. In other words, assume that any of the metal ion that enters the solution from the "insoluble" compound that dissolves is present as the complex ion. For example, suppose we were working a problem involving the solubility of $Cd(OH)_2$ in NaCN solution. If we found that 0.010 mol of $Cd(OH)_2$ dissolved per liter, then the complex concentration would be taken to be 0.010 *M*, and the concentration of uncomplexed Cd^{2+} would be assumed to be very small compared to this value.

Self-Test

15. Write the equilibria that are associated with the equations for K_{form} for each of the following complex ions. Write also the equations for the K_{form} of each.

    (a) $Hg(NH_3)_4{}^{2+}$ ____________________________

    (b) $SnF_6{}^{2-}$ ____________________________

    (c) $Fe(CN)_6{}^{3-}$ ____________________________

16. Cobalt(II) ion, Co^{2+}, forms a complex with ammonia that has the formula $Co(NH_3)_6{}^{2+}$.

    (a) Write a chemical equation involving this complex ion for which the equilibrium constant would be referred to as K_{form}.

    ____

    (b) Write a chemical equation involving this complex for which the equilibrium constant would be referred to as K_{inst}.

    ____

17. Silver ion forms a complex ion with cyanide, $Ag(CN)_2^-$, that has a formation constant equal to 5.3×10^{18}. What is the value of this complex's instability constant?

____

18. (a) Calculate the molar solubility of AgI in 0.010 *M* NaCN solution. (For AgI, $K_{sp} = 8.3 \times 10^{-18}$; for $Ag(CN)_2^-$, $K_{form} = 5.3 \times 10^{18}$.)

____

(b) Calculate the molar solubility of AgI in pure water.

____

(c) By what factor has the solubility been increased by complex formation?

____

New Terms

Write the definitions of the following terms, which were introduced in this section. If necessary, refer to the Glossary at the end of the text.

formation constant
instability constant
stability constant

Answers to Thinking It Through

1 What we need is the value of the ion product, $[Ca^{2+}][SO_4^{2-}]$, for $CaSO_4$. If it *exceeds* the solubility product constant, K_{sp}, for $CaSO_4$, a precipitate must form. To find values for $[Ca^{2+}]$ and $[SO_4^{2-}]$, we have to calculate the ratio of the moles of each ion to the *final* volume , assuming that no precipitate forms. For the calcium ion, the number of moles is found as follows.

$$40 \text{ mL } (CaCl_2 \text{ soln}) \times \frac{2.0 \times 10^{-3} \text{ mol } Ca^{2+}}{1000 \text{ mL } CaCl_2 \text{ soln}} = 8.0 \times 10^{-5} \text{ mol } Ca^{2+}$$

The final volume is 40.0 mL + 60.0 mL = 100.0 mL or 0.100 L, so the concentration of Ca^{2+} in the final solution, assuming no precipitation is

$$[Ca^{2+}] = \frac{8.0 \times 10^{-5} \text{ mol } Ca^{2+}}{0.100 \text{ L}} = 8.0 \times 10^{-4} \, M$$

When we do the same kinds of calculations for the SO_4^{2-}ion its concentration is found as follows.

$$60.0 \text{ mL } (Na_2SO_4 \text{ soln}) \times \frac{3.0 \times 10^{-2} \text{ mol } SO_4^{2-}}{1000 \text{ mL } Na_2SO_4 \text{ soln}} = 1.8 \times 10^{-2}\ M$$

Finally, calculating the ion product, we have

$$[Ca^{2+}][SO_4^{2-}] = (8.0 \times 10^{-4})(1.8 \times 10^{-2}) = 1.4 \times 10^{-5}$$

Because the ion product is *less* than K_{sp}, no precipitate forms.

2 We first must assemble equilibrium equations and their acid solubility products (from Table 17.3, page 756).

For Ag_2S:

$$Ag_2S(s) + 2H^+(aq) \rightleftharpoons 2Ag^+(aq) + H_2S(aq),$$

$$K_{spa} = \frac{[Ag^+]^2[H_2S]}{[H^+]^2} = 6 \times 10^{-30}$$

For CoS:

$$CoS(s) + 2H^+(aq) \rightleftharpoons Co^{2+}(aq) + H_2S(aq)$$

$$K_{spa} = \frac{[Co^{2+}][H_2S]}{[H^+]^2} = 5 \times 10^{-1}$$

Now we check to see if there is an upper limit to the value of $[H^+]$ above which the less soluble sulfide, Ag_2S, would dissolve. The concentration of Ag^+ is given as 0.010 *M*, and from the text the concentration of H_2S in a saturated solution is 0.1 *M*, so

$$K_{spa} = 6 \times 10^{-30} = \frac{(0.010)^2(0.1)}{[H^+]^2}$$

$$[H^+] = 1.3 \times 10^{12} \text{ mol/L}$$

No acid solution can be this concentrated, so there is no upper limit to the acidity; Ag_2S is truly an acid-insoluble sulfide.

Next we check the lower limit; $[Co^{2+}] = 0.010\ M$, so

$$K_{spa} = 5 \times 10^{-1} = \frac{(0.010)(0.1)}{[H^+]^2}$$

$$[H^+] = 0.04 \text{ and pH} = 1.4$$

If the value of $[H^+]$ is made greater than 0.04 mol/L (or the pH is made less than 1.4), then cobalt sulfide will not precipitate.

Answers to Self-Test Questions

1. $H_2CO_3 \rightleftharpoons H^+ + HCO_3^- \quad K_{a_1} = \dfrac{[H^+]\,[HCO_3^-]}{[H_2CO_3]}$

$HCO_3^- \rightleftharpoons H^+ + CO_3^{2-} \qquad K_{a_2} = \dfrac{[H^+][CO_3^{2-}]}{[H_2CO_3]}$

2. $[H^+] = 2.8 \times 10^{-3}$ *M*, pH = 2.55, $[Asc^{2-}] = 1.6 \times 10^{-12}$ *M*
3. pH = 12.95
4. $K_{sp} = 5.0 \times 10^{-10}$
5. $K_{sp} = 2.6 \times 10^{-13}$
6. $K_{sp} = 4.8 \times 10^{-20}$
7. 2.7×10^{-7} *M*
8. 3.7×10^{-12} *M*
9. 4.7×10^{-5} *M*
10. 1.9×10^{-4} *M*
11. Yes. The ion product is equal to 1.8×10^{-5}, which is greater than K_{sp}.
12. No. The ion product is less than K_{sp}.
13. (a) $PbS(s) + H_2O \rightleftharpoons Pb^{2+}(aq) + OH^-(aq) + HS^-(aq)$
    (b) $PbS(s) + 2H^+(aq) \rightleftharpoons Pb^{2+}(aq) + H_2S(aq)$
14. pH = 0.7
15. (a) $Hg^{2+}(aq) + 4NH_3(aq) \rightleftharpoons Hg(NH_3)_4^{2+}(aq) \qquad K_{form} = \dfrac{[Hg(NH_3)_4^{2+}]}{[Hg^{2+}][NH_3]^4}$

    (b) $Sn^{4+}(aq) + 6F^-(aq) \rightleftharpoons SnF_6^{2-}(aq) \qquad K_{form} = \dfrac{[SnF_6^{2-}]}{[Sn^{4+}][F^-]^6}$

    (c) $Fe^{3+}(aq) + 6CN^-(aq) \rightleftharpoons Fe(CN)_6^{3-}(aq) \qquad K_{form} = \dfrac{Fe(CN)_6^{3-}]}{[Fe^{3+}][CN^-]^6}$
16. (a) $Co^{2+}(aq) + 6NH_3(aq) \rightleftharpoons Co(NH_3)_6^{2+}(aq)$
    (b) $Co(NH_3)_6^{2+}(aq) \rightleftharpoons Co^{2+}(aq) + 6NH_3(aq)$
17. $K_{inst} = 1.9 \times 10^{-19}$
18. (a) 1.9×10^{-2} *M*, (b) 2.8×10^{-9} *M*, (c) 6.6×10^{-6} (AgI is 6.6 million times more soluble in the NaCN solution.)

Tools you have learned

Remove this chart from the Study Guide and keep it handy when tackling homework problems.

Tool	*Function*
K_{a_1} of a weak polyprotic acid For the acid H_yA as the sole solute (where $y > 1$): $K_{a_1} = \frac{[H^+][H_{(y-1)}A^-]}{[H_yA]}$	To calculate $[H^+]$ and pH for a solution of a weak polyprotic acid (the second or higher ionization being ignored).
K_{a_2} of a weak diprotic acid For the acid H_2A (as the sole solute): $[A^{2-}] = K_{a_2}$	To calculate the molarity of the anion A^{2-}.
K_b and pH: salt of polyprotic acid For the anion of a weak diprotic acid, A^{2-}, $A^{2-} + H_2O \rightleftharpoons HA^- + OH^-$ $K_b = \frac{[HA^-][OH^-]}{[A^{2-}]}$	Calculate $[OH^-]$, pOH, and pH for a solution of a salt of a weak, diprotic acid.
Solubility product constant, K_{sp} For the sparingly soluble salt, M_mA_a, $K_{sp} = [M]^m[A]^a$	Calculate K_{sp} from the molar solubility of the salt. Calculate a solubility from the K_{sp}. Calculate a solubility when a common ion is present.

<table>
<tr><td>

Predicting precipitation reactions

Ion product > K_{sp}

precipitate will form

Ion product = K_{sp} or

Ion product < K_{sp}

precipitate will not form

</td><td>Using a calculated ion product to predict the likelihood of a precipitate forming when ionic substances are mixed in water.</td></tr>
<tr><td>

Acid Solubility product constant, K_{spa}

For the sparingly soluble sulfide, *M*S, in an acidic medium:

$$MS(s) + 2H^+(aq) \rightleftharpoons M^{2+}(aq) + H_2S(aq)$$

$$K_{spa} = \frac{[M^{2+}][H_2S]}{[H^+]^2}$$

(Sulfides with different formulas require K_{spa} equations modified by exponents that reflect the subscripts.)

</td><td>

Use K_{spa} data to calculate the solubility of a metal sulfide at a given pH.

Use K_{spa} date for two or more metal sulfides to calculate the pH at which one will selectively precipitate from a solution saturated in H_2S.

</td></tr>
<tr><td>

Formation constants (stability constants) of complexes

For the equilibrium involving a central cation, a ligand species and the complex, K_{form} is the regular equilibrium expression. For example, in the equilibrium

$$Cu^{2+}(aq) + 4NH_3(aq) \rightleftharpoons Cu(NH_3)_4(aq)$$

$$K_{form} = \frac{[Cu(NH_3)_4]}{[Cu^{2+}][NH_3]^4}$$

</td><td>Make judgements concerning the relative stabilities of complexes.</td></tr>
</table>

Complex ion formation and salt solubility When a sparingly soluble salt with its own K_{sp} is in a solution where the cation can form a complex ion with having its own K_{form}, the overall equilibrium constant for the two equilibria is $K_c = K_{form}K_{sp}$	Calculate how the solubility of a sparingly soluble salt changes when its cation is able to form a complex ion with a ligand added to the solution.

Summary of Important Equations

Successive acid ionization constants, polyprotic acids (illustrated by H_2A)

$$\text{For } H_2A \rightleftharpoons H^+ + HA^- \qquad K_{a_1} = \frac{[H^+][HA^-]}{[H_2A]}$$

$$\text{For } HA^- \rightleftharpoons H^+ + A^{2-} \qquad K_{a_2} = \frac{[H^+][A^{2-}]}{[HA^-]}$$

Concentration of A^{2-} in a solution of H_2A

$$[A^{2-}] = K_{a_2}$$

Solubility product equilibrium and solubility product constant for the sparingly soluble salt, M_mA_a

$$M_mA_a(s) \rightleftharpoons mM^{a+}(aq) + aA^{m-}$$

$$K_{sp} = [M^{a+}]^m[A^{m-}]^a$$

Solubility product equilibrium and acid solubility product constant for a sparingly soluble sulfide, MS, in an acidic medium. (Sulfides with different formulas require K_{spa} equations modified by exponents that reflect the subscripts.)

$$MS(s) + 2H^+(aq) \rightleftharpoons M^{2+}(aq) + H_2S(aq)$$

$$K_{spa} = \frac{[M^{2+}][H_2S]}{[H^+]^2}$$

Formation constants (stability constants) of complex ions. These are written as ordinary equilibrium constant equations with the complex ion as the *product* and so in the *numerator* of K_{form}.

Instability constants of complex ions.

$$K_{inst} = \frac{1}{K_{form}}$$

Chapter 18

ELECTROCHEMISTRY

Some of the most useful and common practical applications of chemistry involve the use of or production of electricity. Our lives are touched constantly by the ultimate fruits of electrolysis reactions, such as aluminum, bleach, halogenated organic molecules in plastics and insecticides, and soap. Chemical reactions that produce electrical power in batteries start our cars, run electronic calculators and portable radios, keep wristwatches running, and set proper exposures in cameras. Besides all of these things, the relationship between electricity and chemical change has become an extremely useful tool in the laboratory for probing chemical systems of all kinds.

Learning Objectives

As you study of this chapter, keep in mind the following goals:

1 To learn about the kinds of chemical reactions that can be studied electrically.

2 To learn what electrolysis is, to study how an electrolysis apparatus (electrolysis cell) is constructed, and to learn how to write equations for the reactions that take place in an electrolysis cell.

3 To learn how to compute the amount of chemical change caused by the flow of a given amount of electricity.

4 To study various practical applications of electrolysis.

5 To learn how a spontaneous redox reaction can be set up to deliver electrical energy in a galvanic cell.

6 To see how the voltage, or potential, of a galvanic cell can be considered to arise as the difference between the potentials that each half-cell has for reduction.

7 To learn how reduction potentials are measured by comparison to a standard electrode called the hydrogen electrode.

8 To learn how to use standard reduction potentials to predict the spontaneous cell reaction, the cell potential, and whether or not a given reaction will proceed spontaneously.

9 To learn how to calculate ΔG° from a cell potential, and vice versa.

10 To learn how to calculate the effect on the cell potential of changing the concentrations of the ions in a galvanic cell.

11 To learn the chemistry of some common types of batteries and to look at possible future developments.

18.1 Electricity and Chemical Change

Review

Electrical devices operate by the flow of electrons — that's what electricity is. Reactions that produce or consume electrical energy, or that can be studied by electrical measurements involve electron transfer. They are oxidation-reduction reactions. Such reactions are common and the applications of electrochemistry affect our daily lives as well as our activities in the laboratory.

Self-Test

1. What is an *electrochemical change*? ________________________________

   ____

2. What is *electrochemistry*? ________________________________

   ____

New Terms

Write the definitions of the following terms, which were introduced in this section. If necessary, refer to the Glossary at the end of the text.

electrochemical change
electrochemistry

18.2 Electrolysis

Review

When a nonspontaneous reaction is forced to occur by the passage of electricity, the process is called electrolysis. An electrolysis cell (electrolytic cell) consists of a pair of electrodes dipping into a chemical system in which there are mobile ions (formed by melting a salt or by dissolving an electrolyte in water). *When electricity flows, oxidation-reduction reactions occur at the electrodes.*

An important thing to learn in this section is that in *any* cell, regardless of whether it is using or producing electricity, the electrode at which oxidation occurs is called the *anode* and the electrode at which reduction occurs is called the *cathode*. (See the summary on page 770.) Thus, we name an electrode according to the chemical reaction that occurs at it, not according to its charge!

In an electrolysis cell, it happens that the anode is positively charged and the cathode is negatively charged. These are the charges that they *must* have to force oxidation and reduction to occur. The positive charge of the anode pulls electrons from substances and causes them to be oxidized, and the negative charge of the cathode pushes electrons onto other substances and causes them to be reduced.

In a metal, electrical conduction takes place by the transport of electrons through the metal from one place to another. This is called *metallic conduction*. Molten ionic compounds and aqueous solutions conduct by a different mechanism called *electrolytic conduction*. In electrolytic conduction, electrical charge is moved from one place to another by the movement of ions, rather than electrons.

The equation for the cell reaction in an electrolysis is obtained by adding the individual oxidation and reduction half-reactions that occur at the electrodes. Remember to be sure that the electrons in the cell reaction cancel. The procedure for this is the same as in the ion-electron method that you learned to apply in balancing equations for redox reactions. If necessary, review the principles of this method in Chapter 12 (pages 522-526).

For reactions in aqueous solution, the redox of water is possible at the electrodes. You should know the following possible electrode reactions:

Anode (oxidation of H_2O) $\quad 2H_2O(l) \rightarrow O_2(g) + 4H^+(aq) + 4e^-$

Cathode (reduction of H_2O) $\quad 4H_2O(l) + 4e^- \rightarrow 2H_2(g) + 4OH^-(aq)$

Although surface effects at certain electrodes make it difficult to predict whether water will participate in the electrode reactions, by observation we can determine which reactions occur most readily. Thus, the reactions actually observed are the ones that occur most easily.

Thinking It Through

For the following, identify the information needed to solve the problem and show (or explain) what must be done with it.

1 When a solution of NiF_2 is electrolyzed, metallic nickel is deposited on one electrode and O_2 is produced at the other. Which substances are oxidized and reduced, and at which electrodes?

Self-Test

3. At which electrode would these reactions occur?

   (a) $2I^-(aq) \rightarrow I_2(aq) + 2e^-$

   ____

   (b) $2Cr^{3+}(aq) + 7H_2O \rightarrow Cr_2O_7^{2-}(aq) + 14H^+(aq) + 6e^-$

   ____

   (c) $NO_3^-(aq) + 2H_2O + 3e^- \rightarrow NO(g) + 4OH^-(aq)$

   ____

4. Write the equation for the reduction of H_2O at the cathode of an electrolytic cell.

   ____

5. Write the equation for the oxidation of H_2O at the anode of an electrolytic cell.

   ____

6. When an aqueous solution of BaI_2 is electrolyzed, I_2 is formed at the anode and H_2 is formed at the cathode. What are the anode and cathode half-reactions?

   anode: ________________________________

   cathode: ________________________________

New Terms

Write the definitions of the following terms, which were introduced in this section. If necessary, refer to the Glossary at the end of the text.

electrolysis	electrolytic conduction	cathode
electrolysis cell	electrolyze	cell reaction
electrolytic cell	anode	metallic conduction

18.3 Stoichiometric Relationships in Electrolysis

Review

The amount of electricity produced during electrolysis is proportional to the number of moles of electrons (the number of faradays) passed through the electrolysis cell. Important relationships to remember are:

$$1 \text{ mol } e^- = 9.65 \times 10^4 \text{ C} \qquad (\text{C} = \text{coulomb})$$

$$1 \text{ C} = 1 \text{ A} \times \text{s} \qquad (\text{A} = \text{ampere, s} = \text{second})$$

$$1 \mathscr{F} = 1 \text{ mol } e^- \qquad (\mathscr{F} = \text{faraday})$$

We use these relationships, along with balanced half-reactions or a knowledge of the number of electrons transferred in a balanced redox equation, to relate the amount of chemical change to amperes of electrical current and to time. Study Examples 18.2 through 18.4 and then work the Thinking It Through and Self-Test questions below.

Thinking It Through

For the following, identify the information needed to solve the problem and show (or explain) what must be done with it.

2 A solution of NaCl was electrolyzed for 30.0 min, producing Cl_2 at the anode and H_2 at the cathode. The resulting solution after electrolysis was titrated with 0.500 *M* HCl solution and required 22.3 mL of the acid to neutralize the solution. What was the current during the electrolysis.

Self-Test

7. A current of 5.00 A flows for 25.0 minutes. How many moles of electrons does this deliver?

   ____

8. For how many seconds must a current of 6.00 A flow to deliver 0.225 mol of electrons?

   ____

9. What current must be supplied to deliver 0.0165 mol e^- in 155 s?

   ____

10. Calculate the number of moles of electrons that must pass through an electrolysis cell to produce 0.0150 mol $Cr_2O_7^{2-}$ by the reaction,

$$2Cr^{3+} + 7H_2O \rightarrow Cr_2O_7^{2-} + 14H^+ + 6e^-$$

   ____

11. How many minutes are needed to make 0.0225 mol $Cr_2O_7^{2-}$ by the equation in Question 10 if the current is 4.00 A?

____

12. What current will produce 25.3 g Fe in 4.00 hours by reduction of Fe^{2+} in an aqueous solution?

____

New Terms

Write the definitions of the following terms, which were introduced in this section. If necessary, refer to the Glossary at the end of the text.

ampere coulomb faraday

18.4 Industrial Applications of Electrolysis

Review

This section describes electroplating and the methods of producing some important commercial metals and chemicals by electrolysis reactions. In studying this section, you should learn the chemical reactions involved in the various processes and the reasons why the reactions are carried out as they are. When you feel you know the material, try the Self-Test below.

Self-Test

13. What is the purpose of electroplating? _______________

____

14. To which electrode do we connect the object to be electroplated?

____

15. What is the name and formula of the solvent for Al_2O_3 that was originally used in the Hall-Héroult process?

____

16. What is the net cell reaction in the Hall-Héroult process?

____

17. What is the major source of magnesium? _______________ What salt of magnesium is used in the electrolysis reaction that produces the free metal?

____

18. What is the purpose of the special construction of the Downs cell?

____

19. Why is the electrolytic refining of copper so economical?

____

20. Give the cathode, anode, and net cell reaction for the electrolysis of brine.

____

____

____

21. What products are formed if the brine solution is stirred while it is electrolyzed?

____

22. Name one advantage and one disadvantage of using a diaphragm cell in the electrolysis of brine?

____

23. What is an advantage and a disadvantage of using a mercury cell in the electrolysis of brine?

____

____

New Terms

Write the definitions of the following terms, which were introduced in this section. If necessary, refer to the Glossary at the end of the text.

electroplating
diaphragm cell
mercury cell

18.5 Galvanic Cells

Review

When a redox reaction occurs, the energy released is normally lost to the environment as heat. By separating the half-reactions and making oxidation and reduction occur in different places (in different half-cells), we can cause the electron transfer to take place by way of a wire through an external electrical circuit. In this way the energy of the reaction can be harnessed. The apparatus to accomplish this is called a galvanic cell.

In a galvanic cell, the cathode is positive and the anode is negative. (See the summary on page 785.) Electrons flow in the external circuit from anode to cathode; in the solution, cations move toward the cathode and anions move toward the anode. The two compartments, or half-cells, must be connected electrolytically—for example, by a salt bridge. As usual, oxidation occurs at the anode and reduction occurs at the cathode.

Study the way the shorthand description of a galvanic cell is written. This is covered in detail on page 786 and in Example 18.5.

Self-Test

24. The following reaction occurs spontaneously in a galvanic cell:

$$4H^+ + MnO_2 + Fe \rightarrow Fe^{2+} + Mn^{2+} + 2H_2O$$

(a) What half-reaction occurs in the cathode compartment?

____

(b) What electrical charge is carried by the iron electrode?

____

(c) Do electrons flow toward or away from the iron electrode?

____

(d) Using the iron half-cell as an example, explain how a salt bridge containing KNO_3 works.

____

25. Write the standard cell notation for the galvanic cell described in the preceding question.

____

26. What are the anode and cathode half-reactions in the galvanic cell described by the notation

$$Zn(s) \mid Zn^{2+}(aq) \parallel Au^{3+}(aq) \mid Au(s)$$

____

____

New Terms

Write the definitions of the following terms, which were introduced in this section. If necessary, refer to the Glossary at the end of the text.

galvanic cell — half-cell
voltaic cell — salt bridge

17.6 Cell Potentials and Reduction Potentials

Review

Central to the development of this section is the concept that each half-reaction has an intrinsic or innate tendency to proceed as a reduction. The magnitude of this tendency is given by its *reduction potential* (or *standard reduction potential* if the concentrations of all the ions are 1 *M*, the pressure is 1 atm, and the temperature is 25 °C). When two half-cells compete for electrons, as they do in a galvanic cell, the one with the larger reduction potential proceeds as reduction and the other is forced to become an oxidation. The standard cell potential, E°_{cell}, is given by Equation 18.2. This is a very important equation, so be sure you have learned it.

$$E^\circ_{cell} = \begin{pmatrix}\text{standard reduction} \\ \text{potential of the} \\ \text{substance reduced}\end{pmatrix} - \begin{pmatrix}\text{standard reduction} \\ \text{potential of the} \\ \text{substance oxidized}\end{pmatrix} \qquad (18.2)$$

The values of standard reduction potentials are compared to that of a standard electrode called the hydrogen electrode.

$$2H^+(aq,\ 1.00\ M) + 2e^- \rightleftharpoons H_2(g,\ 1\ \text{atm}) \qquad E^\circ = 0.00\ \text{V}$$

Self-Test

27. A lead half-cell was constructed using a lead electrode dipping into a 1.00 *M* $Pb(NO_3)_2$ solution. This was connected to a standard hydrogen electrode. A voltage of 0.13 V was measured for the cell when the positive terminal of the voltmeter was connected to the hydrogen electrode.

    (a) On a sheet of paper, sketch and label a diagram of the galvanic cell.

    (b) What substance is being reduced in the cell? ____________________

    (c) What is $E^\circ_{Pb^{2+}}$ for the half-cell, $Pb^{2+}(aq) + 2e^- \rightleftharpoons Pb(s)$?

    ____

28. Referring to Table 18.1, to which electrode should the negative terminal of a voltmeter be connected in a cell constructed of the half-cells.

$$Au^{3+} + 3e^- \rightleftharpoons Au \quad \text{and} \quad Zn^{2+} + 2e^- \rightleftharpoons Zn?$$

____

29. How is the volt defined in terms of SI units? ____________________

New Terms

Write the definitions of the following terms, which were introduced in this section. If necessary, refer to the Glossary at the end of the text.

electromotive force
volt
cell potential, E_{cell}
standard cell potential, E^{o}_{cell}
reduction potential
standard reduction potential
hydrogen electrode

18.7 Using Standard Reduction Potentials

Review

For a given pair of half-reactions, the one having the higher (more positive) reduction potential occurs spontaneously as reduction; the other is reversed and occurs as oxidation. After setting up the half-reactions in this way, the cell reaction is obtained by adding the reduction and oxidation half-reactions in such a way that all electrons cancel. The procedure is the same as the one you used in the ion-electron method (Chapter 12, pages 522-526). Factors are used to adjust the coefficients so that equal numbers of electrons are gained and lost. *Notice, however, that these factors are not used as multiplying factros when combining the half-reactions*! ***The cell potential is obtained simply by subtracting one reduction potential from the other using Equation 18.2.*** For a spontaneous cell reaction, this difference has a positive algebraic sign.

When asked whether or not a given overall reaction is spontaneous, divide the reaction into its two half-reactions and compute. Find the reduction potential for each half reaction and compute the cell potential using Equation 18.2. If the result is positive, the reaction is spontaneous; if it is negative, however, the reaction is *not* spontaneous in the direction written. (In fact, the reaction is spontaneous in the opposite direction.)

Thinking It Through

For the following, identify the information needed to solve the problem and show (or explain) what must be done with it.

3 What will be the spontaneous reaction if we added nickel and iron filings to a solution that contains $NiCl_2$ and $FeCl_2$?

Self-Test

30. Given the following half-reactions and their reduction potentials,

$ClO_3^-(aq) + 6H^+(aq) + 6e^- \rightleftharpoons Cl^-(aq) + 3H_2O \qquad E° = +1.45\ V$

$Hg_2HPO_4(s) + H^+(aq) + 2e^- \rightleftharpoons 2Hg(l) + H_2PO_4^-(aq) \qquad E° = +0.64\ V$

(a) determine the net spontaneous cell reaction.

____

(b) determine the cell potential. ______________

31. Without actually calculating $E°_{cell}$, determine the spontaneous cell reaction involving the following half-reactions.

$BrO_3^-(aq) + 6H^+(aq) + 6e^- \rightleftharpoons Br^-(aq) + 3H_2O \qquad E° = +1.44\ V$

$H_3AsO_4(aq) + 2H^+(aq) + 2e^- \rightleftharpoons HAsO_2(aq) + 2H_2O \qquad E° = +0.58\ V$

____

32. Will the following reaction occur spontaneously?

$H_2SO_3(aq) + H_2O + Br_2(aq) \rightarrow SO_4^{2-}(aq) + 4H^+(aq) + 2Br^-(aq)$

____

New Terms

none

18.8 Cell Potentials and Thermodynamics

Review

The free energy change, ΔG, and the standard free energy change, $\Delta G°$, for a reaction is related to the cell potential and standard cell potential, respectively, by the equations

$$\Delta G = -n\mathscr{F}E_{cell} \qquad (18.5)$$

$$\Delta G^\circ = -n\mathscr{F}E^\circ_{cell} \qquad (18.6)$$

where n is the number of electrons transferred and $\mathscr{F} = 9.65 \times 10^4$ C/mol e^-. Since E_{cell} and E°_{cell} are in volts, and 1 V = 1 J/C, ΔG and ΔG° are in units of joules.

The standard cell potential is also related to the equilibrium constant for the reaction.

$$E^\circ_{cell} = \frac{0.0592 \text{ V}}{n} \log K_c \qquad (18.7)$$

Thinking It Through

For the following, identify the information needed to solve the problem and show (or explain) what must be done with it.

4 An equilibrium was set up for the reaction

$$Ni(s) + Cd^{2+}(aq) \rightleftharpoons Cd(s) + Ni^{2+}(aq)$$

If the concentration of Ni^{2+} at equilibrium is 3.0×10^{-8} M, what will be the concentration of Cd^{2+}?

Self-Test

33. Calculate ΔG° in kJ for the reaction in Question 30.

____

34. The reaction, $2Al^{3+} + 3Cu \rightarrow 3Cu^{2+} + 2Al$, has $E^\circ_{cell} = -2.00$ V. What is ΔG° in kJ for this reaction?

____

35. The reaction, $2AgBr + Pb \rightarrow PbBr_2 + 2Ag$, has $\Delta G^\circ = -237$ kJ. What is E°_{cell}?

____

36. What is the value of K_c for the reaction in Question 30?

____

37. What is the value of K_c for the reaction in Question 34?

____

38. The reaction, $PbI_2(s) + Zn(s) \rightleftharpoons Pb(s) + 2I^-(aq) + Zn^{2+}(aq)$, has an equilibrium constant, $K_c = 2.2 \times 10^{13}$ at 25 °C. What is the value of $E°_{cell}$ for this reaction?

____

New Terms

none

18.9 The Effect of Concentration on Cell Potential

Review

At 25 °C, the cell potential given by the Nernst equation is

$$E_{cell} = E°_{cell} - \frac{0.0592 \text{ V}}{n} \log Q$$

where n is the number of electrons transferred and Q is the reaction quotient (the value of the mass action expression) for the reaction. In using this equation, there are some important points to note. First, notice that the equation uses *common* logarithms, not natural logarithms. Be sure you use the correct key on your calculator. Second, since many electrochemical reactions are heterogeneous (one or more of the electrode materials are solids), you must be careful to follow the correct procedures for writing the mass action expression. Remember that the concentrations of pure solids and liquids do not appear in the mass action expression. The concentration of water also is omitted because it is essentially a constant, too. (If you need review on this, see Section 15.8, pages 654 to 656.)

Example 18.1 Writing the Nernst Equation for a Reaction

Problem

What is the correct form for the Nernst equation for the reaction

$$3PbSO_4(s) + 2Cr(s) \rightarrow 3Pb(s) + 3SO_4^{2-}(aq) + 2Cr^{3+}(aq)$$

for which $E°_{cell} = 0.38$ V?

Solution

The oxidation of two chromium atoms to chromium(III) ions involves a transfer of $6e^-$, so $n = 6$ for this reaction. Therefore,

$$E_{cell} = 0.38\text{ V} - \frac{0.0592\text{V}}{6} \log Q$$

In constructing Q, we omit the concentrations of the solids. This gives

$$E_{cell} = 0.38\text{ V} - \frac{0.0592\text{ V}}{6} \log ([SO_4^{2-}]^3 [Cr^{3+}]^2)$$

or

$$E_{cell} = 0.38\text{ V} - (0.00989\text{ V}) \log ([SO_4^{2-}]^3 [Cr^{3+}]^2)$$

The measurement of cell potentials provides a means for determining unknown concentrations of ions in a half-cell, as illustrated by Example 18.15 on page 800. Notice that in this calculation we first solve for Q (the reaction quotient). Then we substitute known concentrations and solve for the unknown value.

Thinking It Through

For the following, identify the information needed to solve the problem and show (or explain) what must be done with it.

5 A galvanic cell was set up with a zinc electrode dipping into 100 mL of 1.00 M Zn^{2+} and an iron electrode dipping into 100 mL of 1.00 M Fe^{2+}. If the cell reaction $Zn(s) + Fe^{2+}(aq) \rightarrow Zn^{2+}(aq) + Fe(s)$ delivers a constant current of 0.500 A, what will be the potential of the cell after 1500 min has passed?

Self-Test

39. Write the correct Nernst equation for the reaction

$$MnO_2(s) + 2H^+(aq) + H_3PO_2(aq) \rightarrow Mn^{2+}(aq) + H_3PO_3(aq) + H_2O$$

for which $E^\circ_{cell} = 1.73$ V.

____

40. Write the correct Nernst equation for the reaction

$$Cd(s) + Cr^{3+}(aq) \rightarrow Cd^{2+}(aq) + Cr(s)$$

Use the data in Table 18.1. (Note: The equation is not balanced.)

____

41. What is the cell potential for the reaction in Question 39 if $[H_3PO_2]$ = 0.0010 *M*, $[Mn^{2+}]$ = 5.0×10^{-4} *M*, $[H_3PO_3]$ = 0.15 *M*, and the pH equals 5.0?

---

42. A chemist who wished to monitor the concentration of Cd^{2+} in the waste water leaving a chemical plant set up a galvanic cell consisting of a cadmium electrode that could be dipped into solutions suspected to contain Cd^{2+}, and a silver electrode that was immersed in a 0.100 *M* solution of $AgNO_3$. In a particular analysis, the potential of the cell was determined to be 1.282 V. The standard cell potential for the reaction, $Cd + 2Ag^+ \rightarrow Cd^{2+} + 2Ag$, has been accurately measured to be 1.202 V. What was the Cd^{2+} concentration in the solution that was analyzed?

New Terms

Write the definition of the following term, which was introduced in this section. If necessary, refer to the Glossary at the end of the text.

Nernst equation

18.10 Practical Applications of Galvanic Cells

Review

In this section, the chemical reactions and construction of some common batteries is described. You should be sure you know the chemical reactions that occur at the electrodes and which substances serve as cathode and anode. You should also understand the advantages and disadvantages of the various cells.

Fuel cells offer increased thermodynamic efficiency in converting the energy of chemical reactions into work because they operate under conditions approaching reversibility. Another advantage is that the fuel can be fed to them continuously, so they don't need recharging.

Self-Test

43. What is the cathode reaction in the lead storage battery while it is being discharged?

---

44. What is the cathode reaction in the lead storage battery while it is being charged?

____

45. A lead storage cell produces an emf of about 2 V. How can an automobile battery produce 12 V?

____

46. What substance serves as the anode in the common dry cell?

____

47. What substance serves as the anode in an alkaline battery?

____

48. What is the anode reaction in a nicad battery when it is being discharged?

____

49. Why can a hydrometer be used to test the state of charge of a lead storage battery?

____

____

50. What is the anode material in a "mercury battery"? _______________

What reaction does it undergo during discharge of the cell?

____

51. Write the half-reaction that takes place at the cathode during the discharge of a silver oxide battery.

____

52. What is a disadvantage of present-day fuel cells? _______________

____

New Terms

Write the definitions of the following terms, which were introduced in this section. If necessary, refer to the Glossary at the end of the text.

lead storage battery	nicad battery	silver-oxide battery
zinc-carbon dry cell	mercury battery	fuel cell

Answers to Thinking It through

1 Oxidation and reduction can be identified by changes in oxidation number. In NiF_2, the nickel has an oxidation number of +2 and in the metallic nickel deposited on the electrode has an oxidation number of zero, because it is now a free element.

$$\underset{+2}{NiF_2} \rightarrow \underset{0}{Ni}$$

A decrease in oxidation number from +2 to 0 corresponds to reduction, so the nickel is reduced and therefore must be deposited on the cathode.

The O_2 formed at the "other electrode" must come from H_2O, and the half-reaction that produces O_2 corresponds to oxidation, so the O_2 must be formed at the anode.

2. That the reaction produces H_2 at the cathode means that reduction of water must be taking place. This follows the following half-reaction, as you learned in the preceding section.

$$4H_2O(l) + 4e^- \rightarrow 2H_2(g) + 4OH^-(aq)$$

Therefore, the solution becomes basic as a result of the electrolysis. From the volume and concentration of the HCl solution, we can calculate the number of moles of HCl used. This is equal to the number of moles of OH^- that were formed in the solution (H^+ from the HCl and OH^- react in a 1-to-1 ratio). The half-reaction above tells us that the number of moles of OH^- equals the number of moles of e^- used in the electrolysis. Multiplying the moles of e^- by 9.65×10^4 C/mol e^- gives the number of coulombs used. Dividing the number of coulombs by the time is seconds (30.0 min × 60 s/min = 1800 s) gives the current in amperes. [The answer is 0.598 A]

3 The mixture will contain Ni(*s*), Fe(*s*), $Ni^{2+}(aq)$, and $Fe^{2+}(aq)$. The two possible reduction half-reactions are

$$Ni^{2+}(aq) + 2e^- \rightarrow Ni(s)$$
$$Fe^{2+}(aq) + 2e^- \rightarrow Fe(s)$$

To determine the spontaneous reaction, we find the reduction potentials of Ni^{2+} and Fe^{2+} in Table 18.1. The one with the more positive reduction potential (Ni^{2+}) will be reduced, so we write its half-reaction as reduction. The other half-reaction (that for Fe^{2+}) will be reversed to occur as oxidation. We then add the two half-reactions, making sure the electrons cancel. The spontaneous reaction will be $Ni^{2+} + Fe(s) \rightarrow Ni(s) + Fe^{2+}$.

4 To answer the question, we need to have the equilibrium constant for the reaction. We can look up the reduction potentials of Ni^{2+} and Cd^{2+} in Table 18.1 and from them calculate the value of *E*\s($_{o,cell}$) for the reaction *as writ-*

ten ($E^\circ_{cell} = -0.15$ V). Then, we use this value of E°_{cell} to calculate K_c with Equation 18.7 ($K_c = 8.6 \times 10^{-6}$). Once we have the value of K_c, we write the equilibrium law,

$$K_c = \frac{[Ni^{2+}]}{[Cd^{2+}]}$$

Then we substitute the known Ni^{2+} concentration and solve for the unknown Cd^{2+} concentration. The answer is $[Cd^{2+}] = 3.5 \times 10^{-3}$ *M*.

5 From the current and time, we can calculate the number of coulombs, and from that, the number of moles of electrons that flow in this time period. When one Zn exchanges electrons with one Fe^{2+}, two electrons are exchanged, so the number of moles of Zn that reacts equals the number of moles of e^- divided by 2. The value obtained is the amount by which the Zn^{2+} concentration increases and the amount by which the Fe^{2+} concentration decreases. We then compute the new concentrations, which we can use in the Nernst equation. To set up the Nernst equation, we have to compute the value of E°_{cell} from tabulated reduction potentials of Zn^{2+} and Fe^{2+}. The value of *n* for the cell is 2, and the mass action expression is $[Zn^{2+}]/[Fe^{2+}]$. The value of *Q* is computed by substituting the new calculated values for $[Zn^{2+}]$ and $[Fe^{2+}]$.

$$E_{cell} = E^\circ_{cell} - \frac{0.0592\text{ V}}{2} \log \frac{[Zn^{2+}]}{[Fe^{2+}]}$$

[After 1500 min, $[Zn^{2+}] = 1.23$ *M* and $[Fe^{2+}] = 0.767$ *M*. The change in the cell potential is only 0.006 V. The cell potential becomes smaller by this amount.]

Answers to Self-Test Questions

1. Electrochemical changes produce or are caused by electricity.
2. Electrochemistry is the study of electrochemical changes.
3. (a) anode
   (b) anode
   (c) cathode
4. $2H_2O + 2e^- \rightarrow H_2(g) + 2OH^-(aq)$
5. $2H_2O \rightarrow 4H^+(aq) + O_2(g) + 4e^-$
6. anode: $2I^-(aq) \rightarrow I_2(aq) + 2e^-$
   cathode: $2H_2O + 2e^- \rightarrow H_2(g) + 2OH^-(aq)$
7. 7.77×10^{-2} mol e^-
8. 3.62×10^3 s
9. 10.3 A
10. 0.0900 mol e^-

11. 36.2 minutes
12. 6.08 A
13. To beautify and protect metals.
14. cathode
15. cryolite, Na_3AlF_6
16. $4Al^{3+} + 6O^{2-} \rightarrow 4Al(l) + 3O_2(g)$
17. the ocean; $MgCl_2$
18. To keep the Cl_2 and Na apart so they don't reform NaCl.
19. The anode mud contains precious metals whose value helps pay for the electricity that's used.
20. cathode: $2e^- + 2H_2O \rightarrow H_2(g) + 2OH^-(aq)$
    anode: $2Cl^-(aq) \rightarrow Cl_2(g) + 2e^-$
    net: $2Cl^-(aq) + 2H_2O \rightarrow H_2(g) + Cl_2(g) + 2OH^-aq)$
21. Cl^- is gradually changed to OCl^-.
22. Advantage: No OCl^- is formed by reaction of Cl_2 with OH^-
    Disadvantage: NaOH solution is contaminated by small amounts of unreacted NaCl.
23. Advantage: Very pure NaOH is produced.
    Disadvantage: There is a potential for mercury pollution.
24. (a) $MnO_2 + 4H^+ + 2e^- \rightarrow Mn^{2+} + 2H_2O$
    (b) negative
    (c) away
    (d) NO_3^- ions flow into the iron half-cell compartment to compensate for the charge of the Fe^{2+} ions entering the solution
25. $Fe(s)\,|\,Fe^{2+}(aq)\,||\,Mn^{2+}(aq)\,|\,MnO_2(s)$
26. anode: $Zn(s) \rightarrow Zn^{2+}(aq) + 2e^-$; cathode: $Au^{3+}(aq) + 3e^- \rightarrow Au(s)$
27. (a)

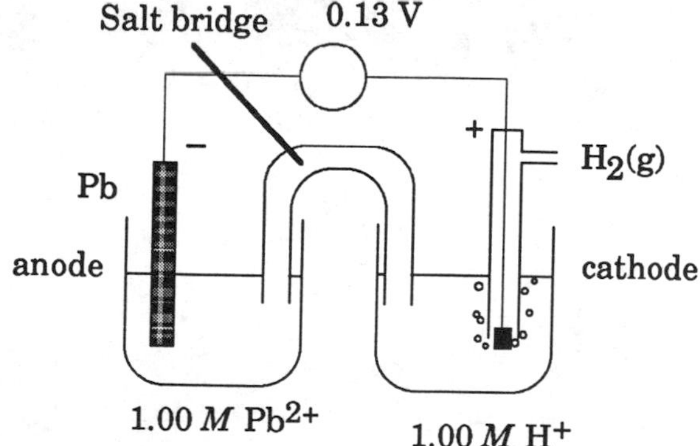

    (b) H^+ (c) −0.13 V
28. zinc
29. 1 V = 1 J/C
30. (a) $ClO_3^-(aq) + 9H^+(aq) + 3Hg_2HPO_4(s) \rightarrow Cl^-(aq) + 3H_2O + 6Hg(l) + 3H_2PO_4^-(aq)$

(b) $E° = 0.81$ V

31. $BrO_3^-(aq) + 3HAsO_2(aq) + 3H_2O \rightarrow Br^-(aq) + 3H_3AsO_4(aq)$
32. yes, $E°_{cell} = +0.90$
33. $\Delta G° = -470$ kJ
34. $\Delta G° = +277$ kcal
35. $E° = 1.23$ V
36. $K_c = 2.3 \times 10^{82}$
37. $K_c = 2 \times 10^{-203}$
38. $E°_{cell} = 0.395$ V
39. $E_{cell} = 1.73\text{ V} - (0.0296\text{ V}) \times \log\left(\frac{[Mn^{2+}][H_3PO_3]}{[H^+]^2[H_2PO_2]}\right)$
40. $E_{cell} = -0.34\text{ V} - (0.00987\text{ V}) \times \log\left(\frac{[Cd^{2+}]^3}{[Cr^{3+}]^2}\right)$
41. $E_{cell} = 1.47$ V
42. 2.0×10^{-5} *M*
43. $PbO_2(s) + 4H^+(aq) + SO_4^{2-}(aq) + 2e^- \rightarrow PbSO_4(s) + 2H_2O$
44. $PbSO_4(s) + 2e^- \rightarrow Pb(s) + SO_4^{2-}(aq)$
45. Six cells are connected in series, so their voltages add.
46. Zinc
47. Zinc
48. $Cd(s) + 2OH^-(aq) \rightarrow Cd(OH)_2(s) + 2e^-$
49. During discharge, H_2SO_4 is used up, and the density of the electrolyte changes (decreases).
50. Zinc; $Zn(s) + 2OH^-(aq) \rightarrow ZnO(s) + H_2O + 2e^-$
51. $Ag_2O(s) + H_2O + 2e^- \rightarrow 2Ag(s) + 2OH^-(aq)$
52. They are expensive and bulky.

Tools you have learned

Remove this chart from the Study Guide and keep it handy when tackling homework problems.

Tool	*Function*
Definition of anode and cathode	Identify an electrode according to the chemical reaction that takes place.
Coulombs = amperes × seconds	Relate current and time to the amount of charge passing through a cell.
1 mol e^- = 9.65×10^4 C	Relate coulombs of charge to moles of electrons.
Signs of electrodes in a galvanic cell	Identifies an electrode as anode or cathode according the the kind of charge the electrode carries.
$E^\circ_{cell} = E^\circ_{reduced} - E^\circ_{oxidized}$	To calculate the potential associated with a cell reaction. ($E^\circ_{reduced}$ is the reduction potential of the substance reduced, $E^\circ_{oxidized}$ is the reduction potential of the substance oxidized.) The spontaneous reaction has a positive E°_{cell}.
E°_{cell}	Calculate ΔG° for a reaction Calculate K_c for a reaction Use with Nernst Equation Predict spontaneity of reaction
Nernst equation $E_{cell} = E^\circ_{cell} - \frac{0.0592\text{ V}}{n}\log Q$	To determine E_{cell} when concentrations of reactants are not 1.00 M; to determine a concentration from measured cell potential.

Summary of Important Equations

Stoichiometric relationships in electrolysis

$$1\ \text{C} = 1\ \text{A} \times 1\ \text{s}$$

$$1\ \text{mol}\ e^- = 9.65 \times 10^4\ \text{C}$$

Using reduction potentials to calculate cell potentials

$$E^\circ_{\text{cell}} = \begin{pmatrix}\text{standard reduction}\\ \text{potential of the}\\ \text{substance reduced}\end{pmatrix} - \begin{pmatrix}\text{standard reduction}\\ \text{potential of the}\\ \text{substance oxidized}\end{pmatrix}$$

Cell potential and free energy change

$$\Delta G = -n\mathcal{F}E_{\text{cell}}$$

$$\Delta G^\circ = -n\mathcal{F}E^\circ_{\text{cell}}$$

Cell potential and K_c

$$E^\circ_{\text{cell}} = \frac{0.0592\ \text{V}}{n}\log K_c$$

Nernst equation

$$E_{\text{cell}} = E^\circ_{\text{cell}} - \frac{0.0592\ \text{V}}{n}\log Q$$

Chapter 19

SIMPLE MOLECULES AND IONS OF NONMETALS: PART I

Nonmetals form compounds with each other in ways denied the metals, and this gives to the nonmetals a special place in the study of chemistry. Without the nonmetals there would be much less to the chemistry of oxidizing and reducing agents; of acids, bases, and salts; and of food and agricultural chemicals—so much less that there would be no living thing around to study them!

This chapter and the next three involve what chemists call "descriptive chemistry," as distinguished from quantitative, theoretical, and mathematical aspects of chemistry. If someone who has studied no chemistry were to ask you, "Do you know any chemistry?" they would mean "Do you know anything about the substances around us, their composition, their physical states, and their chemical properties?" This chapter is, thus, also at the heart of the study of chemistry.

Learning Objectives

As you study of this chapter, keep in mind the following objectives:

1 To learn which kinds of elements dominate the composition of the universe, our solar system, the earth and its crust, and the human body.

2 To learn the preparation, isotopes, chemical properties, and uses of hydrogen and some of its binary hydrides.

3 To learn important physical and chemical properties of oxygen; to study the properties of ozone; to see how the properties of oxides are related to the location of the central atom in the periodic table and its oxidation number; and to survey the properties of the various peroxides and superoxides.

4 To learn the uses of nitrogen in nature and industry; to study the formation and properties of ammonia both as a chemical and as a solvent; and to study other compounds with nitrogen in a negative oxidation state—the nitrides, hydrazine, hydroxylamine, and hydrazoic acid.

5 To learn the preparation and properties of nitric acid and its salts, nitrous acid and its salts, and the oxides of nitrogen.

6 To learn the names, symbols, and some characteristics of the members of the carbon family and some of their binary hydrides; the allotropes of carbon; the formation and properties of carbon monoxide, carbon dioxide, carbonic acid, bicarbonates, carbonates, cyanides, and carbides.

19.1 The Prevalence of the Nonmetallic Elements

Review

According to the Big Band theory, the elements formed by nuclear fusions and transformation from an initial plasma. In composition, the elements hydrogen and helium dominate the universe as a whole as well as the sun. In the earth overall, oxygen leads on an atom basis but iron on a mass basis and air is 79% (v/v) nitrogen and 21% (v/v) oxygen.

Self-Test

1. The substance whose percent in air is 21% is ______________________
2. The seven elements that make up the compounds constituting over 97% of the human body are

 ____

3. The element that provides the majority of the atoms of the earth's crust is

 ____

New Terms

none

19.2 Hydrogen

Review

As you study this section, try to assemble a collection of basic facts about hydrogen. With these facts you should be able to comment on the following.

1. Its physical properties—colorless, odorless, tasteless, gas.

2. How it is prepared, both industrially (action of high-temperature water on a hydrocarbon, like propane) and in the lab (acid on an active metal like zinc, or water on calcium hydride).
3. The names and formulas of its isotopes; that hydrogen is the lightest element; and that it stands first in the periodic table. You should be able to tell what an *isotope effect* is, and what specifically is the *deuterium isotope effect.*
4. Its *hydrides*—ionic, covalent, and transition metal. Which behave as donors of hydride ion (the ionic hydrides), which as donors of hydrogen ion (binary hydrides of several nonmetals—the covalent hydrides), which as a donor of hydrogen atoms (methane), and which as donors of diatomic hydrogen (transition metal hydrides).
5. The major industrial uses of hydrogen—to make ammonia, hydrogenated vegetable oils, and chemicals and as a rocket fuel.

Self-Test

4. Write the equation for the production of hydrogen by the action of high-temperature water on propane (in the presence of a catalyst).

   ____

5. Write net ionic equations of the preparation of hydrogen by the reaction of hydrochloric acid with

   (a) zinc ______________________________

   (b) magnesium ______________________________

6. Give the names, atomic symbols and the numbers of neutrons in the three isotopes of hydrogen.

   ____

   ____

   ____

7. Give an example of the deuterium isotope effect. ______________________________

   ____

8. The following are the formulas of some binary hydrides.

   HCl NH_3 CaH_2 CH_4 UH_3 B_2H_6

Limiting your choices to these, write the formula of a hydride that is able to react in each of the following ways.

(a) When electricity is passed through the hydride in its liquid state, hydrogen is released at the anode. ____________

(b) It spontaneously burns in air. ____________

(c) It dissolves in water to generate H_3O^+ ____________

(d) It tends to donate H_2 ____________

(e) It tends to donate H· ____________

(f) It is an acceptor of H^+ ____________

9. Write the molecular equation for each reaction.

   (a) Potassium hydride with water ____________

   (b) Calcium hydride with water ____________

10. By means of an equation, show how hydrogen is used in the manufacture of hydrogenated vegetable oils.

    ____

New Terms

Write the definitions of the following terms, which were introduced in this section. If necessary, refer to the Glossary at the end of the text.

deuterium isotope effect

isotope effect

hydride

19.3 Oxygen

Review

With respect to diatomic oxygen, you should learn the following.

1. Its physical properties—colorless, odorless, tasteless gas.
2. How it is made industrially (from liquid air) and on a small scale (decomposition of potassium chlorate).
3. Its chief commercial uses (metals industry, manufacture of organic oxygen compounds, water treatment, health industry, and rocketry).

4. Its involvement in the natural processes of the oxygen cycle, particularly in photosynthesis, respiration, decay, and combustion.
5. The names and symbols of the other members of the oxygen family.

♦♦♦

Concerning ozone, you should learn the following.

1. Its formula, color, paramagnetism, and that it is an *allotrope*.
2. The steps in the overall reaction whereby it is made from oxygen.
3. Its very powerful oxidizing properties (making it a dangerous substance, but useful in water treatment).

♦♦♦

Concerning oxides, you should learn the following, including equations that illustrate the properties learned.

1. Metal oxides in which the metal is in a lower oxidation state (+1, +2, and sometimes +3) are basic oxides and ionic in character. Those that dissolve in water do so by a reaction that changes the oxide ion to the hydroxide ion. The ionic oxides include all the oxides of Groups IA and IIA metals (except the oxide of beryllium, which is covalent).

   Be able to write equations that illustrate the reaction of a basic, ionic oxide with water and with hydronium ion.
2. Metal oxides of transition metals in high oxidation states (+4 to +8, and occasionally +3) are covalent oxides (often with low melting points). The high oxidation state so distorts electron clouds of the oxide ion that they are forced into shared electron-pair (covalent) bonds. These oxides are acidic oxides; they react with water to give acids.
3. Metals in a +3 oxidation state (e.g., Al, Cr) often form amphoteric oxides—they react with acids and bases.

   Be able to illustrate by equations the amphoteric nature of aluminum oxide.
4. The oxides of nonmetals are generally covalent and acidic, and become more strongly acidic as the oxidation state of the nonmetal increases.

♦♦♦

Concerning the hydrogen peroxide, you should learn the following.

1. The formula, physical nature, and dangers of hydrogen peroxide.
2. How it decomposes (the equation) and what accelerates this.
3. That it is a strong oxidizing agent, particularly in acid.

♦♦♦

With respect to the oxides and superoxides of metals, you should concentrate on the following.

1. What specifically forms (name and formula) when each member of Group IA reacts with oxygen.
2. The reaction with water of
   (a) sodium peroxide (and the peroxide ion).
   (b) the superoxide ion.
3. The particular danger of superoxides.

Self-Test

11. What is the industrial source of oxygen? ____________________

12. Write the equation for the thermal decomposition of potassium chlorate.

    ____

13. What are five important commercial uses of oxygen?

    ____

    ____

14. What occurs, overall, in photosynthesis? (Write and equation.)

    ____

15. What processes are the largest users of oxygen *in nature?*

    ____

16. How does the thermal decomposition of carbonate rock in volcanic activity interact with the oxygen cycle?

    ____

17. What role do phytoplankton have in the oxygen cycle?

    ____

18. What is "lithifaction," and how does it participate in the control of the oxygen content of the air?

    ____

    ____

19. What are the names and atomic symbols of the members of the oxygen family?

____

20. What is the structure of ozone?

____

21. What are the steps in the formation of ozone from oxygen in the presence of UV radiation?

____

____

22. Why is ozone such a dangerous pollutant in smog? _______________

____

At what concentration of O_3 in air (in ppm) should vigorous activities by children be curtailed?

____

23. The following are the formulas of several oxides.

SO_3 MgO CrO_3 SO_2 Na_2O Al_2O_3

Answer the following questions choosing from these.

(a) This covalent acidic metal oxide that reacts with water as follows:

____

(b) This ionic oxide reacts with water as it dissolves in water. Write the net ionic equation.

____

(c) This amphoteric oxide neutralizes hydrochloric acid as follows: (Write the net ionic equation.)

____

It neutralizes sodium hydroxide by the following net ionic equation.

____

(d) This ionic oxide does not dissolve in water but does dissolve in hydrochloric acid by the following net ionic equation.

____

(e) This gaseous covalent oxide reacts with water to make the solution strongly acidic. The equation is

____

(f) This gaseous covalent oxide has a nonmetal atom in the +4 oxidation state.

____

24. Why is CrO_3 a more acidic oxide than Cr_2O_3?______________________

____

25. What is the equation for the decomposition of hydrogen peroxide?

____

26. Write the half-reaction that illustrates what happens to H_2O_2 when it serves as an oxidizing agent in acidic solution.

____

Signifying its ability to be an oxidizing agent, the standard electrode potential for this would be negative or positive?

____

27. Complete and balance the following equations.

$Li + O_2 \rightarrow$ ______________________________

$Na + O_2 \rightarrow$ ______________________________

$K + O_2 \rightarrow$ ______________________________

$Rb + O_2 \rightarrow$ ______________________________

$Cs + O_2 \rightarrow$ ______________________________

28. Write the net ionic equation for the reaction of the superoxide ion with water.

____

29. How does potassium superoxide react with carbon dioxide? Write the molecular equation.

____

30. Why should one never cut into an encrusted sample of potassium metal?

____

Does an encrusted sample of sodium metal pose the same danger?

__________ Why? ______________________________

____

New Terms

Write the definitions of the following terms, which were introduced in this section. If necessary, refer to the Glossary at the end of the text.

allotrope
oxygen family
peroxide
photosynthesis

19.4 Nitrogen

Review

Concerning elemental nitrogen, the following should be learned.

1. The concentration (% v/v) of nitrogen in air, and its most obvious physical properties—colorless, odorless, tasteless.
2. Its industrial source—liquid air—and its major industrial uses—enhanced oil recovery in oil fields, inert gas atmosphere, inexpensive, unreactive coolant, and the synthesis of nitrogen compounds, particularly ammonia.
3. The oxidation states it can exhibit.
4. Its use in nature in nitrogen fixation, and the importance of this in the nitrogen cycle.

♦♦♦

In your study of ammonia and amides, be sure to check the following, and be able to write equations where relevant.

1. How ammonia is manufactured.
2. The nature of the equilibrium in aqueous ammonia, and why this system is not called ammonium hydroxide.
3. The proton-accepting ability of ammonia.
4. What forms when ammonia burns (a) without a catalyst and (b) with a platinum-rhodium catalyst.

5. How liquid ammonia dissolves Group IA and IIA metals, what an ammoniated electron is, and how metal amides form.
6. The strong basicity of the amide ion (a) in liquid ammonia toward the ammonium ion and (b) toward water.
7. The chemical properties of the ammonium ion

   (a) in water toward strong bases,

   (b) in the thermal decomposition of ammonium salts that do not have an anion that is an oxidizing agent, as well as when such an anion (NO_3^-, or $Cr_2O_7^{2-}$) is present.

♦♦♦

Concerning other compounds in which N is in a negative oxidation state, the following are important areas to study.

1. What the nitrides are—both metal nitrides and others.
2. The reaction of a metal nitride with water. (Think of N^{3-} as a powerful proton-acceptor able to take enough protons from water to form NH_3 and $3OH^-$.)
3. The formula, toxicity, and strong reducing powers of hydrazine.
4. What forms when hydrazine burns.
5. The formula and ammonia-like behavior of hydroxylamine.
6. The formula and weak acidity of hydrazoic acid.
7. The formula and structure of the azide ion and which azides are dangerous.

♦♦♦

Concerning nitric acid, the following features should be learned.

1. The steps in the Ostwald synthesis.
2. The equation by which nitric acid slowly decomposes, particularly in sunlight.
3. The chief industrial uses of nitric acid.
4. The reactions of nitric acid with metal oxides, hydroxides, carbonates and bicarbonates (both molecular and net ionic equations).
5. The half-cell reactions when the nitrate ion functions as an oxidizing agent and its nitrogen changes its oxidation state from +5 to +4, +3, and +2.

♦♦♦

The nitrate salts should be studied so as to learn the following.

1. Nearly all metal nitrates crystallize as hydrates that, when heated, may decompose to nitrites or oxides.
2. Ammonium nitrate can detonate at temperatures above 300 °C (giving nitrogen, oxygen, and water). Be able to write the equation.
3. That a lower temperature decomposition of ammonium nitrate is a synthesis of dinitrogen monoxide (nitrous oxide, laughing gas).

♦♦♦

Concerning the oxides of nitrogen, learn the following.

1. The systematic and common names of each of them.
2. How each is synthesized (using equations).
3. What chemically happens (if anything) when each is heated in air.
4. Which are colored, and which are paramagnetic.
5. How the following react with water and with aqueous strong base: N_2O_3, NO_2, N_2O_5.
6. How nitrous acid and salts of nitrous acid are made.

♦♦♦

The flow chart on page 394 summarizes the chemistry of nitrogen compounds.

Self-Test

31. What is the source of nitrogen, industrially? ______________________

32. What are three major uses of nitrogen in industry? ________________

____

33. What is "nitrogen fixation?" ______________________________

____

34. What is the Haber-Bosch process? (Give an overall equation.)

____

35. Write in the species that is the strongest acid that can exist in liquid ammonia. (Write its formula.) ______________________

Do the same for the strongest base. ____________________

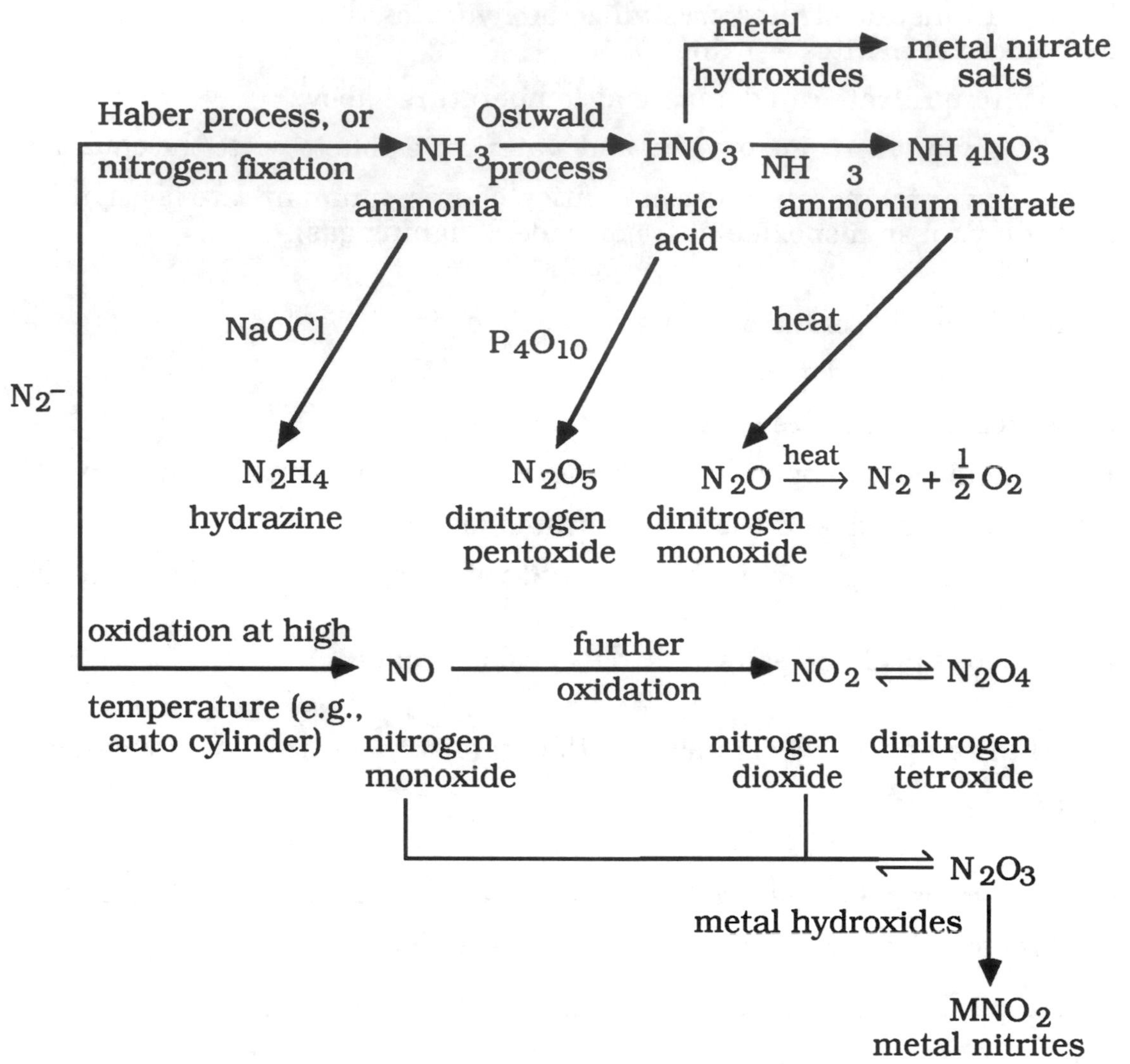

Flow chart showing the chemistry of nitrogen and some of its compounds

36. Write the equilibrium present in aqueous ammonia.

____

37. What is the net ionic equation for the reaction of hydrochloric acid with aqueous ammonia.

____

38. What is the oxidation number of N in NH_3? ______________

In NH_4^+ ________ In Na_3N? ______ In N^{3-}? ______________

39. What is an ammoniated electron and how does it form?

____

____

40. How is sodium amide made? (Write the equation.)

____

41. Write the equations for the combustion of ammonia,

without a catalyst: ______________________________

with a Pt/Rh catalyst: ____________________________

42. Complete and balance the following equations. (Write molecular equations when reactants are shown as whole formula units, and write net ionic equations where one or more reactant is given as an ionic species.)

(a) $NaNH_2(s) + H_2O \rightarrow$ ______________________________

(b) $NH_4Cl(aq) + NaOH(aq) \rightarrow$ ______________________________

(c) $NH_4Cl(s) \xrightarrow{\text{heat}}$ ______________________________

(d) $NH_4NO_3(s) \xrightarrow{\text{heat}}$ ______________________________

(e) $(NH_4)_2Cr_2O_7(s) \xrightarrow{\text{heat}}$ ______________________________

(f) $NH_2^-(s) + H_2O \rightarrow$ ______________________________

(g) $NH_4^+(am) + NH_2^-(am) \rightarrow$ ______________________________

(h) $Mg(s) + NH_3(g) \xrightarrow{\text{heat}}$ ______________________________

(i) $Li_3N(s) + H_2O \rightarrow$ ______________________________

(j) $N_2H_4(l) + O_2(g) \rightarrow$ ______________________________

(k) $Ca(s) + NH_3(g) \xrightarrow{\text{heat}}$ ______________________________

(l) $N^{3-}(s) + H_2O \rightarrow$ ______________________________

43. Write the structures of the following.

(a) hydrazine ______________________

(b) hydroxylamine ______________________

(c) azide ion ______________

44. Write balanced molecular equations for the steps in the Ostwald process for the synthesis of nitric acid.

____

____

____

45. Write the equation for the decomposition of concentrated nitric acid as it remains in a laboratory bottle exposed to sunlight.

____

46. What is the largest single industrial use of nitric acid?

____

47. Write molecular and net ionic equations for the following reactions.

(a) Nitric acid and sodium hydroxide.

____

____

(b) Nitric acid and potassium carbonate

____

____

(c) Nitric acid and sodium bicarbonate

____

____

(d) Nitric acid and calcium oxide

____

____

(e) Concentrated nitric acid and copper

____

____

(f) Dilute nitric acid and copper

____

____

48. Ammonium nitrate thermally decomposes in different ways according to the temperature. Write the molecular equations for these.

    (a) At above 300 °C ________________________________

    (b) Between 200 and 260 °C ________________________________

49. Write equations for the synthesis of each of the following in the lab.

    (a) N_2O ________________________________

    (b) NO ________________________________

    (c) NO_2 ________________________________

    (d) N_2O_5 ________________________________

50. Write the formulas and formal names for each of the following.

    (a) laughing gas ______ ________________________________

    (b) nitric oxide ______ ________________________________

    (c) nitrous oxide ______ ________________________________

    (d) nitric anhydride ______ ________________________________

    (e) nitrous anhydride ______ ________________________________

51. What two oxides of nitrogen give the equivalent of dinitrogen trioxide? (Give their formulas.) ________________________________

52. Write the formulas of the following nitrogen oxides.

    (a) It reacts with water to give nitric acid. ______

    (b) It forms in an automobile cylinder. ______

    (c) It decomposes when heated to nitrogen and oxygen ______

    (d) It is further oxidized by oxygen in air. ______

    (e) It is colorless and paramagnetic. ______

    (f) It is red and paramagnetic. ______

(g) It forms when another oxide of nitrogen is cooled. ____________

53. How can a solution of nitrous acid be made from a salt? (Write the equation.)

____

54. How does aqueous nitrous acid decompose? (Write the equation.)

____

55. What oxide of nitrogen contributes to acid rain? ____________

Write the equation for its reaction with water.

____

How does this nitrogen oxide become an air pollutant?

____

New Terms

Write the definitions of the following terms, which were introduced in this section. If necessary, refer to the Glossary at the end of the text.

Haber-Bosch process	nitrogen fixation
nitrogen family	Ostwald process

19.5 Carbon

Review

The most notable trends in the carbon family are that its members become more metallic as the family is descended and their binary hydrides become less stable and much less numerous. Carbon is unique among all elements in its ability to form strong single, double, or triple covalent bonds to its own kind while also being able to form strong bonds to other nonmetals.

The two chief allotropes of carbon—graphite and diamond—differ markedly in properties because they differ so much in structure. Charcoal, coke, and carbon black are mostly graphite. The fullerenes consist of C_n molecules, where n is 60, 70, or higher numbers, and whose molecules resemble soccer balls or geodesic domes in shape.

With respect to the oxides of carbon and related compounds, be sure to learn the following.

1. The monoxide is a colorless, odorless, poisonous gas that is present when any carbon compound (or carbon) burns in a supply of oxygen insufficient to form carbon dioxide entirely. It is itself a fuel and is used as a reducing agent in the minerals industry. It is also a ligand in coordination chemistry.
2. The dioxide is a colorless, odorless gas that, although not poisonous, cannot support life and so can cause suffocation. It is the major end product of the combustion of carbon compounds. It forms when many metal carbonates and bicarbonates are heated strongly. It is used in the synthesis of urea, as a fire extinguisher, and it is the dissolved gas in many beverages. When bubbled into aqueous bases, it neutralizes them, forming bicarbonates and carbonates.
3. Carbonic acid forms in an equilibrium with aqueous carbon dioxide. It is an unstable acid, so it cannot be isolated. It is a weak acid, and it is the parent of two families of salts—bicarbonates and carbonates.
4. Metal carbonates and bicarbonates react with acids to generate carbonic acid from which, by prompt decomposition, carbon dioxide is released.

From the standpoint of future studies in chemistry, the properties of carbon dioxide, carbonic acid, and its salts are by far the most important topics in this section.

♦♦♦

Concerning the greenhouse effect, learn the following.

1. What the term "greenhouse effect" means.
2. What are the chief greenhouse gases and how they cause the greenhouse effect.
3. The sources of the chief greenhouse gas, CO_2, and the sources of the slow but steady increase in the level of this gas in air over the last century.
4. What the CFCs are (in general terms) and how their involvement in ozone depletion in the stratosphere contributes to global warming.
5. The effects of volcanic emissions on global warming.

♦♦♦

Concerning other inorganic compounds of carbon, learn the following.

1. The formula of its sulfide, and that this compound is poisonous and can explode when heated.
2. The cyanides are salts of hydrogen cyanide, a poisonous gas that is a weak acid in water, and the cyanide ion is a moderately strong Brønsted base.
3. The binary carbides occur as salt-like compounds of the C_2^{2-} ion (e.g., CaC_2), or of the C^{4-} ion (as Mg_2C). Both of these anions are powerful

Brønsted bases that take protons from water to form hydrides. The C_2^{2-} ion changes to acetylene and the C^{4-} ion to methane. Some carbides are covalent (e.g., SiC) and some are interstitial (e.g., WC).

Self-Test

56. What are the names of the two chief allotropes of carbon?

____

57. In which allotrope of carbon is there a tetrahedral array of carbon atoms? _______________

58. Which allotrope of carbon has carbon atoms in sheets of hexagonal arrays of atoms? _______________

59. Which form of carbon consists of molecules resembling soccer balls in shape and form? _______________

60. Water gas is a mixture of what two combustible substances? (Write their formulas.) _______________

61. Carbon monoxide can be used to reduce CuO. Write the equation.

____

62. Compounds in which CO occurs as a ligand are called what?

____

63. Industrially, what is the chief source of carbon dioxide?

____

64. Name four important uses of CO_2 _______________

____

65. Name each of the following.
    (a) H_2CO_3 _______________
    (b) HCN _______________
    (c) CS_2 _______________
    (d) K_2CO_3 _______________
    (e) $NaHCO_3$ _______________

(f) CaC_2 ________________

(g) KCN ________________

66. Complete and balance the following equations.

(a) $NaOH(aq) + CO_2(g) \rightarrow$ ________________

(b) $NaHCO_3(aq) + HNO_3(aq) \rightarrow$ ________________

(c) $K_2CO_3(aq) + HBr(aq) \rightarrow$ ________________

(d) $CaC_2(s) + H_2O \rightarrow$ ________________

(e) $CN^-(aq) + H_2O \rightarrow$ ________________

(f) $Mg_2C(s) + H_2O \rightarrow$ ________________

(g) $C + S \xrightarrow{heat}$ ________________

(h) $CO_2(aq) + H_2O \rightleftharpoons$ ________________

67. What is the "greenhouse effect?" ________________

____

68. What are the names of four greenhouse gases? ________________

____

69. Which green house gas is totally of human origin? ________________

70. What is a necessary function of the greenhouse gases from a human point of view?

____

71. Which greenhouse gas contributes most to global warming?

____

New Term

Write the definition of the following term, which was introduced in this section. If necessary, refer to the Glossary at the end of the text.

carbon family

Answers to Self-Test Questions

1. oxygen
2. carbon, hydrogen, oxygen, nitrogen sulfur, phosphorus, and chlorine
3. oxygen

4. $C_3H_8 + 3H_2O \xrightarrow[\text{catalyst}]{900\ ^\circ C} 3CO + 7H_2$

5. (a) $Zn(s) + 2H^+(aq) \rightarrow Zn^{2+}(aq) + H_2(g)$
   (b) $Mg(s) + 2H^+(aq) \rightarrow Mg^{2+}(aq) + H_2(g)$
6. protium, H, 0 neutrons; deuterium, D, 1 neutron; tritium, T, 2 neutrons
7. D_2 reacts more slowly with Cl_2 than H_2.
8. (a) CaH_2 (b) B_2H_6 (c) HCl (d) UH_3 (e) CH_4 (f) NH_3
9. (a) $KH(s) + H_2O \rightarrow KOH(aq) + H_2(g)$
   (b) $CaH_2(s) + 2H_2O \rightarrow Ca(OH)_2(s) + 2H_2(g)$
10. Under heat and pressure and in the presence of a catalyst, hydrogen adds to carbon-carbon double bonds as follows (where only partial structures are shown).

$$>C{=}C< \quad + \quad H_2 \longrightarrow -\overset{|}{C}H-\overset{|}{C}H-$$

11. Distillation of liquid air.
12. $2KClO_3(s) \rightarrow 2KCl(s) + 3O_2(g)$
13. Metal industry (as in the basic oxygen process for iron), synthesis of organic oxygen compounds, water treatment, health industry, and rocketry.
14. $nCO_2 + nH_2O \xrightarrow[\text{(several steps)}]{\text{photosynthesis}} (CH_2O)_n + nO_2$
15. Combustion, decay, and respiration.
16. It releases CO_2, a raw material for photosynthesis
17. Phytoplankton are photosynthesizers
18. Lithifaction is the use of CO_2 to make carbonate rocks, like limestone, and this removes some CO_2 from its place as a raw material for photosynthesis.
19. Oxygen, O; Sulfur, S; Selenium, Se; Tellurium, Te; Polonium, Po
20. It is a bent molecule that is a hybrid of the following resonance structures.

$$:\ddot{O}{=}\ddot{O}-\ddot{O}:^{-} \longleftrightarrow {}^{-}:\ddot{O}-\ddot{O}{=}\ddot{O}:$$

21. $O_2 \xrightarrow{\text{UV radiation}} 2\,O$
    $O + O_2 + M \rightarrow O_3 + M$
22. It is a powerful oxidizing agent toward most substances in cells and it attacks lung tissue. 0.5 ppm.
23. (a) $CrO_3(s) + H_2O \rightarrow H_2CrO_4(aq)$
    (b) $Na_2O(s) + H_2O \rightarrow 2NaOH(aq)$

(c) $Al_2O_3(s) + 6H^+(aq) \rightarrow 2Al^{3+}(aq) + 3H_2O$

$Al_2O_3(s) + 2OH^-(aq) + 7H_2O \rightarrow 2[Al(H_2O)_2(OH)_4]^{2-}(aq)$

(d) $MgO(s) + 2H^+(aq) \rightarrow Mg^{2+}(aq) + H_2O$

(e) $SO_3(g) + H_2O \rightarrow H_2SO_4(aq)$

(f) SO_2

24. In CrO_3, Cr is in the +6 oxidation state and it strongly attracts electron density from the oxygens making them unable to function as basic oxide ions. The Cr center is more able to accept another oxygen (by accepting a water molecule) and change into an acid.
25. $2H_2O_2(l) \rightarrow 2H_2O + O_2(g)$
26. $H_2O_2(aq) + 2H^+(aq) + 2e^- \rightleftharpoons 2H_2O$
27. $4Li(s) + O_2(g) \rightarrow 2Li_2O(s)$
    $2Na(s) + O_2(g) \rightarrow Na_2O_2(s)$
    $K(s) + O_2(g) \rightarrow KO_2(s)$
    $Rb(s) + O_2(g) \rightarrow RbO_2(s)$
    $Cs(s) + O_2(g) \rightarrow CsO_2(s)$
28. $2O_2^-(s) + H_2O \rightarrow O_2(g) + HO_2^-(aq) + OH^-(aq)$
29. $4KO_2(s) + 2CO_2(g) \rightarrow 2K_2CO_3(s) + 3O_2(g)$
30. The superoxide coating in the crust, KO_2, can detonate when mechanically cut. The crust on sodium is not a superoxide, and it does not detonate when cut.
31. Liquid air.
32. Enhanced oil recovery, inert blanketing gas, ammonia synthesis.
33. Reactions used by microorganisms that convert nitrogen into ammonia and other nitrogen compounds in soil or water.
34. It is the industrial synthesis of ammonia. $N_2 + 3H_2 \rightarrow 2NH_3$
35. NH_4^+. NH_2^-
36. $NH_3(aq) + H_2O \rightleftharpoons NH_4^+(aq) + OH^-(aq)$
37. $NH_3(aq) + H^+(aq) \rightarrow NH_4^+(aq)$
38. It is –3 in all of these.
39. It is an electron associated with an ammonia molecule in liquid ammonia which forms when a Group IA metal dissolves in liquid ammonia.
40. $2Na(s) + 2NH_3(l) \rightarrow 2NaNH_2(am) + H_2(g)$
41. $4NH_3(g) + 3O_2(g) \rightarrow 2N_2(g) + 6H_2O$
    $4NH_3(g) + 5O_2(g) \xrightarrow{Pt/Rh} 4NO(g) + 6H_2O$
42. (a) $NaNH_2(s) + H_2O \rightarrow NaOH(aq) + NH_3(aq)$
    (b) $NH_4Cl(aq) + NaOH(aq) \rightarrow NH_3(aq) + H_2O + NaCl(aq)$
    (c) $NH_4Cl(s) \xrightarrow{heat} NH_3(g) + HCl(g)$

(d) $NH_4NO_3(s) \xrightarrow{\text{heat}} N_2O(g) + 2H_2O(g)$
(e) $(NH_4)_2Cr_2O_7(s) \xrightarrow{\text{heat}} N_2(g) + Cr_2O_3(s) + 4H_2O(g)$
(f) $NH_2^-(s) + H_2O \rightarrow NH_3(aq) + OH^-(aq)$
(g) $NH_4^+(am) + NH_2^-(am) \rightarrow 2NH_3(am)$
(h) $3Mg(s) + 2NH_3(g) \xrightarrow{\text{heat}} Mg_3N_2(s) + 3H_2(g)$
(i) $Li_3N(s) + 3H_2O \rightarrow 3LiOH(aq) + NH_3(aq)$
(j) $N_2H_4(l) + O_2(g) \rightarrow N_2(g) + 2H_2O(g)$
(k) $3Ca(s) + 2NH_3(g) \xrightarrow{\text{heat}} Ca_3N_2(s) + 3H_2(g)$
(l) $N^{3-}(s) + 3H_2O \rightarrow NH_3(aq) + 3OH^-(aq)$

43. (a)
```
  H  H
  |  |
H-N--N-H
```
(b)
```
  H
  |
H-N-O-H
```
(c) $[N{\equiv}N{-}N]^-$

44. $4NH_3(g) + 5O_2(g) \rightarrow 4NO(g) + 6H_2O$
$2NO(g) + O_2(g) \rightarrow 2NO_2(g)$
$3NO_2(g) + H_2O \rightarrow 2HNO_3(aq) + NO(g)$
45. $4HNO_3 \rightarrow 4NO_2 + 2H_2O + O_2$
46. The manufacture of ammonium nitrate.
47. (a) $HNO_3(aq) + NaOH(aq) \rightarrow NaNO_3(aq) + H_2O$
$H^+(aq) + OH^-(aq) \rightarrow H_2O$
(b) $2HNO_3(aq) + K_2CO_3(aq) \rightarrow 2KNO_3(aq) + CO_2(g) + H_2O$
$2H^+(aq) + CO_3^{2-}(aq) \rightarrow CO_2(g) + H_2O$
(c) $HNO_3(aq) + NaHCO_3(aq) \rightarrow NaNO_3(aq) + CO_2(g) + H_2O$
$H^+(aq) + HCO_3^-(aq) \rightarrow CO_2(g) + H_2O$
(d) $2HNO_3(aq) + CaO(s) \rightarrow Ca(NO_3)_2(aq) + H_2O$
$2H^+(aq) + CaO(s) \rightarrow Ca^{2+}(aq) + H_2O$
(e) $Cu(s) + 4HNO_3(aq) \rightarrow Cu(NO_3)_2(aq) + 2NO_2(g) + 2H_2O$
$Cu(s) + 2NO_3^-(aq) + 4H^+(aq) \rightarrow Cu^{2+}(aq) + 2NO_2(g) + 2H_2O$
(f) $3Cu(s) + 8HNO_3(aq) \rightarrow 3Cu(NO_3)_2(aq) + 2NO(g) + 4H_2O$
$3Cu(s) + 2NO_3^-(aq) + 8H^+(aq) \rightarrow 3Cu^{2+}(aq) + 2NO(g) + 4H_2O$
48. (a) $2NH_4NO_3(s) \rightarrow 2N_2(g) + O_2(g) + 4H_2O(g)$
(b) $NH_4NO_3(s) \rightarrow N_2O(g) + 2H_2O(g)$
49. (a) $NH_4NO_3(s) \xrightarrow{\text{heat}} N_2O(g) + 2H_2O(g)$
(b) $N_2(g) + O_2(g) \xrightarrow{\text{heat}} 2NO$
(c) $2NO(g) + O_2(g) \rightarrow 2NO_2(g)$
(d) $4HNO_3(aq) + P_4O_{10}(s) \rightarrow 2N_2O_5(s) + 4HPO_3(l)$
50. (a) N_2O, dinitrogen monoxide
(b) NO, nitrogen monoxide

(c) N_2O, dinitrogen monoxide
(d) N_2O_5, dinitrogen pentoxide
(e) N_2O_3, dinitrogen trioxide

51. $NO + NO_2$
52. (a) N_2O_5 (b) NO (c) N_2O (d) NO (e) NO (f) NO_2 (g) N_2O_4
53. $NaNO_2(s) + HCl(aq) \rightarrow HNO_2(aq) + NaCl(aq)$
54. $3HNO_2(aq) \rightarrow HNO_3(aq) + 2NO(g) + H_2O$
55. NO_2. $2NO_2 + H_2O \rightarrow HNO_3 + HNO_2$. NO_2 forms when NO in auto exhaust reacts with oxygen in air: $2NO + O_2 \rightarrow 2NO_2$
56. graphite and diamond
57. diamond
58. graphite
59. the fullerenes
60. CO and H_2
61. $CuO(s) + CO(g) \rightarrow Cu(s) + CO_2(g)$
62. metal carbonyls
63. From the Haber process.
64. Manufacture of urea; refrigerant; fire extinguishers; production of carbonated beverages.
65. (a) carbonic acid (b) hydrogen cyanide (c) carbon disulfide (d) potassium carbonate (e) sodium bicarbonate (f) calcium carbide (g) potassium cyanide
66. (a) $NaOH(aq) + CO_2(g) \rightarrow NaHCO_3(aq)$
    (b) $NaHCO_3(aq) + HNO_3(aq) \rightarrow NaNO_3(aq) + CO_2(g) + H_2O$
    (c) $K_2CO_3(aq) + 2HBr(aq) \rightarrow 2KBr(aq) + CO_2(g) + H_2O$
    (d) $CaC_2(s) + 2H_2O \rightarrow C_2H_2(g) + Ca(OH)_2(s)$
    (e) $CN^-(aq) + H_2O \rightleftharpoons HCN(aq) + OH^-(aq)$
    (f) $Mg_2C(s) + 4H_2O \rightarrow CH_4(g) + 2Mg(OH)_2(s)$
    (g) $C + 2S \rightarrow CS_2$
    (h) $CO_2(aq) + H_2O \rightleftharpoons H_2CO_3(aq)$
67. The insulating effect of the earth's atmosphere caused by the presence of the greenhouse gases.
68. carbon dioxide, methane, dinitrogen monoxide, and the chlorofluorocarbons
69. the CFCs (chlorofluorocarbons)
70. Without the insulating action of the greenhouse effect, the earth could cool to a temperature too low to sustain human life.
71. carbon dioxide

Chapter 20

SIMPLE MOLECULES AND IONS OF NONMETALS: PART II

In this chapter we conclude what we started in Chapter 19—a survey of the preparation, properties, and reactions of important nonmetals. We study additional members of the families of these elements and point out some general trends within families and periodic relationships.

Learning Objectives

Throughout your study of this chapter, keep in mind the following objectives:

1 To learn how sulfur occurs in nature; and to study its principal allotropes.

2 To learn the preparation, properties, and interrelationships among sulfur dioxide, sulfurous acid, and the salts of this acid.

3 To learn the preparation and the chief properties and uses of sulfur trioxide, sulfuric acid, and the salts of this acid; and to survey briefly the chemistry of the thiosulfate ion.

4 To learn some of the properties of hydrogen sulfide and sulfide compounds.

5 To survey some trends in the sulfur family.

6 To learn of the two chief allotropes of phosphorus; to study the preparation and properties of the chief halides and hydrides of phosphorus.

7 To learn the formation, the various forms, and the main properties of phosphoric acid; but also to learn something about the other oxoacids of phosphorus and to survey some trends in the nitrogen family of elements.

8 To learn the sources of the halogens, and how they react (if at all) with water; to learn the preparations and properties of the binary hydrides and the more stable oxides of the halogens; to study the major halogen oxoacids and salts; and (as important as anything else in the section), to learn the major periodic trends in the family.

9 To learn the names and symbols of the noble gases and some of their properties.

20.1 Sulfur

Review

Sulfur occurs as the free element, as inorganic sulfates and sulfides, and in organic substances. Its two major sources are from underground sulfur deposits (obtained via the Frasch process) and from the conversion of the hydrogen sulfide in natural gas or petroleum to sulfur. You should learn the reaction whereby SO_2 and H_2S react to form sulfur.

The most stable allotrope of sulfur is S_α or orthorhombic sulfur. It consists of S_8 molecules in a cyclic, crown configuration. The same molecules occur in another allotrope, S_β or monoclinic sulfur, but they are stacked differently in the crystal. S_β, on long standing, gradually rearranges to S_α. Other allotropes—plastic sulfur, for example—are mixtures of molecules of sulfur that form when S_8 rings open on heating and then form into chains of varying length on cooling. Plastic sulfur also reverts to S_α on standing.

Sulfur forms compounds with most other elements, including all but three metals. Its compounds with nonmetals are restricted largely to those with high electron affinities—oxygen, fluorine, and chlorine, for the most part. This is because sulfur itself has a high electron affinity.

Concerning sulfur dioxide, you should learn the following.

1. It is a colorless gas with a sharp odor and it forms when sulfur burns. (Know the equation.)
2. It is quite soluble in water, and the solution is acidic.
3. It is a major air pollutant because it forms when any sulfur-bearing fuel is burned, and it is a major cause of acid rain.

♦♦♦

With respect to sulfurous acid, you should know the following:

1. "Sulfurous acid" is simply a solution of sulfur dioxide in water in which the SO_2 molecules are hydrated. When the water is removed from such a solution, everything goes into the gaseous state. In other words, "sulfurous acid" is unstable.
2. A molecule with the traditional formula for sulfurous acid, H_2SO_3, does not exist; this formula, however, is a convenient way by which to discuss the acidic nature of aqueous sulfur dioxide. And it is acceptable to use the formula H_2SO_3 in such discussions.

3. The hydrogen sulfite anion, HSO_3^-, does exist, and its salts are known in solution. However, when such solutions are heated to drive off the water and give the solid hydrogen sulfite salts, complex changes occur. The hydrogen sulfite salts do not form.
4. When excess SO_2 is present as water is driven off by heat from a solution of sodium hydrogen sulfite, sodium metabisulfite forms, $Na_2S_2O_5$. When this is redissolved in water, the following equilibrium forms, and this gives us a source of HSO_3^-.

$$2HSO_3^-(aq) \rightleftharpoons S_2O_5^{2-}(aq) + H_2O$$

5. Sodium sulfite can be obtained by driving the water off a solution of SO_2 in water that has been neutralized by the appropriate amount of sodium carbonate.
6. The sulfite ion and the hydrogen sulfite ion are Brønsted bases; their conjugate acids are relatively weak acids.
7. All systems—aqueous SO_2, aqueous $Na_2S_2O_5$ (which contains the bisulfite ion, HSO_3^-) and sulfite salts—are reducing agents.
8. If you acidify a solution of a sulfite or hydrogen sulfite salt, SO_2 will form, and much of it will leave the solution as a gas.

♦♦♦

Concerning acid rain or acid deposition, you should know the following.

1. Oxides of sulfur and oxides of nitrogen are the important agents that make rain acidic. They dissolve in the rain water to generate hydrogen ions.
2. The sulfur oxides—mostly SO_2—come largely from the combustion of sulfur-containing fossil fuels. The nitrogen oxide that directly participates in acid rain is NO_2, which forms from nitrogen monoxide, NO, made from atmospheric nitrogen and oxygen at high temperatures in the cylinders of vehicles or in power plant furnaces.
3. Acid deposition corrodes metals, causes building stones (especially limestone) to deteriorate, and makes bodies of water unfit for fish.
4. The technology for removing sulfur and nitrogen oxides from exhaust gases exists.

♦♦♦

Concerning sulfur trioxide, you should learn the following.

1. It is a colorless gas with an extremely sharp odor and is very damaging to most surfaces in the presence of water.
2. It forms from the oxidation of SO_2 only slowly.
3. It is a highly acidic nonmetal oxide that gives sulfuric acid in water and that directly neutralizes metal hydroxides, carbonates, and bicarbonates.

4. Virtually its only use is in the manufacture of sulfuric acid.

♦♦♦

About sulfuric acid you should learn the following.

1. The individual steps in the contact process for making it.
2. How to handle its concentrated form safely.
3. It is a strong, diprotic acid.
4. Its chief industrial use is to make phosphate fertilizers from phosphate rock.
5. You should review the reactions of all strong acids with metal oxides, hydroxides, carbonates, and bicarbonates.

♦♦♦

Concerning the salts of sulfuric acid, you should know the following.

1. The hydrogen sulfate salts are confined largely to those of the Group IA metals. Sodium hydrogen sulfate, the most common of these kinds of salts, is often used when a solid inorganic acid is needed, because the hydrogen sulfate ion is a moderately strong acid.
2. The Group IA sulfates are readily soluble in water and form neutral solutions.
3. The Group IIA sulfates from calcium and below are insoluble in water. Anhydrous forms are good drying agents.

♦♦♦

About the thiosulfate ion, learn the following.

1. How it can be made (writing the equation), and that its parent acid is unstable and cannot be isolated.
2. It is a moderately good reducing agent (forming the tetrathionate ion or the sulfate ion, depending on the oxidizing agent used).
3. It is the active species in photographer's hypo being able to form water-soluble complexes with the silver ion.

♦♦♦

Concerning hydrogen sulfide, you should know the following.

1. It is a dangerous, poisonous gas whose rotten-eggs odor can desensitize the nose to its presence.
2. It can evolve from any solution of a metal sulfide when acid is added.

♦♦♦

With respect to other sulfur compounds, we noted the following.

1. Sulfur forms several compounds with fluorine and chlorine and a few with the other halogens. Sulfur hexafluoride is particularly stable.

◆◆◆

As for the other members of the oxygen family, the following generalizations are useful.

1. They become more metallic as the family is descended.
2. All members—O, S, Se, Te, and Po—form binary hydrides of the type H_2Z. All but H_2O and H_2Po are gases at room temperature.
3. O, S, Se, and Te form salts with the Group IA metals—M_2Z—and with the Group IIA metals—MZ.
4. Dihalides, dioxides, and trioxides of S, Se, and Te are known.
5. The oxides of selenium form aqueous acids similar to those formed from the oxides of sulfur.

◆◆◆

The flow-sheet on the next page displays all of the chief sulfur compounds studied and how they are made. Consider M to be either Na or K.

Self-Test

1. How is sulfur dioxide used to make sulfur as a sideline to petroleum refining? (Write the equation.)

   ____

2. How is sulfur dioxide made from hydrogen sulfide? (Write the equation.)

3. When a sulfide ore, like PbS, is heated in air, what happens?

   (Write an equation.)

   ____

4. Briefly describe the structural features in each of the following.

   (a) Orthorhombic sulfur ______________________

   (b) Monoclinic sulfur ______________________

   (c) Plastic sulfur ______________________

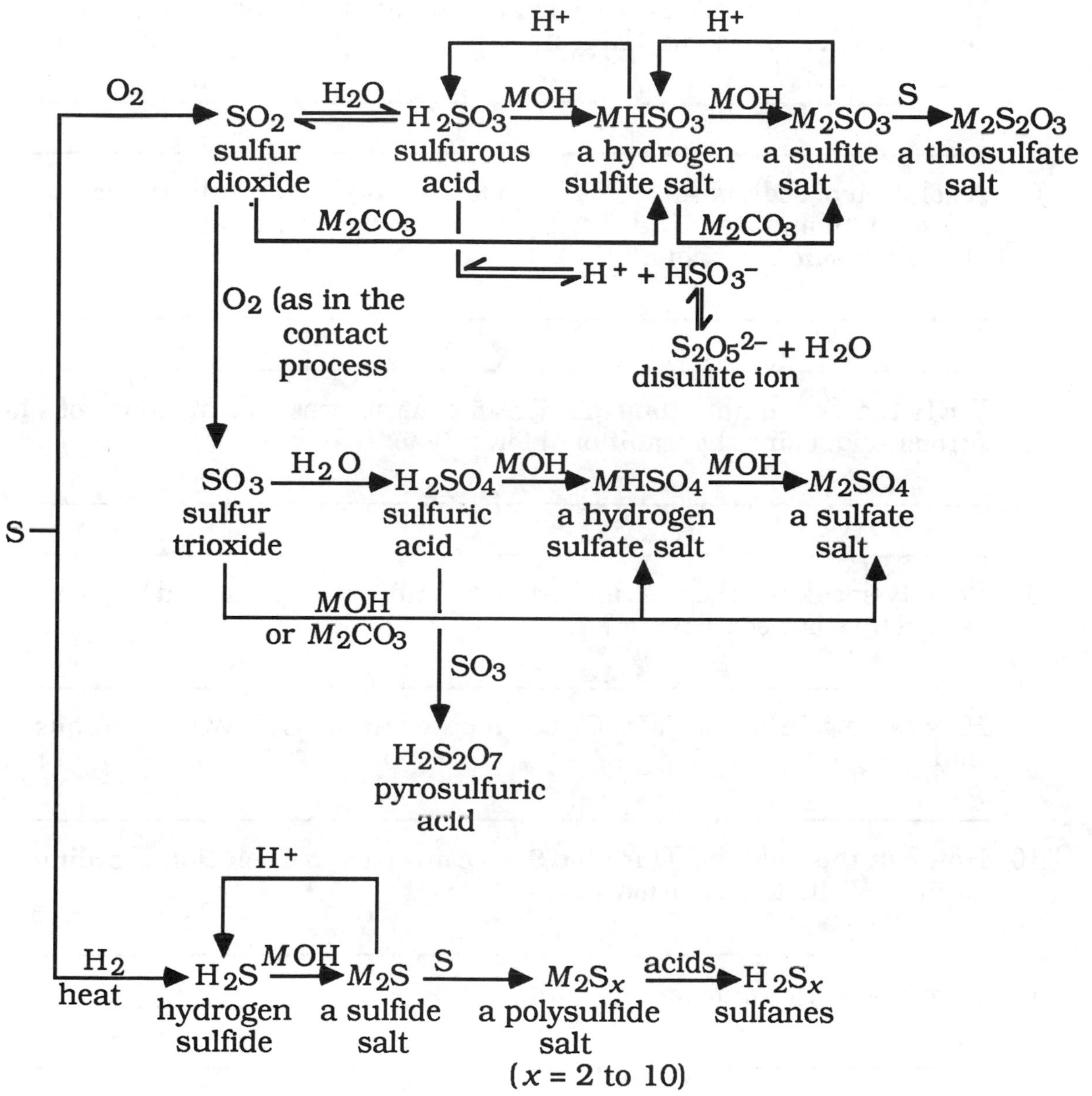

Flow chart showing the chemistry of sulfur and its compounds.

5. What happens structurally when sulfur is heated to a temperature where the molten sulfur is extremely viscous?

____

When very viscous sulfur is heated to still higher temperatures, it becomes much less viscous. Explain.

____

____

6. A solution of sodium sulfite has no odor, but if you add sulfuric acid to the solution and make it definitely acidic, a sharp odor can be noticed. Explain, using a net equation.

____

____

7. Write the two equilibrium equations for the successive ionizations of sulfurous acid, using the traditional formula for it.

____

____

8. Strictly speaking, the first ionization of sulfurous acid should be represented by what equilibrium?

____

9. How can a solution of $NaHSO_3$ be prepared from SO_2? Write the equation.

____

10. How can this solution (Question 9) be converted to a solution of sodium sulfite? Write the equation.

____

11. Describe how sodium metabisulfite is made? ______________

____

12. How can sodium metabisulfite serve as a source of sodium hydrogen sulfite? (Write an equilibrium as part of your answer.)

____

____

13. When sulfurous acid functions as a reducing agent, to what is it oxidized? Write the half-cell reaction when the system is acidic.

____

14. What are the chief causes of acid rain? ____________________

____

15. How can sulfur dioxide be removed from smokestack gases? Write an equation as part of your answer.

____

____

16. Write the equations for the steps in the contact process.

____

____

____

____

17. Complete and balance the following equations.

(a) ________ + $KOH(aq) \rightarrow KHSO_4(aq)$ + ________

(b) $SO_3(g)$ + ________ $\rightarrow Na_2SO_4(aq)$ + ________

(c) ________ + $K_2CO_3(aq) \rightarrow K_2SO_4(aq) + CO_2(g)$

(d) ________ + $NaHSO_4(aq) \rightarrow Na_2SO_4(aq) + H_2O$

(e) $SO_3(g) + H_2SO_4(l) \rightarrow$ ________

18. The acid ionization constant of HSO_4^- is on the order of

(a) 10^2 (b) 10 (c) 10^{-2} (d) 10^{-5} ________

____

19. List some reasons why concentrated sulfuric acid is dangerous.

____

20. When the thiosulfate ion acts as a reducing agent there are principally two species to which it is thereby oxidized.

(a) A relatively weak oxidizing agent changes the thiosulfate ion to what? (Give its name and formula.)

____

(b) A strong oxidizing agent changes the thiosulfate ion to what? (Give its name and formula.)

---

21. When the thiosulfate ion acts to dissolve AgBr during the "fixing" of a photographic negative, how does it work?

---

22. Complete and balance the following equations.

    (a) $FeS(s) + HCl(aq) \rightarrow$ __________

    (b) $K_2S(aq) + H_2SO_4(aq) \rightarrow$ __________

    (c) $CH_3{-}\overset{\overset{\displaystyle S}{\|}}{C}{-}CH_3 \;+\; H_2O \longrightarrow$ __________

    (d) $NaHS(aq) + HCl(aq) \rightarrow$ __________

23. Why doesn't the sulfide ion, S^{2-}, exist as such in watesr? Include an equation as part of your answer. __________

---

24. How could it happen that a sulfide mineral would not give the rotten egg smell when treated with hydrochlorid acid?

---

25. What is the formula of the most stable sulfur-halogen compound?

____

26. Write formulas for each.

    (a) The possible hydrides of Se, Te, and Po

    ____

    (b) The possible oxides of selenium and tellurium

    ____

    (c) The oxoacids of Se and Te in which these elements are in their +4 oxidation states. Give their names, too.

---

27. Which would be the stronger acid, H_2SeO_4 or H_2SeO_3, and what is the basis for making a judgement?

New Terms

Write the definition of the following term, which was introduced in this section. If necessary, refer to the Glossary at the end of the text.

acid raid

contact process

20.2 Phosphorus

Review

With respect to elemental phosphorus, its hydride and halides, learn the following.

1. The names and structural natures of white and red phosphorus and which is the more stable allotrope.
2. In sufficient air, phosphorus burns to give P_4O_{10}. (In insufficient air, P_4O_6 can form.) Nearly all P_4O_{10} is used to make phosphoric acid by reacting with water.
3. To learn that the phosphorus analog of ammonia is PH_3, phosphine, and that it is very unlike ammonia otherwise (being very poisonous, quite insoluble in water, and easily burned).
4. Although several phosphorus-halogen compounds are known—PX_3, P_2X_4, and PX_5—the ones of widest use are the trichloride and pentachloride of phosphorus. These two are made by direct chlorination, and both react readily with water to give oxoacids of phosphorus—phosphorous acid from PCl_3 and phosphoric acid from PCl_5.
5. Phosphorus oxychloride, $POCl_3$, forms when PCl_3 reacts with oxygen, and $POCl_3$ also reacts readily with water to give phosphoric acid.

♦♦♦

With respect to phosphoric acid, learn the following (being able to write equations where relevant).

1. In most situations, its formula is H_3PO_4, and unless something is said to the contrary, use this formula for it.
2. It is a moderately strong triprotic acid, and three families of salts are known, those of the $H_2PO_4^-$ ion, those of the HPO_4^{2-} ion, and those of the PO_4^{3-} ion.
3. These anions hydrolyze, their order of basicity toward water being:

$$PO_4^{3-} > HPO_4^{2-} > H_2PO_4^-$$

(An aqueous solution of sodium phosphate can be quite alkaline.)

4. Salts of pairs of these ions serve a buffers.
5. Salts of $H_2PO_4^-$ are acid salts and used as acidulants.

♦♦♦

Phosphoric acid can exist in several structural forms of which we are concerned largely with the following, which have P—O—P networks.

1. Diphosphoric acid, $H_4P_2O_7$, is a tetraprotic acid with one

   P—O—P unit, and some of its salts are acidulants.
2. The polyphosphoric acids are like diphosphoric acid except that there are more P atoms in the P—O—P chain.
3. The tripolyphosphate ion, $P_3O_{10}^{5-}$, is the active "phosphate" in phosphate detergents, and it forms water-soluble complexes with the ions that cause hardness in hard water.
4. P—O—P linkages can occur in cyclic systems—in cyclotriphosphoric acid (metaphosphoric acid).

♦♦♦

Phosphorous acid, H_3PO_3, and hypophosphorous acid, H_3PO_2, are two other oxoacids of phosphorus.

1. Phosphorous acid is a moderately strong, *di*protic acid. One of the H atoms in H_3PO_3 is bound directly to P, and one oxygen is a lone oxygen. Salts of this acid are dihydrogen phosphites and monohydrogen phosphites.
2. Hypophosphorous acid is a moderately strong, *mono*protic acid. Two of the H atoms in H_3PO_2 are bound directly to P, and one oxygen is a lone oxygen. Its salts are hypophosphites.

♦♦♦

Concerning trends in the nitrogen family, the following were noted.

1. As the family is descended:

   The elements become more metallic.

   The metal oxides become more basic.

   Their reactivity toward other elements or water decreases.
2. All members of the family form two kinds of oxides—M_2O_3 and M_4O_{10}.
3. All form binary hydrides, MH_3, which are all very toxic. Their stability decreases sharply as the family is descended.

4. All form trihalides with all four halogens. All form pentafluorides, but their stability decreases sharply as the family is descended. All form sulfides.

The flow chart on the next page ties together the reactions of phosphorus and its principal compounds. Consider *M* to be either the sodium or the potassium ion. When phosphoric acid occurs in the chart, which specific form—meta-, di- or ortho-—depends on the relative quantities of water used. Not shown are salts of diphosphoric acid or tripolyphosphoric acid.

Self-Test

28. What is it about the structure of red phosphorus that makes it so much more stable than white phosphorus?

____

____

29. What happens to white phosphorus when it is exposed to air? (Write an equation.)

____

30. Write equations for the preparation of the following.

(a) PH_3 _______________

(b) PCl_3 _______________

(c) PCl_5 _______________

(d) $POCl_3$ _______________

31. Complete and balance the following equations.

(a) $PH_3(g) + O_2(g) \rightarrow$ _______________

(b) $PCl_3(l) + H_2O \rightarrow$ _______________

(c) $PCl_5(s) + H_2O \rightarrow$ _______________

(d) $POCl_3(l) + H_2O \rightarrow$ _______________

32. Name each of the following compounds.

(a) Na_2HPO_4 _______________

(b) KH_2PO_4 _______________

(c) Na_3PO_4 _______________

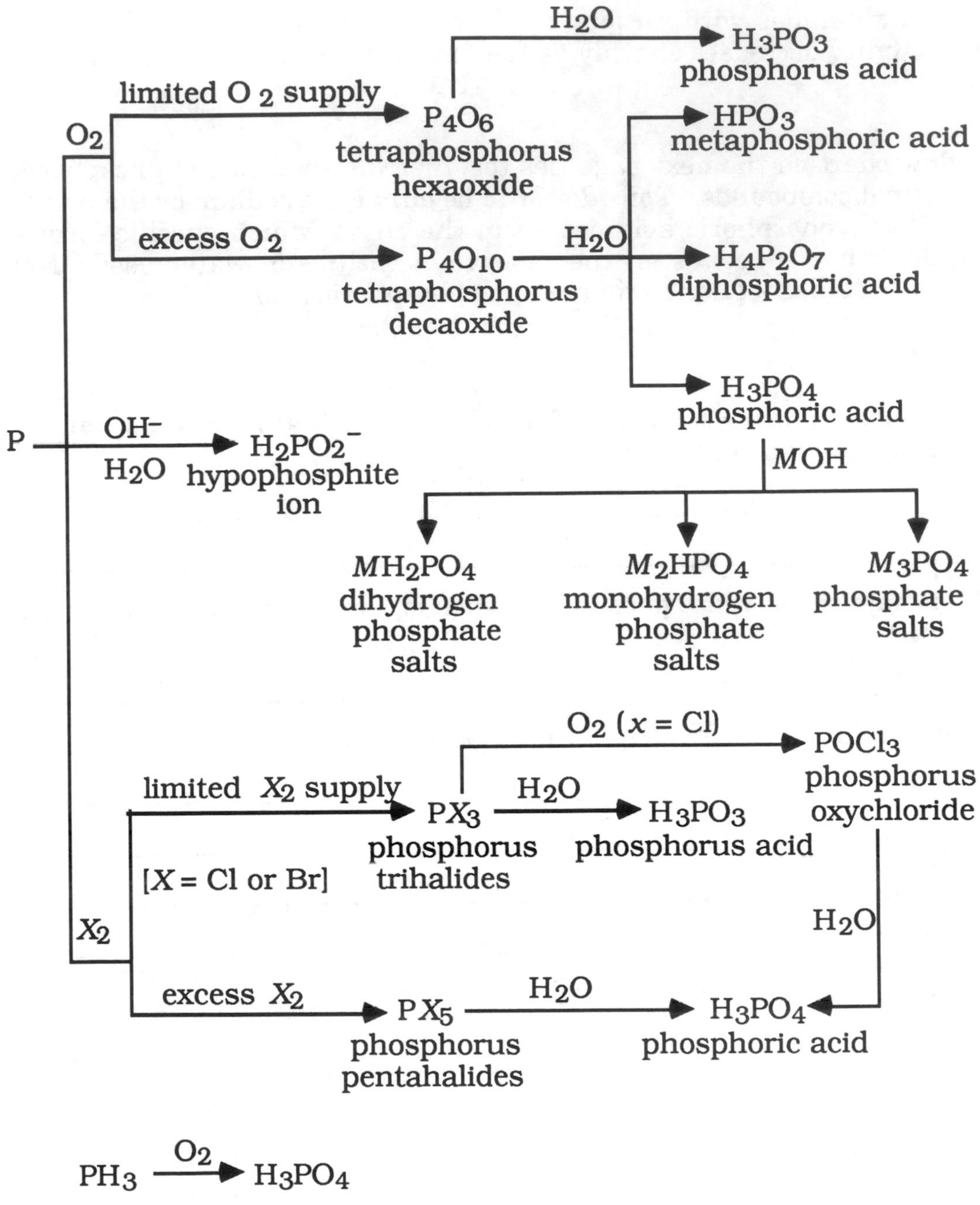

A flow chart showing the chemistry of phosphorus and its principal compounds

(d) $H_4P_2O_7$ __________

(e) P_4O_{10} __________

(f) $Na_5P_3O_{10}$ __________

(g) KH_2PO_2 __________

(h) Na_2HPO_3 __________

33. Write equations for the steps in the synthesis of each of the following from phosphorus.

(a) H_3PO_4 __________

____

(b) H_3PO_3 __________

____

34. In which solution would the pH be higher, 0.1 *M* NaH_2PO_4 or 0.1 *M* Na_2HPO_4? __________ Explain. __________

____

35. A solution made 0.1 *M* in both NaH_2PO_4 and Na_2HPO_4 is buffered. Explain, writing net ionic equations.

____

____

36. The tripolyphosphate ion forms a water-soluble complex with the magnesium ion. Write its formula.

____

New Terms

none

20.3 The Halogens

Review

With respect to fluorine, you should learn the following.

1. It is a pale yellow gas that reacts violently with almost all other elements.

2. It burns in water to give oxygen and HF.
3. Its binary acid, HF, is a weak acid, but very dangerous to flesh and capable of dissolving glass.

The following flow sheet outlines the principal reactions of fluorine that were studied.

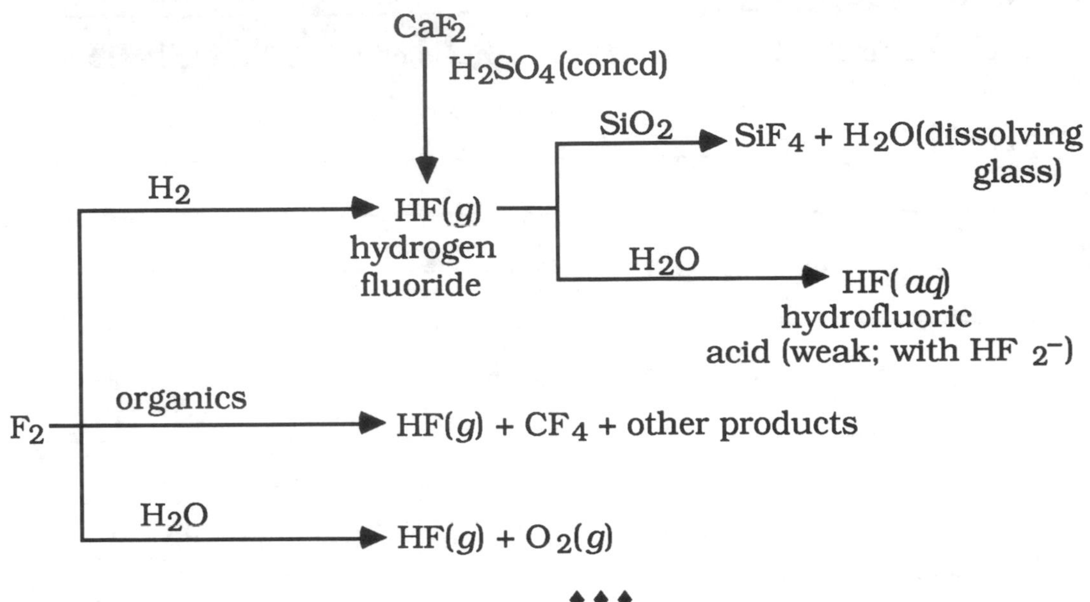

◆◆◆

Concerning chlorine, you should learn the following.

1. It is a yellowish-green poisonous gas made by the electrolysis of aqueous sodium chloride (discussed in Section 18.4).
2. Its binary hydride, HCl, is a gas made by the direct combination of the elements or as a by-product in the manufacture of organic chlorine compounds. (In the lab it can be made by heating a mixture of sodium chloride and sulfuric acid.)
3. A solution of its binary hydride in water is called hydrochloric acid, a common strong acid.
4. It forms oxides—Cl_2O, ClO_2, and Cl_2O_7—of which ClO_2 is the most important (being a bleaching agent), and Cl_2O_7 is explosive.
5. It reacts with water to give an equilibrium mixture of HCl and HOCl.
6. The names and formulas of the four oxoacids—HOCl, $HClO_2$, $HClO_3$, and $HClO_4$— should be learned together with their stabilities and the equations for their decompositions.
7. Cl_2 as well as several oxo compounds of chlorine are bleaches.

The following flow chart outlines the principal reactions of chlorine that we studied.

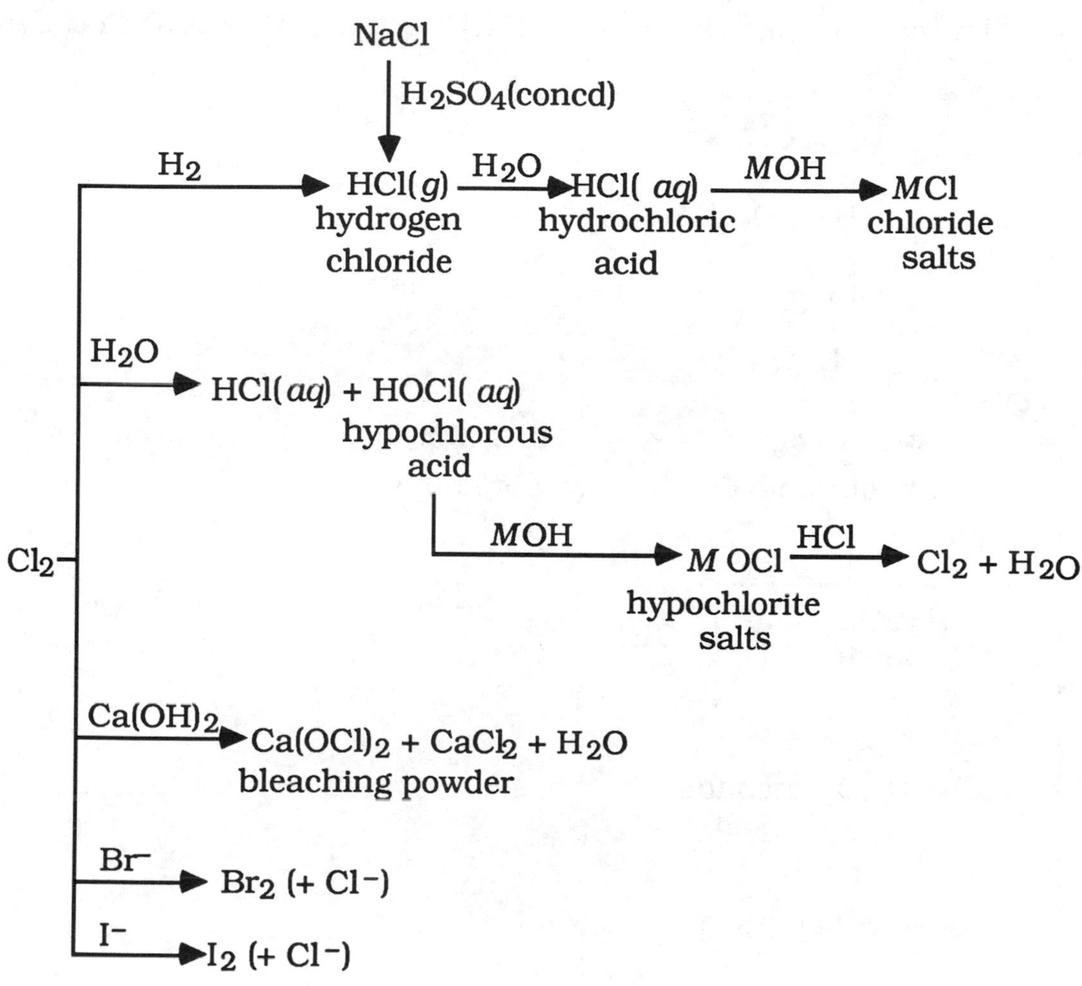

♦♦♦

About bromine, learn the following.

1. It is a dark brown, corrosive liquid whose vapors are poisonous and can scar tissue.
2. It is made by the oxidation of the bromide ion (using Cl_2 as the oxidizing agent.
3. It is less reactive than chlorine, but it forms compounds of the same type as chlorine, including a binary hydride that, in aqueous solution, is hydrobromic acid.

4. It is more soluble in water than chlorine and, unlike chlorine, does not react with water to any appreciable extent. (Know, however, the equilibrium that is present that involves HBr and HOBr.)

The following flow chart displays the chief reactions of bromine that were studied.

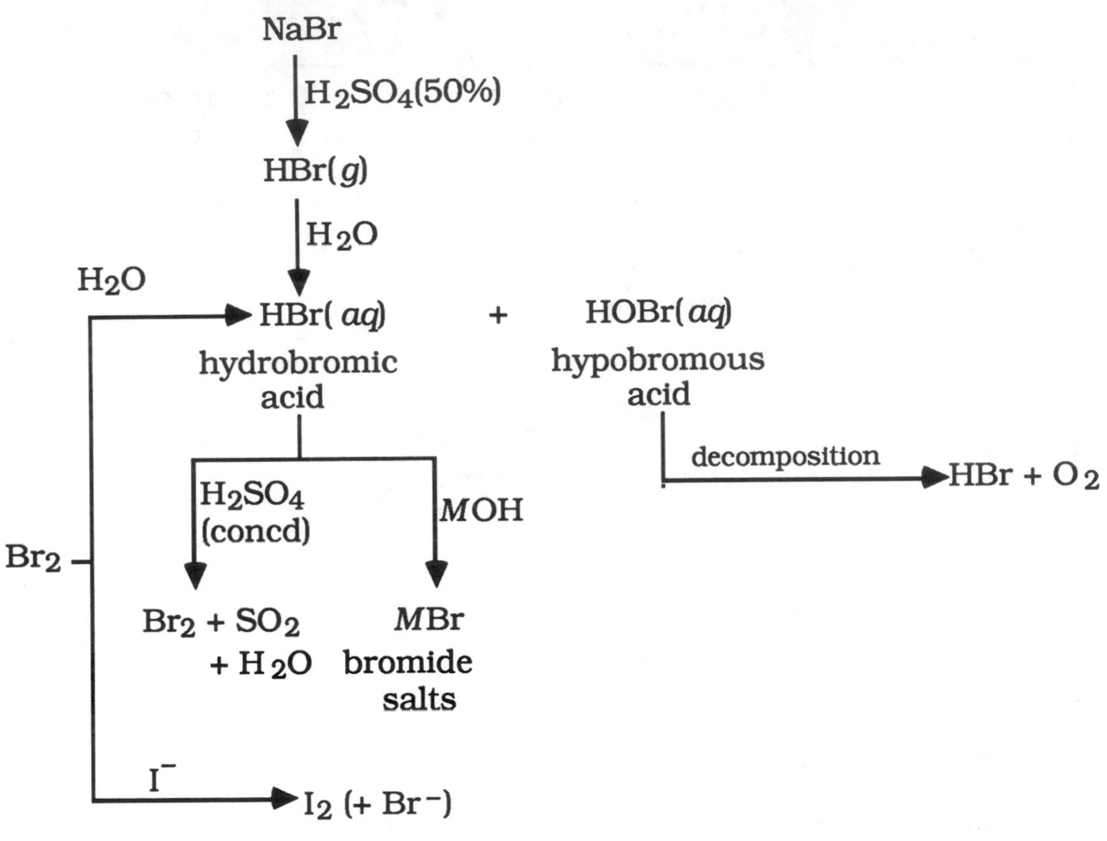

♦♦♦

As for the least reactive halogen, iodine, know the following.

1. It is made by the oxidation of the iodide ion in brines.
2. It is a deep purple (almost black) nonmetallic solid that is relatively insoluble in water.
3. Its binary hydride dissolves in water to give the strongest of the hydrohalogen acids, HI, hydriodic acid.
4. It forms a few oxides and oxoacids similar to those of chlorine, with iodic acid (HIO_3) and periodic acid (HIO_4) being the most important (and are strong oxidizing agents).

The chief reactions of iodine that we studied are outlined in the following flow chart.

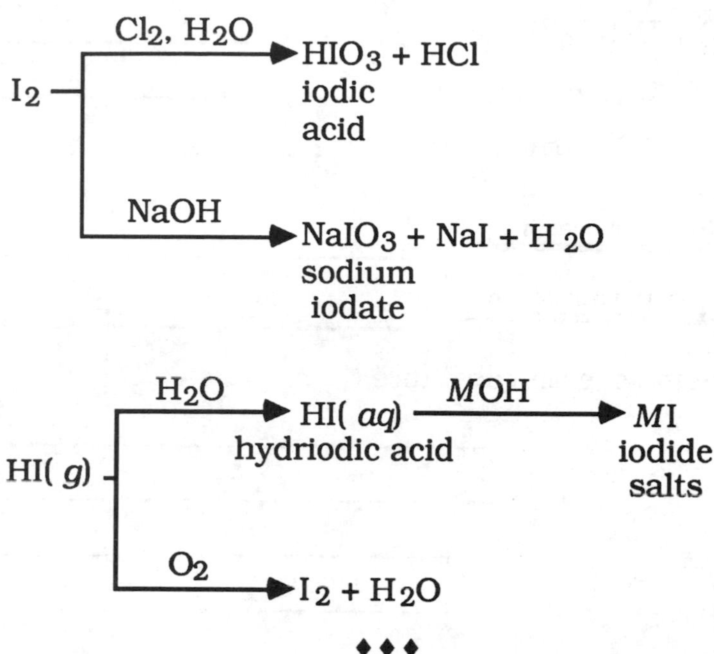

♦♦♦

The periodic trends in the halogen family are particularly important. Be sure to study the summary discussion of these that is found at the end of Section 20.3. Pay particular attention to the trends in reactivity toward other elements, relative electronegativities, strengths of the H*X* types of acids, abilities of the diatomic forms, X_2, to function as oxidizing agents (and the abilities of the anions, X^-, to function as reducing agents—an opposite trend that you should notice). All these trends are in line with broad generalizations that we have made throughout the text.

Self-Test

37. Complete each of the following equations by writing the formulas of the missing substances and then balancing. If no reaction occurs, write NR.

    (a) $H_2(g) + Cl_2(g) \rightarrow$ ____________________

    (b) $F_2(g) + H_2O \rightarrow$ ____________________

    (c) $Ca(OH)_2(s) + Cl_2(g) \rightarrow$ ____________________

    (d) $Br_2(l) + Cl^-(aq) \rightarrow$ ____________________

    (e) $CaF_2(s) + H_2SO_4(l) \rightarrow$ ____________________

(f) $NaCl(aq) + H_2O \rightarrow$ ______________________

(g) $Cl_2(g) + H_2O \rightarrow$ ______________________

(h) $NaCl(s) + H_2SO_4(l) \rightarrow$ ______________________

(i) $I_2(s) + Br^-(aq) \rightarrow$ ______________________

(j) $HOCl(aq) + NaOH(aq) \rightarrow$ ______________________

______

(k) $ClO^- \xrightleftharpoons{\text{disproportionation}}$ ______________________

(l) $HClO_2 \xrightleftharpoons{\text{disproportionation}}$ ______________________

38. **Write the name of each substance.**

(a) $NaIO_3$ ______________________

(b) $HF(g)$ ______________________

(c) HOBr ______________________

(d) KOCl ______________________

(e) $HI(aq)$ ______________________

(f) Cl_2O_7 ______________________

(g) H_5IO_6 ______________________

(h) HIO_3 ______________________

(i) HOCl ______________________

(j) $HClO_2$ ______________________

(k) $HClO_3$ ______________________

(l) ClO_2 ______________________

(m) $NaClO_3$ ______________________

(n) $KClO_4$ ______________________

(o) $NaIO_4$ ______________________

(p) $KBrO_3$ ______________________

(q) $NaClO_2$ ______________________

39. Arrange the halide ions in their order of increasing ability to work as reducing agents. ______________________

40. Arrange the halogen atoms in their order of increasing ionization energy. ______________________

41. Arrange the diatomic halogens in their order of increasing ability to work as oxidizing agents. ______________________

42. Arrange the halide ions in their order of increasing basicity.

____

43. Arrange the halogens in their order of increasing electronegativities.

____

44. Arrange the halide ions in their order of increasing ease of oxidation.

____

45. Arrange the halide ions in their order of increasing enthalpies of hydration.

____

46. Arrange the hydrohalogen acids, $HX(aq)$, in their order of increasing acidity.

____

New Terms

Write the definitions of the following terms, which were introduced in this section. If necessary, refer to the Glossary at the end of the text.

disproportionation
halogen family

20.4 The Noble Gas Elements

Review

The almost total absence of any chemical properties of the noble gases is their most significant chemical feature (and one that makes this section particularly easy). Only xenon (and rarely krypton) gives compounds and then only with the most reactive, electronegative elements, fluorine and oxygen.

Self-Test

47. What noble gas has the highest concentration in the atmosphere?

    ____

48. Give three uses of helium.

    ____

    ____

49. Give the formulas of at least three compounds of xenon.

    ____

50. What is meant by the term "superconductor" and how can helium be used to make one?

    ____

    ____

New Terms

Write the definition of the following term, which was introduced in this section. If necessary, refer to the Glossary at the end of the text.

noble gases

Answers to Self-Test Questions

1. $2H_2S(g) + SO_2(g) \rightarrow 3S(s) + 2H_2O$
2. $2H_2S(g) + 3O_2(g) \rightarrow 2SO_2(g) + 2H_2O$
3. $2PbS + 3O_2(g) \rightarrow 2PbO(s) + 2SO_2(g)$
4. (a) Cyclic S_8 molecules
   (b) Cyclic S_8 molecules
   (c) An amorphous tangle of open-chain molecules of varying lengths.
5. Cyclic molecules open and join end to end to make a mixture of open chain systems of varying but usually very long lengths. These cannot slip by each other readily, so the molecten mixture is viscous. At higher temperatures, these very long molecules break up.
6. The sulfite ion is protonated to give sulfurous acid most of which decomposes to release sulfur dioxide.

   $HSO_3^-(aq) + H^+ \rightarrow SO_2(g) + H_2O$
7. $H_2SO_3(aq) \rightleftharpoons H^+(aq) + HSO_3^-(aq)$

   $HSO_3^-(aq) \rightleftharpoons H^+(aq) + SO_3^{2-}(aq)$

8. $SO_2 \cdot nH_2O(aq) \rightleftharpoons H_3O^+(aq) + HSO_3^-(aq)$
9. By bubbling sulfur dioxide through a sodium carbonate solution.
   $2SO_2(g) + Na_2CO_3(aq) + H_2O \rightarrow 2NaHSO_3(aq) + CO_2(g)$
10. By adding sodium carbonate.
   $2NaHSO_3(aq) + Na_2CO_3(aq) \rightarrow 2Na_2SO_3(aq) + CO_2(g) + H_2O$
11. By the evaporation of aqueous $NaHSO_3$ in the presence of SO_2.
12. In solution the metabisulfite ion exists in equilibrium with the hydrogen sulfite ion. $2HSO_3^-(aq) \rightleftharpoons S_2O_5^{2-}(aq) + H_2O$
13. To SO_4^{2-}. $SO_4^{2-}(aq) + 4H^+(aq) + 2e^- \rightleftharpoons H_2SO_3(aq) + H_2O$
14. The oxides of sulfur and nitrogen, particularly SO_2 and NO_2. These form when sulfur-containing fuels are burned.
15. By absorbing it chemically with calcium hydroxide.
   $SO_2(g) + Ca(OH)_2(s) \rightarrow CaSO_3(s) + H_2O$
16. $S(s) + O_2(g) \rightarrow SO_2(g)$
   $2SO_2(g) + O_2(g) \xrightarrow{V_2O_5 \text{ catalyst}} 2SO_3(g)$
   $SO_3(g) + H_2SO_4(l) \rightarrow H_2S_2O_7(l)$
   $H_2S_2O_7(l) + H_2O \rightarrow 2H_2SO_4(l)$
17. (a) $H_2SO_4(aq) + KOH(aq) \rightarrow KHSO_4(aq) + H_2O$
   (b) $SO_3(g) + 2NaOH(aq) \rightarrow Na_2SO_4(aq) + H_2O$
   (c) $SO_3(g) + K_2CO_3(aq) \rightarrow K_2SO_4(aq) + CO_2(g)$
   (d) $NaOH(aq) + NaHSO_4(aq) \rightarrow Na_2SO_4(aq) + H_2O$
   (e) $SO_3(g) + H_2SO_4(l) \rightarrow H_2S_2O_7(l)$
18. (c)
19. It is viscous and sticks to surfaces; it is strongly acidic; its reaction with moisture is very exothermic.
20. (a) Tetrathionate ion, $S_4O_6^{2-}$.
   (b) Hydrogen sulfate ion, HSO_4^-. (Giving the sulfate ion would also be acceptable; this depends on the pH of the solution.)
21. It forms a soluble complex with the silver ion; $Ag(S_2O_3)_2^{3-}$.
22. (a) $FeS(s) + 2HCl(aq) \rightarrow FeCl_2(aq) + H_2S(g)$
   (b) $K_2S(aq) + H_2SO_4(aq) \rightarrow K_2SO_4(aq) + H_2S(g)$
   (c) $C_3H_6S + H_2O \rightarrow C_3H_6O + H_2S(g)$
   (d) $NaHS(aq) + HCl(aq) \rightarrow NaCl(aq) + H_2S(g)$
23. It is too strong a base; it reacts with water: $S^{2-} + H_2O \rightarrow OH^- + HS^-$
24. The mineral could be an acid-insoluble sulfide, like CuS or HgS.
25. SF_6
26. (a) H_2Se, H_2Te, H_2Po
   (b) SeO_2, SeO_3, TeO_2, TeO_3

(c) H_2SeO_3 (selenous acid), H_2TeO_3 (tellurous acid)

27. H_2SeO_4, because it has an extra oxygen that makes the O—H bond more polar.
28. It has normal, unstrained bond angles.
29. It burns. $4P + 5O_2 \rightarrow P_4O_{10}$
30. (a) $Ca_3P_2(s) + 6H_2O \rightarrow 2PH_3(g) + 3Ca(OH)_2(s)$
    (b) $2P(s) + 3Cl_2(g) \rightarrow 2PCl_3(l)$
    (c) $2P(s) + 5Cl_2(g) \rightarrow 2PCl_5(s)$ [or $(PCl_4)(PCl_6)$]
    (d) $2PCl_3(l) + O_2(g) \rightarrow 2POCl_3(l)$
31. (a) $PH_3(g) + 2O_2(g) \rightarrow H_3PO_4(l)$
    (b) $2PCl_3(l) + 6H_2O \rightarrow 2H_3PO_3(aq) + 6HCl(aq)$
    (c) $PCl_5(s) + 4H_2O \rightarrow H_3PO_4(aq) + 5HCl(aq)$
    (d) $POCl_3(l) + 3H_2O \rightarrow H_3PO_4(aq) + 3HCl(aq)$
32. (a) sodium monohydrogen phosphate
    (b) potassium dihydrogen phosphate
    (c) sodium phosphate
    (d) diphosphoric acid
    (e) tetraphosphorus decaoxide
    (f) sodium tripolyphosphate
    (g) potassium hypophosphite
    (h) sodium hydrogen phosphite
33. (a) $4P(s) + 5O_2(g) \rightarrow P_4O_{10}(s)$
    $P_4O_{10}(s) + 6H_2O \rightarrow 4H_3PO_4(aq)$
    (b) $4P(s) + 3O_2(g) \rightarrow P_4O_6(s)$
    $P_4O_6(s) + 6H_2O \rightarrow 4H_3PO_3(aq)$
34. In Na_2HPO_4, because the HPO_4^{2-} is a stronger Brønsted base than $H_2PO_4^{2-}$.
35. It can neutralize H^+ as follows.
    $H^+(aq) + HPO_4^{2-}(aq) \rightarrow H_2PO_4^-(aq)$
    And it can neutralize OH^- as follows.
    $OH^-(aq) + H_2PO_4^-(aq) \rightarrow H_2O + HPO_4^{2-}(aq)$
36. $CaP_3O_{10}^{3-}$
37. (a) $H_2(g) + Cl_2(g) \rightarrow 2HCl(g)$
    (b) $2F_2(g) + 2H_2O \rightarrow 4HF(g) + O_2(g)$
    (c) $2Ca(OH)_2(s) + 2Cl_2(g) \rightarrow Ca(OCl)_2(s) + CaCl_2(s) + 2H_2O$
    (d) NR
    (e) $CaF_2(s) + H_2SO_4(l) \rightarrow CaSO_4(s) + 2HF(g)$
    (f) NR

(g) $Cl_2(g) + H_2O \rightarrow HOCl(aq) + HCl(aq)$
(h) $2NaCl(s) + H_2SO_4(l) \rightarrow 2HCl(g) + Na_2SO_4(s)$
(i) NR
(j) $HOCl(aq) + NaOH(aq) \rightarrow NaOCl(aq) + H_2O$
(k) $3ClO^-(aq) \rightleftharpoons 2Cl^-(aq) + ClO_3^-(aq)$
(l) $5HClO_2 \rightarrow 4ClO_2 + HCl + 2H_2O$

38. (a) sodium iodate (b) hydrogen fluoride (c) hypobromous acid
(d) potassium hypochlorite (e) hydriodic acid
(f) dichlorine heptoxide (g) periodic acid (h) iodic acid
(i) hypochlorous acid (j) chlorous acid (k) chloric acid
(l) chlorine dioxide (m) sodium chlorate
(n) potassium perchlorate (o) sodium periodate
(p) potassium bromate (q) sodium chlorite
39. $F^- < Cl^- < Br^- < I^-$
40. I < Br < Cl < F
41. $I_2 < Br_2 < Cl_2 < F_2$
42. $I^- < Br^- < Cl^- < F^-$
43. I < Br < Cl < F
44. $F^- < Cl^- < Br^- < I^-$
45. $I^- < Br^- < Cl^- < F^-$
46. HF < HCl < HBr < HI
47. argon
48. to fill lighter-than-air balloons, as a liquid coolant, and as an inert environment gas.
49. Any of the following: XeF_2, XeF_4, XeF_6, XeO_3, $XeOF_4$, XeO_2F_2
50. A conductor that has no resistance to the flow of electricity; some metals become superconductors when cooled with liquid helium.

Chapter 21

METALLURGY AND THE REPRESENTATIVE METALS

Approximately three-fourths of all the elements are metals—they are literally everywhere. Because their properties are so varied, we find them used in a myriad of applications, both as free elements and in their combined states.

In this chapter we study the properties of metals and how these properties vary within the periodic table. We examine how metals are extracted from their compounds and how they are treated to make them ready for practical uses. We also discuss specific chemical and physical properties of the representative metals—those in the A-groups of the periodic table. One the goals that you should have is to learn how these properties relate to where the metals are found in nature, how the metals are obtained from their compounds, and how the metals and their compounds are put to use.

Learning Objectives

As you study of this chapter, keep in mind the following goals:

1 To learn what controls the metallic character of an element and how metallic character varies within the periodic table.

2 To learn where metals are found, how their ores are treated, and how the metals that are obtained from them are purified.

3 To learn about the properties and reactions of the elements belonging to Group IA of the periodic table.

4 To learn about the chemical and physical properties of the elements of Group IIA.

5 To learn the chemical and physical properties and important compounds of aluminum and the post-transition metals.

21.1 Metallic Character and the Periodic Table

Review

Metals are characterized by a *metallic lattice* in which metal atoms lose valence electrons to the lattice as a whole. This happens because there are too few valence electrons to complete the valence shells by electron sharing. The tendency to form ionic compounds—a characteristic of metals—varies with the metal's electronegativity. Metallic character *increases* from top to bottom in a group and *decreases* from left to right in a period (see the margin figure on page 909).

Self-Test

1. For each pair, choose the more metallic element.

   (a) Be and Mg ________________

   (b) Ca and Ge ________________

   (c) Ga and Ba ________________

2. Which element in each pair will have the more basic oxide?

   (a) Be or K ________________

   (b) K or Ga ________________

   (c) B or Ga ________________

New Terms

none

21.2 Metallurgy

Review

Obtaining metals for practical applications can be divided into a number of steps: (1) mining, (2) pretreatment of ores, (3) reduction of metal compounds, and (4) purification.

Sources of metals are the sea and the earth. You should learn which metals are extracted from the sea and how they are obtained. When metal ores are dug from the ground, they must be enriched; dirt and other useless gangue has to be removed to make recovery of the metal economical. Washing with water and flotation are two physical methods of enriching ores.

Some ores, especially sulfide ores, must be roasted in air to convert metal sulfides to metal oxides, which are more easily reduced. You should study the reactions for the chemical purification of bauxite, the ore of aluminum.

Once an ore has been enriched and pretreated (if necessary), it is reduced. The strength of the reducing agent needed depends on the chemical activity of the metal. The most active metals, such as sodium, magnesium, and aluminum, must be obtained by electrolysis. Many metals can be obtained by reduction of the oxide with carbon. Study especially the reactions that take place in the blast furnace.

After the free metal is obtained from its ore, it usually must be purified further before it can be used commercially. The open hearth process, and especially the basic oxygen process, are used to convert pig iron from the blast furnace to steel. Study the reactions that take place during the operation of the basic oxygen furnace.

Self-Test

3. Which two metals are extracted in quantity from the ocean?

   ____

4. Give the chemical equations for the separation of magnesium from sea water and its conversion to magnesium chloride.

   ____

   ____

5. Where are *manganese nodules* found?____________________

6. What kinds of ores are treated by the flotation process?

   ____

7. Give the chemical equation for the roasting of a lead sulfide ore.

   ____

8. What property of aluminum oxide is made use of in the purification of bauxite?

   ____

   Give the chemical equations for the reactions used to purify the bauxite ore.

   ____

____

____

9. What is the active reducing agent in the blast furnace?

____

Give the *overall* equation for the reduction of hematite, Fe_2O_3, using this reducing agent.

____

10. Write equations that illustrates the reactions that $CaCO_3$ undergoes when it is added to the charge in a blast furnace in the reduction of iron ore.

____

____

11. Why has the basic oxygen process become the principle steel-making process?

____

New Terms

Write the definitions of the following terms, which were introduced in this section. If necessary, refer to the Glossary at the end of the text.

metallurgy	coke
manganese nodules	blast furnace
gangue	slag
flotation	open hearth furnace
roasting	basic oxygen process

21.3 The Metals of Group IA: The Alkali Metals

Review

By now you should be familiar with the location of the alkali metals in the periodic table. Of course, even though hydrogen is placed at the top of the group, it is not a metal and has properties that are distinctly different from those elements below it.

In nature, the alkali metals always occur in compounds. They are too reactive to be found as free elements. Because almost all their compounds are water-soluble, they occur in many natural waters, including seawater and underground brines. Where the weather is hot and dry, they are found in deposits on the surface, and there are large deposits of NaCl underground in various locations. Some important facts you should know are:

1. Sodium and potassium are the most abundant of the alkali metals.

2. Potassium ions are essential to both plants and animals, but with few exceptions, only animals need sodium ions.

3. All isotopes of francium are radioactive.

4. In general, the alkali metals are soft, have low melting points, have low ionization energies and low densities, and have very negative reduction potentials. Review the discussion on pages 918 and 919 to be sure you know why.

5. Lithium and sodium are prepared by electrolysis of their molten chlorides; the other alkali metals can be prepared by the general reaction

$$MCl(l) + Na(g) \rightleftharpoons M(g) + NaCl(l)$$

   which takes advantage of the greater volatility of potassium, rubidium, and cesium.

The principal chemical properties of the alkali metals that you should know are the following:

1. They all react with water to generate hydrogen gas by the reaction

$$2M(s) + 2H_2O \rightarrow H_2(g) + 2M^+(aq) + 2OH^-(aq)$$

2. Only lithium reacts with elemental nitrogen to form a nitride, Li_3N.

3. Lithium reacts with O_2 to give the normal oxide, Li_2O; sodium reacts with O_2 to give the peroxide, Na_2O_2; and the rest of the alkali metals react with O_2 to give superoxides, MO_2. (These were discussed in Chapter 19.)

Among the important compounds of the alkali metals are:

1. *Sodium chloride.* This is the chief raw material for other sodium compounds as well as being an important chemical itself. Much NaCl is mined from underground deposits; much is also harvested from the sea.

2. *Sodium hydroxide.* Obtained from NaCl by electrolysis of aqueous solutions. This reaction was described in Chapter 17. You should be able to write equations for the chemical reactions involved.

3. *Sodium carbonate* and *sodium bicarbonate.* Na_2CO_3 is used to make detergents, glass, paper, and other chemicals. It is obtained from the mineral trona, $Na_2CO_3 \cdot NaHCO_3 \cdot 2H_2O$, and can also be manufactured by the Solvay process. Be sure to learn the chemistry of the Solvay process on pages 921 and 922. Sodium bicarbonate, the intermediate in the Solvay process, is used in baking, as a buffer, and in antacid products.

4. *Potassium carbonate.* A chemical found in plant ashes. It is used to make soaps, special glass products, and other compounds of potassium.

5. *Lithium carbonate.* A chemical that has been used to treat manic depression.

A useful fact to remember is that the sodium salts of strong acids are often more soluble in water than the corresponding potassium salts. On the other hand, potassium salts of weak acids are often more soluble in water than the sodium salts.

You should know the colors of the flame tests for sodium, potassium, and lithium. These are shown in photos on page 923.

Self-Test

12. Why are the alkali metals soft? ______________________________

____

13. Why are the alkali metals so reactive? ________________________

____

14. Why can sodium metal be used to reduce CsCl in the production of metallic cesium?

____

15. Why does lithium have such a negative standard reduction potential?

____

____

16. Which compound of the alkali metals has shown promise in the treat-

ment of mental disorders?

____

17. Complete and balance the following equations.

    (a) $Na + O_2 \rightarrow$ ____________________

    (b) $Li + O_2 \rightarrow$ ____________________

    (c) $Cs + O_2 \rightarrow$ ____________________

    (d) $NaHCO_3 \xrightarrow{\text{heat}}$ ____________________

    (e) $KO_2 + H_2O \rightarrow$ ____________________

    (f) $Na_2O_2 + H_2O \rightarrow$ ____________________

18. What color flame is produced when salts of the following elements are introduced into a Bunsen burner flame?

    (a) lithium ____________________

    (b) sodium ____________________

    (c) potassium ____________________

19. The only alkali metal to react with elemental nitrogen is __________

    The product of the reaction is ________________

20. What is the net reaction involving sodium chloride in the Solvay process?

    ____

21. Which compound would you expect to be more soluble in water (on a mole basis)?

    (a) NaF or KF _______________

    (b) $NaClO_3$ or $KClO_3$ _______________

New Terms

Write the definitions of the following terms, which were introduced in this section. If necessary, refer to the Glossary at the end of the text.

Solvay process

flame test

21.4 The Metals of Group IIA: The Alkaline Earth Metals

Review

From past discussions, you should be familiar with the location of the alkaline earth metals in the periodic table, and some of the elements are familiar to you—calcium and magnesium, for example.

The following are the key points to remember concerning these elements. Where appropriate, learn to write chemical equations to describe reactions.

1. Like the elements of Group IA, those in this group are chemically quite reactive, so they are never found as free elements in nature. Calcium and magnesium occur at appreciable concentrations in sea water, and this is one of the principal sources of magnesium. Since most compounds of the Group IIA elements are insoluble in water, many of them are found in mineral deposits such as limestone ($CaCO_3$), gypsum ($CaSO_4 \cdot 2H_2O$), and dolomite (mixed $CaCO_3$ and $MgCO_3$, represented by the approximate formula $CaCO_3 \cdot MgCO_3$). Beryllium is found in beryl, a silicate mineral with the formula $Be_3Al_2(SiO_3)_6$.
2. Compared to the Group IA elements, the alkaline earth metals are not as soft, nor as low melting. They have higher ionization energies and are not quite as strong as reducing agents. Learn the explanations given for this on pages 924 and 925.
3. Only beryllium and magnesium are able to withstand oxygen and moisture sufficiently to make them useful as free metals. Beryllium is obtained by electrolysis of $BeCl_2$, but NaCl must be added as an electrolyte because $BeCl_2$ is molecular and a poor conductor. Magnesium is also obtained by electrolysis of its chloride, which is made from seawater or from dolomite. Learn the chemistry of the recovery of magnesium from these sources as given on page 926.
4. Beryllium and magnesium form oxides that are sufficiently insoluble in water to prevent attack on the metal. The remaining metals react readily with water as follows:

$$M(s) + 2H_2O \rightarrow M^{2+}(aq) + 2OH^-(aq) + H_2(g)$$

   Magnesium reacts similarly with hot water and steam.
5. All the Group IIA metals react readily with oxygen. Combustion of magnesium gives an intense white light useful in flashbulbs, flares, and fireworks.
6. Magnesium is the only Group IIA metal that reacts directly with elemental nitrogen. Note the similarity to lithium, often referred to as a

"diagonal similarity" because of the relative positions of the elements in the periodic table.

7. Beryllium compounds in general tend to be covalent rather than ionic because of the small size of the Be^{2+} ion. Study the molecular structure of $BeCl_2$.
8. Beryllium is amphoteric. The metal reacts with acid and with base to liberate H_2. BeO dissolves in both acid and base. Study the equations near the bottom of page 928.

Among the interesting and important compounds of the alkaline earth metals are the following.

1. The oxides. They can be made by direct reaction of the metal with O_2 and by decomposition of the carbonate. Note especially that:

   Calcium oxide reacts with water to form the hydroxide, $Ca(OH)_2$, in a process called slaking.

   Magnesium oxide is much less soluble in water than calcium oxide and is used to make refractory bricks.

2. The hydroxides. For the elements below Mg, they can be made by reaction of the metal oxide with water. $Mg(OH)_2$ can be made by reacting $Mg^{2+}(aq)$ with base.
3. As mentioned earlier, $BeCl_2$ is covalent with a chain structure in the solid state. $CaCl_2$ is ionic and the solid is deliquescent.
4. The sulfates. The solubilities decrease going down Group IIA. Epsom salts is magnesium sulfate, $MgSO_4 \cdot 7H_2O$, and is quite soluble in water. Much less soluble is calcium sulfate in the form of gypsum, which when partially dehydrated gives plaster of Paris. Barium sulfate is very insoluble.

Flame test can be used to confirm the presence of calcium, strontium, and barium in chemical analyses. Observe in Figure 21.15 the colors of the flames produced by these ions.

Self-Test

22. Why are the alkaline earth metals less reactive than the alkali metals?

    ____

23. Why does $BeCl_2$ form a polymer in the solid state?

    ____

24. Why aren't common objects ever made from calcium?

___

25. Give the chemical formula for
    (a) limestone ___
    (b) Epsom salts ___
    (c) milk of magnesia ___
    (d) gypsum ___
    (e) dolomite ___
    (f) beryl ___
    (g) plaster of Paris ___
    (h) slaked lime ___

26. Complete and balance the following equations:
    (a) $Ba + H_2O \rightarrow$ ___
    (b) $CaO + H_2O \rightarrow$ ___
    (c) $Mg + O_2 \rightarrow$ ___
    (d) $CaCO_3 \xrightarrow{\text{heat}}$ ___
    (e) $Mg(OH)_2 \xrightarrow{\text{heat}}$ ___

27. Give the formula of an alkaline earth metal compound that is deliquescent.

___

28. What color is given to a flame by ions of
    (a) barium ___
    (b) calcium ___
    (c) strontium ___

New Terms

Write the definitions of the following terms, which were introduced in this section. If necessary, refer to the Glossary at the end of the text.

alkaline earth metals

deliquescence

21.5 The Metals of Groups IIIA, IVA, and VA

Review

The metals that are found to the right of the transition elements are all considerably less reactive and less metallic than those in Groups IA and IIA. They are nearly all amphoteric and their oxides are much less basic than the oxides of the alkali and alkaline earth metals. The heavier members of the groups also show two oxidation states, with the lower oxidation state becoming increasingly more stable relative to the higher one as the group is descended.

Group IIIA Metals

Aluminum is the most important member of this family. It shows only a +3 oxidation state. The metals below it also have a +1 state that becomes increasingly more stable going down the group. The ore of aluminum is bauxite and the metal is prepared by electrolysis in the Hall-Héroult process. Aluminum is quite reactive, but is protected by a tough oxide coating. Some special points to note about the chemistry of aluminum are as follows:

1. The metal dissolves in both acid and base with the evolution of H_2. You should be able to write equations for the reactions. (See page 933.)

2. The formation of Al_2O_3 is very exothermic, a fact that is exploited in the thermite reaction described on page 932.

3. In acidic solutions, Al^{3+} exists as $Al(H_2O)_6{}^{3+}$. Polarization of the water molecules by the small, highly charge Al^{3+} ion makes the hydrogens acidic and causes solutions of aluminum salts in water to be acidic. Study Figure 21.20 on page 934. Be sure you can write chemical equations that describe what happens as a solution of $Al(H_2O)_6{}^{3+}$ is gradually made basic.

4. Aluminum oxide occurs in two forms, α-Al_2O_3 and γ-Al_2O_3. The alpha form is called corundum, and ruby and sapphire are gems composed primarily of this form of aluminum oxide.

5. Aluminum sulfate, prepared from Al_2O_3 by reaction with H_2SO_4, has many uses (page 935).

6. Aluminum sulfate forms alums, which are double salts with the general formula $M^+M^{3+}(SO_4)_2{\cdot}12H_2O$. Become familiar with some of the uses of potassium alum, $KAl(SO_4)_2{\cdot}12H_2O$.

7. The aluminum halides tend to be molecular in their anhydrous forms. They exist as dimers with the general formula Al_2X_6. Note the similarities between the structures of the Al_2X_6 dimer and the $BeCl_2$ polymer.

Metals of Group IVA—Tin and Lead

Tin and lead both have two oxidation states, +2 and +4. For tin, both are important and there are many Sn^{II} and Sn^{IV} compounds. For lead, the most stable state is +2; Pb^{IV} compounds tend to be strong oxidizing agents.

The following are special points of interest about these elements and their compounds.

1. Tin occurs as its oxide, which is reduced to the free metal with carbon. Lead occurs as PbS, which is changed to the oxide by roasting in air; the oxide is reduced with carbon. Both metals are purified electrolytically (following a procedure similar to that for copper—see page 780).

2. Tin occurs in two allotropic forms. White tin has mostly metallic properties. Gray tin has the same crystal structure as diamond and has nonmetallic properties. Conversion of white tin to gray tin occurs spontaneously at low temperatures.

3. Tin dissolves in HCl(*aq*) with the evolution of H_2, but reacts with HNO_3 to give SnO_2. On the other hand, lead reacts with acids to give only the +2 oxidation state.

4. Tin and lead are amphoteric. The metals dissolve in base with the evolution of H_2. Learn to write the equations for the reactions.

5. Tin(IV) halides are covalent. Polarization of the halide ions by the very small, highly positive Sn^{4+} cation produces essentially covalent bonds between the Sn and the halogen.

6. Important oxygen compounds of tin are SnO_2, used to make electrically conducting glass, and bis(tributyltin) oxide, $[(C_4H_9)_3Sn]_2O$, which prevents wood rot and the growth of marine organisms such as barnacles.

7. Lead(IV) halides are very unstable because of the tendency of the Pb^{IV} to oxidize the halide ion to the free halogen.

8. Important lead compounds are PbO(litharge), Pb_3O_4 (red lead), PbO_2 (lead(IV) oxide or lead dioxide), $PbCrO_4$ (lead chromate), and $Pb_3(OH)_2(CO_3)_2$ (white lead). These are discussed on page 940.

Bismuth—the only metal in Group VA

Bismuth occurs in nature as its oxide or sulfide. In preparing the metal, the sulfide is roasted to give the oxide. Reduction of the oxide with carbon gives the free metal. Bismuth is used in making Wood's metal—a special low-melting alloy. In compounds, bismuth occurs in two oxidation states, +3 and +5. The lower oxidation state is more stable, and compounds containing Bi^{V} are powerful oxidizing agents.

BiF_3 is ionic, but the other halides are covalent. $BiCl_3$ hydrolyzes in water to give BiOCl. Other Bi^{III} compounds also hydrolyze to give the BiO^+ ion.

Sodium bismuthate, $NaBiO_3$, is a powerful oxidizing agent and is used to test for the presence of Mn^{2+} ion in a solution.

Self-Test

29. What is bauxite? ______________________________

30. Write equations for the reaction of metallic aluminum with an acid and with a base.

____

____

31. What is the thermite reaction?

____

32. (a) What are two forms of aluminum oxide? ____________

(b) Which form occurs in ruby and sapphire? ____________

(c) What ion gives ruby its color? ____________

33. What would be the formula of an alum formed from $Al_2(SO_4)_3$ and $(NH_4)_2SO_4$?

____

34. Sketch the structure of Al_2Br_6?

35. Write a chemical equation that illustrates why solutions of aluminum salts in water are acidic.

____

36. Write an equation for the reaction of gelatinous aluminum hydroxide with excess base.

____

37. Write equations for

    (a) roasting of galena in air

    ____

    (b) the reduction of SnO_2 by carbon

    ____

    (c) the reduction of PbO by carbon

    ____

38. How are tin and lead metals refined? Explain the process.

____

____

39. Give the products of the following reactions.

    (a) $Pb + O_2 \rightarrow$ ____________________

    (b) $Sn + O_2 \rightarrow$ ____________________

40. Which is the better oxidizing agent, SnO_2 or PbO_2? ____________________

41. Write chemical equations for

    (a) the reaction of tin with nitric acid.

    ____

    (b) the reaction of lead with nitric acid.

    ____

42. Write a general equation that illustrates the reactions of tin and lead with concentrated base.

____

43. What type of solid (ionic, covalent, molecular or metallic) would be expected to be formed by tin(IV) bromide?

____

44. What reaction occurs if $PbCl_4$ is heated?

____

45. What is Wood's metal? ______________________________

46. What is the formula for

    (a) bismuthyl ion ____________________

    (b) sodium bismuthate ____________________

47. What chemical reaction is used in the analytical test to determine the presence of Mn^{2+} in a solution?

____

New Terms

Write the definitions of the following terms, which were introduced in this section. If necessary, refer to the Glossary at the end of the text.

thermite reaction

double salt

alum

Answers to Self-Test Questions

1. (a) Mg, (b) Ca, (c) Ba
2. (a) K, (b) K, (c) Ga
3. magnesium and sodium
4. $Mg^{2+} + 2OH^- \rightarrow Mg(OH)_2(s)$
   $Mg(OH)_2(s) + 2H^+ + 2Cl^- \rightarrow Mg^{2+} + 2Cl^- + 2H_2O$
5. On the ocean floor
6. sulfide ores of copper and lead
7. $2PbS + 3O_2 \rightarrow 2PbO + 2SO_2$
8. Amphoteric behavior; $Al_2O_3(s) + 2OH^- \rightarrow 2AlO_2^- + H_2O$
   $AlO_2^- + H^+ + H_2O \rightarrow Al(OH)_3(s)$
   $2Al(OH)_3 \xrightarrow{\text{heat}} Al_2O_3 + 3H_2O$
9. carbon monoxide; $Fe_2O_3 + 3CO \rightarrow 2Fe + 3CO_2$
10. $CaCO_3 \rightarrow CaO + CO_2$
    $CaO + SiO_2 \rightarrow CaSiO_3$ (slag)
11. It is fast and produces good-quality steel.
12. There are ions with only a 1+ charge in the metallic lattice.
13. They have very low ionization energies and therefore lose electrons very easily.

14. In the reaction, $CsCl(l) + Na(g) \rightleftharpoons Cs(g) + NaCl(l)$, Cs is more volatile than Na.
15. The Li^+ ion is very small and has an exceptionally large hydration energy, which must be overcome when Li^+ is reduced in the presence of water.
16. Li_2CO_3
17. (a) $2Na + O_2 \rightarrow Na_2O_2$
    (b) $4Li + O_2 \rightarrow 2Li_2O$
    (c) $Cs + O_2 \rightarrow CsO_2$
    (d) $2NaHCO_3 \xrightarrow{heat} Na_2CO_3 + H_2O + CO_2$
    (e) $2KO_2 + 2H_2O \rightarrow 2KOH + O_2 + H_2O_2$
    (f) $Na_2O_2 + 2H_2O \rightarrow 2NaOH + H_2O_2$
18. (a) red (b) yellow (c) pale-violet
19. lithium, Li_3N
20. $NaCl + NH_3 + CO_2 + H_2O \rightarrow NaHCO_3 + NH_4Cl$
21. (a) KF, (b) $NaClO_3$
22. They have larger ionization energies.
23. By forming coordinate covalent bonds from the chlorine atoms of one $BeCl_2$ unit to the Be atoms of other $BeCl_2$ units, each Be achieves an octet of electrons.
24. Calcium reacts readily with both O_2 and moisture.
25. (a) $CaCO_3$, (b) $MgSO_4{\cdot}7H_2O$, (c) $Mg(OH)_2$, (d) $CaSO_4{\cdot}2H_2O$
    (e) $CaCO_3{\cdot}MgCO_3$, (f) $Be_3Al_2(SiO_3)_6$, (g) $CaSO_4{\cdot}\frac{1}{2}H_2O$, (h) $Ca(OH)_2$
26. (a) $Ba + 2H_2O \rightarrow Ba(OH)_2 + H_2$
    (b) $CaO + H_2O \rightarrow Ca(OH)_2$
    (c) $2Mg + O_2 \rightarrow 2MgO$
    (d) $CaCO_3 \xrightarrow{heat} CaO + CO_2$
    (e) $Mg(OH)_2 \xrightarrow{heat} MgO + H_2O$
27. $CaCl_2$
28. (a) pale yellow-green, (b) orange-red, (c) bright red
29. aluminum ore, $Al_2O_3{\cdot}xH_2O$
30. $2Al + 6H^+ \rightarrow 2Al^{3+} + 3H_2$
    $2Al + 2OH^- + 2H_2O \rightarrow 2AlO_2^- + 3H_2$
31. $Fe_2O_3 + 2Al \rightarrow 2Fe + Al_2O_3$
32. (a) α-Al_2O_3 and γ-Al_2O_3, (b) corundum, α-Al_2O_3, (c) Cr^{3+}
33. $(NH_4)Al(SO_4)_2{\cdot}12H_2O$
34. See Figure 21.24, page 936.
35. $Al(H_2O)_6^{3+} + H_2O \rightleftharpoons Al(H_2O)_5(OH)^{2+} + H_3O^+$
36. $Al(OH)_3(H_2O)_3(s) + OH^-(aq) \rightarrow Al(OH)_4(H_2O)_2^-(aq) + H_2O$
37. (a) $2PbS + 3O_2 \rightarrow 2PbO + 2SO_2$
    (b) $SnO_2 + C \rightarrow Sn + CO_2$

(c) $2PbO + C \rightarrow 2Pb + CO_2$

38. Electrolytically. Impure tin (or lead) anodes are dipped into a solution of the metal salt and when electrolysis is carried out the anodes dissolve and the pure metal deposits on the cathode.
39. (a) PbO (b) SnO_2
40. PbO_2
41. (a) $Sn(s) + 4HNO_3(aq) \rightarrow SnO_2(s) + 4NO_2(g) + 2H_2O$
    (b) $3Pb(s) + 8HNO_3(aq) \rightarrow 3Pb(NO_3)_2(aq) + 2NO(g) + 4H_2O$
42. $M(s) + 2OH^-(aq) + 2H_2O \rightarrow M(OH)_4^{2-}(aq) + H_2(g)$
43. molecular ($SnBr_4$ should be a covalently bonded molecule, similar to $SnCl_4$.)
44. $PbCl_4 \rightarrow PbCl_2 + Cl_2$
45. A low-melting alloy of bismuth, lead, tin, and cadmium.
46. (a) BiO^+ (b) $NaBiO_3$
47. $14H^+ + 5BiO_3^- + 2Mn^{2+} \rightarrow 2MnO_4^- + 5Bi^{3+} + 7H_2O$

Chapter 22

TRANSITION METALS AND THEIR COMPLEXES

The transition elements (transition metals) are located in the main body of the periodic table between Groups IIA and IIIA. They include some of our most familiar and useful metals. In this chapter we explore their physical and chemical properties, and we study in greater depth the structures and bonding in complex ions formed by the transition metals.

Learning Objectives

As you study of this chapter, keep in mind the following goals:

1 To learn how the transition elements are categorized, what some of their general properties are.

2 To learn what the lanthanide contraction is and how it affects the properties of the elements in period 6.

3. To learn the origin of ferromagnetism, the kind of magnetic phenomenon that we associate with the element iron.

4 To learn about the major chemical and physical properties of the most important transition metals.

5 To learn some of the practical applications of the transition metals and their compounds.

6 To learn about the kinds of structures formed by complexes when the metal ion is surrounded by various numbers of ligands.

7 To learn how the formation of ring structures within complexes leads to an extra stability of the complex.

8 To learn how more than one compound can have the same chemical formula—a phenomenon called isomerism—and to learn to identify different kinds of isomers.

9 To learn how the colors, magnetic properties, and the stabilities of the oxidation states of metal ions in complexes can be explained by considering how the ligands influence the energies of the *d* orbitals of the central metal ion.

22.1 The Transition Metals: General Characteristics and Periodic Trends

Review

The transition elements, located in the center region of the periodic table, are referred to as the *d*-block elements. The inner transition elements (lanthanides and actinides) are found in the two long rows beneath the main body of the table.

General properties of the transition elements are as follows:

1. They tend to be hard and high-melting.

2. Most of them exhibit two or more oxidation states (besides zero for the free element).

3. Their ions form many complex ions with neutral molecules and anions.

4. Many of their compounds are colored.

The sizes of the transition metals decrease from left to right across a period, and show a minimum near the center of each row. Within a group, size increases from period 4 to period 5, but hardly any change occurs from period 5 to period 6. This is because of the lanthanide contraction, which occurs because of the filling of the 4f subshell in elements 58–71. The lanthanide contraction causes the elements that immediately follow lanthanum in period 6 to be exceptionally dense and resistant to oxidation.

Ferromagnetism—the strong type of magnetic behavior shown by the elements iron, cobalt, and nickel—appears to result from the magnetic alignment of enormous numbers of individual paramagnetic atoms within crystalline regions called domains. These domains become aligned when the metal is placed in a magnetic field and the magnetic attraction is very strong.

Self-Test

1. Among the elements Sc, Ag, Gd, Np, Zr, Cs, Ge, and U, which ones are

   (a) *d*-block elements? ________________

   (b) inner transition elements? ________________

   (c) lanthanides? ________________

   (d) actinides? ________________

2. What would be the charge on the complex ion, $Cr(NH_3)_4(CN)_2$ if it is formed from Cr^{3+}, NH_3, and CN^-?

____

3. What do the trends in melting points suggest about the solids formed by the transition metals?

____

4. Is the trend in melting points across a period consistent with the trend in atomic sizes? Explain.

____

5. Give the electron configuration of
   (a) titanium _______________
   (b) nickel _______________
   (c) copper _______________

6. Why is the density of tungsten so much larger than that of molybdenum?

____

7. Iron is ferromagnetic, but ruthenium (just below iron in the periodic table) is not. What is a reasonable explanation for this?

____

____

8. Which pair of ions should be more chemically alike, Ti^{4+} and Zr^{4+}, or Zr^{4+} and Hf^{4+}? Justify your answer.

____

New Terms

Write the definitions of the following terms, which were introduced in this section. If necessary, refer to the Glossary at the end of the text.

transition elements (transition metals)
inner transition elements
lanthanide elements
actinide elements
complex ion
lanthanide contraction
ferromagnetism
domains

22.2 Properties of Some Important Transition Elements

Review

In this section, we examine some of the chemical and physical properties of the major transition metals. These are the ones you are most likely to come into contact with personally or read about elsewhere. They include the transition metals that are most important to our modern society, and along with learning their chemistry, you should also become aware of their practical applications.

Titanium

This is a strong, lightweight (meaning *low density*), corrosion resistant metal important in the aircraft and aerospace industry. The most important oxidation state of titanium is +4, and its most important compound is TiO_2. It occurs in the titanium ore called rutile, and is a white substance that is the most common white pigment in modern paints. Titanium is extracted from its ore by reaction of TiO_2 with carbon and chlorine, which converts it to titanium tetrachloride, $TiCl_4$, a molecular substance that's a liquid at room temperature. Be sure to review the reasons for its covalence. Reaction of $TiCl_4$ with magnesium produces metallic titanium. Titanium tetrachloride reacts with water to give TiO_2 and HCl.

Vanadium

Vanadium is used mostly in steel alloys to give them ductility and shock resistance. Although it exhibits a variety of oxidation states (+2, +3, +4, and +5), the most important is +5. Vanadium pentoxide, V_2O_5, is important because it is a catalyst for the oxidation of SO_2 to SO_3. When V_2O_5 is dissolved in water, no simple V^{5+} cation is formed because of extensive hydrolysis, as depicted in Figure 22.6. In an acidic solution, vanadium(V) exists as the VO_2^+ ion; in base it exists as the VO_4^{3-} ion.

Chromium

This very hard, white, corrosion-resistant metal is familiar to everyone in the form of chrome plate, normally applied electrolytically to steel or other metals. It is also one of the principal ingredients in stainless steel alloys and it is found in nichrome wire, which is used for heating elements in toasters and the like.

All chromium compounds are colored, and pigments are among the principal uses for these substances. (Note that the oxide Cr_2O_3 is a common, stable green pigment.) There are three main oxidation states of chromium, +2, +3, and +6. The +2 state is very easily oxidized to the +3 state, which is the most stable. In water, Cr^{3+} exists as the complex ion, $Cr(H_2O)_6^{3+}$, which is slightly acidic. Review the equation for the reaction of this ion as a weak acid and study the equations for the reactions that take place when solutions containing the amphoteric $Cr(H_2O)_6^{3+}$ ion are gradually made basic. Note, too, the color changes that take place.

In the +6 oxidation state, the most important species are CrO_3, H_2CrO_4 (formed when CrO_3 is dissolved in water), and the ions CrO_4^{2-} and $Cr_2O_7^{2-}$. Chromate ion (yellow) exists in basic solutions and is changed to dichromate ion (reddish-orange) in acidic solutions. Study the equilibria described on pages 955 and 956. In the +6 oxidation state, chromium is a powerful oxidizing agent, especially in acidic solutions.

Manganese

Manganese is more reactive than chromium; its properties are close to those of iron. Its principal use as a free metal is in alloys. The most important oxidation states of Mn are +2, +3, +4, +6, and +7. The +2 state is most stable, and in solution the manganese(II) ion exists as the pale pink $Mn(H_2O)_6^{2+}$ ion.

Addition of base to $Mn(H_2O)_6^{2+}$ precipitates $Mn(OH)_2$ which is easily oxidized to MnO(OH), a compound of manganese(III). Oxidation of the metal also gives a manganese(III) compound, Mn_2O_3. Manganese dioxide, MnO_2, contains manganese(IV) and is used in making dry cell batteries and as a starting material in the synthesis of other manganese compounds. MnO_2 is nonstoichiometric, which means that its composition is somewhat variable and there isn't an exact 2-to-1 ratio of oxygen to manganese. Oxidation of MnO_2 in a basic medium gives the green manganate ion, MnO_4^{2-}.

The violet permanganate ion, MnO_4^-, is a very powerful oxidizing agent and is a frequently used reagent for redox titrations in the laboratory. Learn to write equations for its reduction reactions in acidic solutions and in basic solutions.

Iron

This is certainly one of the most common transition metals, and its alloys—steels—have numerous practical, structural applications. Pure iron is not very hard and is quite reactive. It dissolves readily in nonoxidizing acids with the evolution of H_2, and it corrodes in an atmosphere of moist oxygen. However, iron is made *passive* in concentrated nitric acid.

There are only two important oxidation states of iron, +2 and +3. Both are common, but the +2 state is rather easily oxidized to +3, especially in a basic medium. The oxides of iron are FeO, Fe_3O_4 (magnetic), and Fe_2O_3.

In water, iron(II) exists as the pale green $Fe(H_2O)_6^{2+}$ ion. Many iron(II) salts have a green color because they contain this ion. Iron(III) exists as $Fe(H_2O)_6^{3+}$ in water. It is weak acid and its reaction with water gives a yellow ion, $Fe(H_2O)_5(OH)^{2+}$. Addition of base to $Fe(H_2O)_6^{2+}$ precipitates $Fe(OH)_2$, which is rapidly oxidized in air to $Fe_2O_3{\cdot}xH_2O$. Addition of base to $Fe(H_2O)_6^{3+}$ precipitates gelatinous $Fe(H_2O)_3(OH)_3$, which is not amphoteric.

Complex ions mentioned in this section are the red $Fe(SCN)_6^{3-}$ ion and the ferrocyanide and ferricyanide ions. Reaction of Fe^{2+} with $Fe(CN)_6^{3-}$, and reaction of Fe^{3+} with $Fe(CN)_6^{4-}$ give the same deep-blue compound, $Fe_4[Fe(CN)_6]_3 \cdot 16H_2O$.

Cobalt

This is a hard metal, somewhat less reactive than iron, that is used in alloys and in catalysts. It has two oxidation states, +2 and +3. The $Co(H_2O)_6^{2+}$ ion is pink and gives its color to many cobalt(II) salts. The $Co(H_2O)_6^{3+}$ ion oxidizes water to O_2, but most complex ions of Co^{III} are stable in water. When heated, the pink compound $[Co(H_2O)_6]Cl_2$ loses water and turns blue as the compound $[Co(H_2O)_4]Cl_2$ is formed. The reaction is reversible, and gradual absorption of water from the air yields the original pink salt.

Nickel

This metal is corrosion resistant and is applied over other metals as protective coatings. It is also widely used to make alloys. Nickel steel has many industrial applications In combination with chromium and iron, nickel forms stainless steel, and when combined with copper, it gives an alloy called monel.

After extraction from its ores, nickel metal is refined, either electrolytically or by the Mond process. The latter makes use of the ease of formation of nickel carbonyl, $Ni(CO)_4$ from nickel and CO. Nickel dissolves in nonoxidizing acid with the evolution of H_2. The chief oxidation state is +2 and in water the nickel ion is green owing to the presence of $Ni(H_2O)_6^{2+}$. Nickel forms many complexes, including the blue $Ni(NH_3)_6^{2+}$ ion. Nickel(IV) oxide (NiO_2) is the anode material found in nickel-cadmium batteries.

Copper, Silver, and Gold

These are known as the coinage metals. Review their applications on page 961.

The common oxidation states of the coinage metals are:

Copper	+1, +2
Silver	+1
Gold	+1, +3

None of them dissolves in nonoxidizing acids. Copper and silver dissolve in nitric acid and hot concentrated H_2SO_4 (which is a moderately strong oxidizing agent). You should be able to write equations for these reactions.

Compounds and Reactions of Copper When metallic copper is heated in air, it forms the black oxide, CuO, but slow corrosion in the atmosphere produces instead a green coating of the "basic carbonate" $Cu_2(OH)_2CO_3$. Oxidizing acids yield copper(II) salts when they dissolve the metal. Copper(I) ion disproportionates in water, but is stabilized when isolated in an insoluble salt (see the reaction for the formation of CuCl on page 962). One of the most important compounds of copper is copper sulfate, which crystallizes from water as the blue hydrate, $CuSO_4{\cdot}5H_2O$. This compound loses water when heated strongly, but gradually absorbs water from the atmosphere to re-form the hydrate. Copper(II) ion forms many complexes. Three with which you should be familiar are the pale blue $Cu(H_2O)_4^{2+}$ ion, the yellow $CuCl_4^{2-}$ ion, and the deep blue $Cu(NH_3)_4^{2+}$ ion.

Compounds and Reactions of Silver Silver is not easily oxidized by air, but when H_2S is present, it forms a black film of Ag_2S. You should be able to write the equation for this reaction, as well as the reaction of silver with both concentrated and dilute nitric acid. When made basic, solutions containing Ag^+ yield brown Ag_2O. Except for AgF, the silver halides have very low solubilities in water, and they become progressively less soluble from AgCl to AgI. The halides are photosensitive and tend to undergo photodecomposition, a property that permits compounds such as AgBr to be used in photographic film and paper. Silver ion forms many complexes. You should study the reactions involved in the qualitative analysis of silver ion, which are given on page 963.

Reactions of Gold Gold is unreactive toward both nonoxidizing and oxidizing acids. However, in the presence of chloride ion, it is oxidized by HNO_3 to give a complex ion $AuCl_4^-$. Apparently the tendency for gold to form complexes with Cl^- combined with the oxidizing power of the HNO_3 are sufficient together to cause the gold to react. The mixture that accomplishes this is called aqua regia and consists of 1 part concentrated HNO_3 and 3 parts concentrated HCl, by volume. In general, gold compounds are quite unstable and tend to be easily decomposed by heat.

Zinc, Cadmium, and Mercury

Zinc is an extremely important metal. It is used to coat steel (galvanizing) and to make alloys such as brass and bronze. It is present in the various dry cell batteries we use, and its compounds ($ZnCl_2$, ZnO, and ZnS, for example) find many applications.

Zinc is easily oxidized and gives only one oxidation state, +2. The metal dissolves in acids such as HCl with the evolution of H_2. It is amphoteric and also dissolves in base. (You should be able to write equations for both reactions.) In air, the metal corrodes slowly with the formation of a coating of a basic carbonate $Zn_2(OH)_2CO_3$ (similar to that formed by copper). Zinc hydroxide, which precipitates when solutions of Zn^{2+} are made basic, dissolves in excess base. As with the other transition metals, complex ions of zinc are common.

Cadmium is similar in reactivity to zinc and also gives just one oxidation state (+2). However, cadmium is not amphoteric. Neither the metal nor its hydroxide dissolves in concentrated base. Cadmium ions are toxic and cadmium is one of the "heavy metals" to be avoided.

Mercury, as you know, is a liquid at room temperature and one of its common uses is as the fluid in thermometers. Another common use is in making amalgams for dental fillings. The ability of mercury to dissolve gold is used in gold mining operations. Mercury does not dissolve in nonoxidizing acids, but it does dissolve in nitric acid. Two oxidation states are known. Mercury(II) is Hg^{2+}, but mercury(I) is Hg_2^{2+}. This latter ion consists of two mercury atoms joined by a covalent bond, with the pair of atoms carrying a total charge of 2+. Many mercury(II) "salts" are predominantly covalent and dissociate to only a limited degree in water. Two important compounds of mercury are $HgCl_2$ and Hg_2Cl_2. Be sure you know the reaction that occurs when solid Hg_2Cl_2 is treated with aqueous ammonia. You should also know what happens when solutions that contain mercury(II) are treated with H_2S.

Self-Test

9. What is the most important compound of titanium? ____________

10. Write the chemical equation for the reaction of $TiCl_4$ with water.

    ____

11. What are the principal uses of vanadium?

    ____

12. Why can't a simple V^{5+} ion exist in water?

    ____

13. What is the formula for the catalyst in the contact process for the production of H_2SO_4?

    ____

14. Give the formula for:

    (a) chromate ion ____________________

    (b) dichromate ion ____________________

    (c) chromic acid ____________________

15. What compound is the acid anhydride of chromic acid?

    ____

16. Write the equation for the equilibrium between chromate ion and dichromate ion in water.

    ____

17. The most stable oxidation state of chromium is__________ and in aqueous solution this ion has the formula ____________________

18. What are the oxidation states of manganese?

____

19. Why is permanganate a useful titrant for redox titrations in acidic solutions?

____

____

20. What is the manganese-containing product when MnO_4^- is reduced
    (a) in acidic solution? ______________
    (b) in basic solution? ______________

21. What is a disproportionation reaction? ______________

____

22. What are the common oxidation states of cobalt?

____

What is the formula for the pink ion that gives many cobalt salts their color?

____

23. What is a nonstoichiometric compound? ______________

____

____

Give two examples. ______________

24. What are the chief components of stainless steel?

____

25. What are the formulas of the two iron ions that exist in aqueous solutions of iron salts?

____

26. Write an equation for the reaction of aqueous iron(III) ion with base.

____

27. What net ionic reaction occurs between solutions of ferric chloride and potassium ferrocyanide?

____

28. What compound of nickel is formed in the Mond process?

____

29. What is the most stable oxidation state of nickel? ____________

30. Which are the coinage metals? ______________________

What are the formulas of their chlorides? ________________

31. Write the net ionic equations for the reactions of copper and silver with concentrated nitric acid.

____

____

32. What is aqua regia? ______________________________

33. What is the formula for the complex ion formed during the test for silver?

____

34. What advantage does cadmium have over zinc as a protective coating on metals such as iron?

____

35. A solution of a metal in mercury is called ____________________

36. The two ions of mercury are __________ and __________.

New Terms

Write the definitions of the following terms, which were introduced in this section. If necessary, refer to the Glossary at the end of the text.

nonstoichiometric compound
disproportionation
amalgam
coinage metals
aqua regia

22.3 Complexes of the Transition Metals

Review

The structures of complexes are most easily classified according to the number of ligand atoms bonded to the metal ion. This number is called the coordination number. Polydentate ligands are able to form ring structures with the metal ion. Such complexes, called chelates, are more stable than similar complexes formed by monodentate ligands because when one end of a ring lets go of the metal, the other end is still firmly attached. There is a large probability that the ring will reform before this other end also becomes detached. Because of this, it is more difficult to cause a polydentate ligand to dissociate from the metal ion than it is to cause a similar dissociation of monodentate ligands, so the complex appears more stable. This phenomenon is called the chelate effect.

Common coordination numbers in complexes include 2, 4, and 6. For coordination number 2, the complex usually has a linear geometry. For coordination number 4, both tetrahedral and square planar structures are observed. For coordination number 6, almost all complexes are octahedral. Such octahedral complexes are formed with monodentate ligands as well as polydentate ligands. Study Figures 22.20 and 22.21. Be sure you can sketch an octahedral complex; review Figure 22.22.

Self-Test

37. Sketch chelate rings formed by

    (a) ethylenediamine

    (b) oxalate ion

38. Which complex would you expect to be more stable, $[Cr(en)_3]^{3+}$ or $[Cr(NH_3)_6]^{3+}$?

    ____

39. Name two biologically important complexes that contain the porphyrin structure.

    ____

40. What property do metal ions that form tetrahedral complexes usually have?

____

41. What is the coordination number of the metal ion in each of the following?

    (a) $[Cu(NH_3)_4]^{2+}$ ____________

    (b) $[Co(NH_3)_4Cl_2]^{+}$ ____________

    (c) $[Ni(en)_3]^{2+}$ ____________

New Terms

Write the definitions of the following terms, which were introduced in this section. If necessary, refer to the Glossary at the end of the text.

coordination number

chelate effect

22.4 Isomers of Coordination Compounds

Review

When two or more different compounds have the same chemical formula, they are said to be isomers of each other. For coordination compounds, there are several ways for this to occur, but the most important one for you to learn about is the type of isomerism called stereoisomerism. This occurs when two or more compounds have the same formula, but differ in the way their atoms are arranged in space.

Geometric isomers of square planar complexes such as $[Pt(NH_3)_2Cl_2]$, or octahedral complexes such as $[Cr(H_2O)_4Cl_2]^{+}$, exist in *cis* and *trans* forms. In the *cis* isomer, the chloride ions are next to each other on *the same side* of the metal ion. In the *trans* isomer, the chloride ions are opposite each other. In general, *cis* and *trans* isomers can exist for square planar complexes with the general formula Ma_2b_2 (where M is a metal ion and a and b are monodentate ligands). *Cis* and *trans* isomers also exist for octahedral complexes with the general formula Ma_2b_4, and for octahedral complexes with the general formula MA_2b_2 (where A is a bidentate ligand and b is a monodentate ligand). These are illustrated on page 970 of the text.

Chiral isomers occur when two structures differ only in that one is the nonsuperimposable mirror image of the other—that is, when the mirror image of one isomer looks exactly like the other isomer, but the two isomers themselves do not match exactly when one is place over the other. Chiral isomers exist for complexes with the general formula MA_3 and *cis*-MA_2b_2 (where *A* stands for a bidentate ligand and *b* stands for a monodentate ligand). Chiral isomers are also known as optical isomers because the two isomers affect polarized light in opposite ways.

Self-Test

42. Is *cis-trans* isomerism possible for tetrahedral complexes? Explain.

____

____

43. How many different isomers exist for the complex $[Co(en)_2Br_2]^+$?

____

44. Is the *trans* isomer of $[Ni(C_2O_4)_2(CN)_2]^{4-}$ chiral? (Note: $C_2O_4{}^{2-}$ is oxalate ion, a bidentate ligand.)

____

New Terms

Write the definitions of the following terms, which were introduced in this section. If necessary, refer to the Glossary at the end of the text.

isomerism	*cis*-isomer	superimposability
stereoisomerism	*trans*-isomer	enantiomers
geometric isomerism	chirality	optical isomers

22.5 Bonding in Complexes

Review

The crystal field theory is used to explain the properties of complexes in which the metal ion has a partially filled *d* subshell. To understand the theory, it is necessary to know the shapes and directional properties of the *d* orbitals. Study Figure 22.28.

In an octahedral complex, we can imagine the ligands to lie along the *x*, *y*, and *z* axes of a Cartesian coordinate system with the metal ion in the center (at the origin). The negative charges of the ligands (either anions or the negative ends of ligand dipoles) point directly at the metal ion's $d_{x^2-y^2}$ and d_{z^2} orbitals, but they point between the d_{xy},, d_{xz},, and d_{yz}, orbitals. Because of this, the ligands repel electrons in the $d_{x^2-y^2}$ and d_{z^2} orbitals more than they repel electron in the other three. This raises the energies of the $d_{x^2-y^2}$ and d_{z^2} orbitals above the energies of the d_{xy},, d_{xz},, and d_{yz}, orbitals. The net result is an energy level diagram like that shown in Figure 22.30 on page 974. The energy difference between the two energy levels is called the crystal field splitting and is given the symbol Δ.

The magnitude of Δ depends on the nature of the ligands attached to the metal ion, the oxidation state of the metal, and the period in which the metal occurs. In general, as the oxidation state increases, other things being equal, the size of Δ becomes larger. Going down a group, Δ becomes larger, too.

In this section, the usefulness of the crystal field theory is illustrated by considering three phenomena—the stabilities of certain oxidation states of metal ions in complexes, the origin of the colors of complexes, and the magnetic properties of complexes.

For chromium(II) ion, you see that the removal of a high-energy electron, along with an increase in the magnitude of Δ that accompanies the increase in oxidation state, helps to make the oxidation of $[Cr(H_2O)_6]^{2+}$ to $[Cr(H_2O)_6]^{3+}$ energetically favorable. Stated in another way, in water, chromium(III) ion is the more stable oxidation state, because chromium(II) ion is so easily oxidized to it.

According to crystal field theory, the color of a complex arises from the absorption of a photon that has an energy equal to Δ. For transition metal complexes, this photon has a frequency that places it in the visible region of the spectrum. The color observed for the complex corresponds to the color of the light that *isn't* absorbed.

Some ligands always produce a large Δ, regardless of the metal ion, and some ligands always produce a small Δ. The list of ligands arranged in order of their ability to produce a large Δ is called the spectrochemical series. This is given on page 977.

The amount of energy needed to cause two electrons to become paired in the same orbital is called the pairing energy, to which we have given the symbol *P*. For certain numbers of *d* electrons, there is a choice as to how the electrons are to be distributed among the higher and lower *d*-orbital energy levels. Pairing an electron with another in a low-energy *d* orbital costs energy equal to the pairing energy, but it saves an energy equal to Δ. On the other hand, placing the electron in the higher energy orbital cost an energy equal to Δ, but saves energy equal to the pairing energy. Which energy distribution prevails depends on how the magnitudes of Δ and *P* compare. When $\Delta > P$, pairing of electrons in the lower-energy level is preferred; when $\Delta < P$, then spreading the electrons out as much as possible is preferred.

Self-Test

45. Which of the *d* orbitals point directly along the *x, y,* and *z* axes?

____

46. Which complex has the larger Δ, $[CrCl_6]^{4-}$ or $[CrCl_6]^{3-}$?

____

47. Which complex has the larger Δ, $Ni(CN)_4{}^{2-}$ or $Pt(CN)_4{}^{2-}$?

____

48. Cyanide ion produces a very large crystal field splitting. Should it be easy or difficult to oxidize $[Co(CN)_6]^{4-}$ to $[Co(CN)_6]^{3-}$? Explain your answer in terms of the populations of the *d* orbitals of the metal ion.

49. Which complex would be expected to absorb light of longer wavelength?

(a) $[Ti(H_2O)_6]^{3+}$ or $[Ti(H_2O)_6]^{2+}$ _______________

(b) $[Ni(H_2O)_6]^{2+}$ or $[Ni(CN)_6]^{4-}$ _______________

50. Which complex has a larger Δ, one that absorbs red light or one that absorbs blue light?

____

51. How many unpaired electrons would you expect to find in each of the following?

(a) $[CrI_6]^{4-}$ _______________ (c) $[Fe(H_2O)_6]^{3+}$ _______________

(b) $[Cr(CN)_6]^{4-}$ _______________ (d) $[Fe(CN)_6]^{3-}$ _______________

New Terms

Write the definitions of the following terms, which were introduced in this section. If necessary, refer to the Glossary at the end of the text.

crystal field theory
crystal field splitting
spectrochemical series
high-spin complex
low-spin complex
pairing energy

Answers to Self-Test Questions

1. (a) Sc, Ag, Zr (b) Gd, Np, U (c) Gd (d) Np, U
2. 1+, the formula is $Cr(NH_3)_4(CN)_2^+$
3. Some covalent bonding exists between atoms of elements located around the center of a row of transition elements.
4. Yes. Attractive forces that produce high melting points would also be expected to draw the atoms closer together, thereby yielding smaller atomic radii.
5. (a) [Ar] $3d^24s^2$ (b) [Ar] $3d^84s^2$ (c) [Ar] $3d^{10}4s^1$
6. The lanthanide contraction causes tungsten to be about the same size as the atom above it in the periodic table, but the atomic mass of tungsten is about 88 u larger. This larger mass is packed into the same volume, so the density is much larger.
7. The Ru atoms are not spaced just right to cause them to lock onto each other in magnetic domains, which is required for ferromagnetism.
8. Zr^{4+} and Hf^{4+}. The ions are very nearly the same size (because of the lanthanide contraction) and they have the same charge, so they behave in nearly identical ways in chemical reactions.
9. TiO_2
10. $TiCl_4 + 2H_2O \rightarrow TiO_2 + 4HCl$
11. In alloys.
12. A highly charged V^{5+} ion would polarize the surrounding H_2O molecules to such a large extent that the hydrogens attached to the water-oxygens would transfer as H^+ to other water molecules in the solvent. As a result, ions such as VO_2^+ and VO_4^{3-} are formed.
13. V_2O_5
14. (a) CrO_4^{2-} (b) $Cr_2O_7^{2-}$ (c) H_2CrO_4
15. CrO_3
16. $CrO_4^{2-} + 2H^+ \rightleftharpoons Cr_2O_7^{2-} + H_2O$
17. +3, $[Cr(H_2O)_6]^{3+}$
18. +2, +3, +4, +6, +7
19. The MnO_4^- ion has a deep purple color, but its reduction product, Mn^{2+} is very pale pink. As MnO_4^- is added during a titration, the color fades as it reacts. When the equivalence point is reached, the solution becomes colored by the next drop of titrant because an excess of MnO_4^- now exists, and this signals the end point.
20. (a) Mn^{2+} [Actually, $Mn(H_2O)_6^{2+}$] (b) $MnO_2(s)$

21. A reaction in which a portion of the reactant becomes oxidized while the rest of it becomes reduced. The reactant is simultaneously an oxidizing agent and a reducing agent.
22. +2, +3; $[Co(H_2O)_6]^{2+}$
23. A compound in which the elements are not combined in an exactly whole number ratio by moles, and in which the mole ratio is variable from sample to sample. Some examples are MnO_2, FeO, Fe_2O_3, and Fe_3O_4.
24. Iron, chromium, and nickel.
25. $[Fe(H_2O)_6]^{2+}$ and $[Fe(H_2O)_6]^{3+}$
26. $[Fe(H_2O)_6]^{3+}(aq) + 3OH^-(aq) \rightarrow Fe(H_2O)_3(OH)_3(s) + 3H_2O$
27 $3[Fe(H_2O)_6]^{3+}(aq) + [Fe(CN)_6]^{4-}(aq) \rightarrow Fe_4[Fe(CN)_6]_3 \cdot 16H_2O + 2H_2O$
28. $Ni(CO)_4$
29. +2
30. Cu, Ag, Au; CuCl, $CuCl_2$, AgCl, AuCl, $AuCl_3$
31. $Cu + 2NO_3^- + 4H^+ \rightarrow Cu^{2+} + 2NO_2 + 2H_2O$
    $Ag + NO_3^- + 2H^+ \rightarrow Ag^+ + NO_2 + H_2O$
32. A 1:3 mixture (by volume) of concentrated HNO_3 and concentrated HCl.
33. $[Ag(NH_3)_2]^+$
34. It is not amphoteric, so it doesn't corrode in a basic environment as would zinc.
35. amalgam
36. Hg^{2+}; Hg_2^{2+}
37. (a) 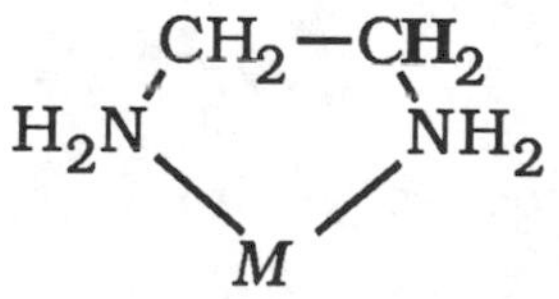 CH_2-CH_2 / H_2N NH_2 / *M* (b) 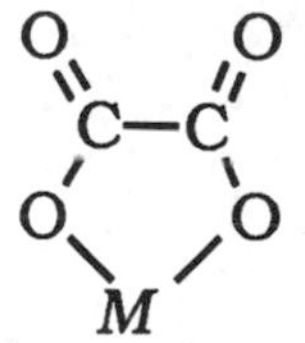 O O / C—C / O O / *M*
38. $[Cr(en)_3]^{3+}$ (the chelate effect)
39. hemoglobin, myoglobin, Vitamin B_{12}
40. They usually have filled *d* subshells (i.e., they don't have partially filled *d* subshells).
41. (a) 4 (b) 6 (c) 6
42. No. For complexes of the type Ma_2b_2, only one tetrahedral structure can be constructed.
43. Three; two *cis* isomers that are enantiomers, and one *trans* isomer.
44. No. See Figure 22.26.
45. $d_{x^2-y^2}$ and d_{z^2}.
46. $[CrCl_6]^{3-}$, because it has chromium in the higher oxidation state.
47. $Pt(CN)_4^{2-}$, because Pt is below Ni in its group.

48.

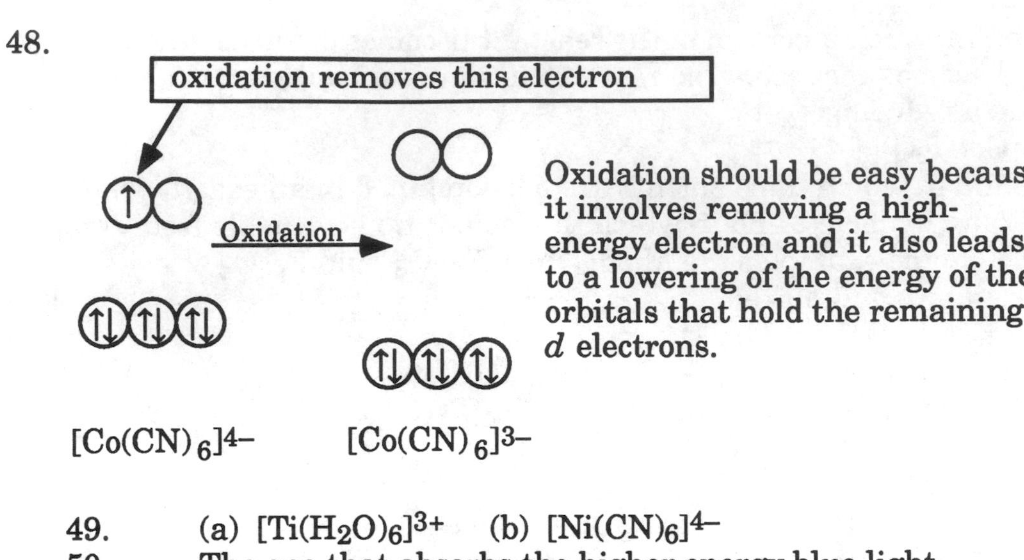

Oxidation should be easy because it involves removing a high-energy electron and it also leads to a lowering of the energy of the orbitals that hold the remaining *d* electrons.

49. (a) $[Ti(H_2O)_6]^{3+}$ (b) $[Ni(CN)_6]^{4-}$

50. The one that absorbs the higher energy blue light.

51. (a) 4 (b) 2 (c) 5 (d) 1

Chapter 23

NUCLEAR REACTIONS AND THEIR ROLE IN CHEMISTRY

The unstable nuclei of many naturally occurring and synthetic isotopes present both risks and opportunities. To understand them, we have to learn what radiations are, their energies and penetrating abilities, how to detect and measure them, and both the dangers and benefits they make possible.

Learning Objectives

Throughout your study of this chapter, keep in mind the following objectives:

1. To learn the circumstances that allow us to use the law of conservation of mass and the law of conservation of energy as independent laws; and to learn the combined law of conservation of mass-energy and when it must be used.
2. To learn how nuclear binding energy is calculated and how the nuclear binding energy per nucleon is a measure of nuclear stability.
3. To learn how the strong force and the electrostatic force are involved in holding nucleons together; to learn the modes of radioactive decay and the associated radiations; and to learn how to balance nuclear equations.
4. To study factors that are associated with the stability of nuclei, such as the odd-even rule, the existence of "magic numbers," and how the ratio of neutrons to protons correlates with the stability of a nucleus.
5. To learn how the bombardment of the atoms of specific isotopes by various particles cause transmutations.
6. To learn how radiations are detected and how radioactive materials or their radiations are quantitatively described; to learn about the background radiation; and to see how protections against radiations can be achieved.
7. To see how radiations from radionuclides can be used to locate (trace) other objects; determine the kinds and quantities of impurities; and to date ancient artifacts or geological strata.

8 To learn about the problems and prospects for controlled nuclear fusion as a source of useful energy.

9 To learn how fission occurs in a nuclear chain reaction; to see how fission gives energy and how this energy can be used to make electricity; and to study various aspects of safety—radioactive wastes and their storage, and loss-of-coolant emergencies.

10 To learn how a non-fissile isotope can be changed to one that is fissile and what this might mean both to nuclear energy and to the availability of atomic bomb material.

23.1 Conservation of Mass-Energy

Review

In the study of nuclear binding energies in the next Section, we will refer to the "rest mass" of a nuclear particle. The distinction between a rest mass and the mass of a particle in motion is important only in nuclear sciences. Chemists can ignore the distinction in all situations involving the stoichiometry of reactions. But Einstein had to postulate a relationship between the mass of a particle and its velocity in order to develop the important relationship between mass and energy and the *law of conservation of mass-energy*. The *Einstein equation*, $\Delta E = \Delta mc^2$, is essential to a discussion of nuclear stability because it lets us calculate nuclear binding energies (Section 23.2). It also lets us understand the huge energy yields from small quantities of "fuel" in such *nuclear reactions* as nuclear fusion (Section 23.8) and nuclear fission (Section 23.9).

A calculation in this Section using the Einstein equation illustrates how extremely small the error is when we ignore the mass-energy equivalence in matters of enthalpy changes in chemical reactions.

Self-Test

1. The enthalpy of combustion of acetylene is -1.30×10^3 kJ/mol. When 1.00 mol of acetylene burns, how much mass (in nanograms) changes to energy?

____

New Terms

Write the definitions of the following terms, which were introduced in this section. If necessary, refer to the Glossary at the end of the text.

Einstein equation

nuclear reactions

law of conservation of mass-energy

23.2 Nuclear Binding Energies

Review

The energy that leaves the system when nucleons come together to form a nucleus would be the energy required to break up the nucleus. This is why this energy that leaves the system is called the nuclear *binding energy*. The greater this binding energy is, the more stable is the nucleus. In another sense, the nuclear binding energy is the energy the nucleus does not have because some of the mass of the nucleons changed to energy and left the system as the nucleus formed. Without this energy, the nucleus is more stable than it could have been with this energy.

When binding energies per nucleon are plotted against atomic number (Figure 23.1 in the text), the curve rises rapidly from the least stable nuclei to peak in the vicinity of isotopes of atomic number 26 (iron). Then the curve drops slowly as the highest atomic numbers are approached. In other words, on strictly the grounds of net energy changes, remembering that nature tends to favor events that are exothermic, the fusion of small nuclei into larger ones should release energy. And nuclear fusion does this. Likewise, the breaking up of very large nuclei into those of intermediate atomic numbers, should also give an overall gain in nuclear stability and the release of energy. Nuclear fission does this. Many nuclei change in the direction of greater stability by less drastic events. They emit radiations that transport energy out of their nuclei.

Self-Test

2. Why do we call the nuclear binding energy the energy that a nucleus does not have? ______________________________

   ____

3. How do we explain the fact that the total mass of the nucleons in helium-4 is less than the actual mass of its nucleus?

   ____

   ____

   ____

   ____

4. In the plot of Figure 23.1 in the text, what does the maximum point of the curve correspond to, to a point of high stability or a point of low stability for a nucleus at or near it?

   ____

New Term

Write the definition of the following term, which was introduced in this section. If necessary, refer to the Glossary at the end of the text.

binding energy

23.3 Radioactivity

Review

Within a nucleus, the electrostatic force causes protons to repel each other, which lessens nuclear stability. But the strong force, which causes nucleons to attract each other, acts to overcome the electrostatic force. The electrostatic force, however, is able to act over longer distances than the strong force, so if a nucleus does not have enough neutrons to "dilute" the electrostatic force, it is unstable. A common consequence of such instability is *radioactive decay.*

Various *radionuclides* decay by the following modes.

Decay Mode	Change to Nucleus	Change in mass no.	Change in atomic no.
alpha emission	loss of $^{4}_{2}He$ (and usually also $^{0}_{0}\gamma$)	–4	–2
beta emission	loss of $^{0}_{-1}e$ (and usually also $^{0}_{0}\gamma$)	none	+1
gamma emission	loss of $^{0}_{0}\gamma$ in the 1 MeV energy range	none	none
positron emission	loss of $^{0}_{1}e$ (followed by an annihilation collision producing gamma radiation)	none	–1
neutron emission	loss of $^{1}_{0}n$	–1	none
electron capture	gain of one $^{0}_{-1}e$ followed by X-ray loss	none	–1

The energy of a radiation is usually described by some multiple of the *electron-volt* (eV), and the relative instability of a radionuclide is described by its half-life.

When we write *nuclear equations,* the sums of the mass numbers on each side of the arrow must be equal as well as the sums of the atomic numbers on each side.

The most penetrating radiations are those with neither mass nor charge (gamma and X rays) or with mass but no charge (neutrons).

Several radionuclides of high mass number do not achieve stable nuclei by one nuclear change. Additional changes take place as a *radioactive disintegration series* is descended to a stable isotope.

Self-Test

5. Consider the natures of the electrostatic force and the strong force in an atomic nucleus.

   (a) Which acts between both protons and neutrons? ________________

   (b) Which acts only between protons? ________________

   (c) Which destabilizes nuclei? ________________

   (d) Which acts over the shorter distance? ________________

   (e) Which is a force of attraction? ________________

6. What is present in a nucleus, besides the strong force, that helps to lessen repulsions between protons? ________________

7. Write the nuclear equations for the decay of a hypothetical isotope, ${}^{279}_{111}X$, by each process. (Use Z as the atomic symbol for any new nuclide that forms from each process.)

   (a) by beta and gamma emission ________________

   (b) by alpha and gamma emission ________________

   (c) by positron emission ________________

   (d) by neutron emission ________________

   (e) by electron capture and X-ray emission ________________

8. State what kind of particle or radiation is *emitted* when

(a) a neutron changes to a proton. ____________

(b) an electron capture takes place. ____________

(c) the radionuclide's atomic number increases by 1. ____________

(d) the atomic number decreases by 2. ____________

(e) 2 photons of gamma radiation are produced following decay. ____________

(f) the mass number decreases by 1. ____________

(g) the atomic number decreases by 1. ____________

(h) the mass number decreases by 4. ____________

(i) no change to a different element occurs. ____________

9. Gamma rays have energies on the order of

(a) 0.1 MeV (b) 1.0 MeV (c) 10 MeV (d) 1.0 keV ____________

10. The radiation with the best ability to penetrate lead is

(a) alpha radiation (c) gamma radiation
(b) beta radiation (d) positron radiation

____

11. Annihilation radiation photons result from the collision of an electron with

(a) a positron (c) a neutron
(b) another electron (d) a proton ____________

12. The net effect of electron capture is the conversion of

(a) an electron into a proton
(b) a proton into a neutron
(c) a neutron into a proton
(d) a positron into an electron ____________

New Terms

Write the definitions of the following terms, which were introduced in this section. If necessary, refer to the Glossary at the end of the text.

alpha particle

alpha radiation

neutron emission

nuclear equation

antimatter
beta particle
beta radiation
electron capture
electron-volt (eV)
gamma radiation
K-capture
positron
radioactive
radioactive decay
radioactive disintegration series
radioactivity
radionuclide
X ray

23.4 The Band of Stability

Review

Nuclear instability is prevalent among nuclides having odd numbers for either the mass number or the atomic number, and particularly when both are odd and when both make the isotope lie outside the band of stability. When one or both numbers is a magic number (2, 8, 20, 28, 50, 82 or 126), the nuclide is more stable than those nearby in the band of stability. Among the elements below atomic number 83, radionuclides with too high a neutron/proton ratio tend to be beta emitters. Those with too low a value of this ratio tend to emit positrons. Radionuclides with atomic numbers above 83 are most often alpha emitters.

Self-Test

13. The most stable isotope of the following four isotopes (where we use *hypothetical* atomic symbols) is

    (a) ${}^{15}_{8}X$ (b) ${}^{131}_{53}Y$ (c) ${}^{16}_{8}Z$ (d) ${}^{32}_{15}A$ __________

14. At which atomic number is the nuclide most likely to be both stable and have a neutron to proton ratio very nearly equal to 1?

    (a) 10 (b) 40 (c) 80 (d) 106 __________

15. An isotope of atomic number 65 and mass number 140

    (a) lies below the band of stability.
    (b) lies within the band of stability.
    (c) lies above the band of stability.
    (d) has one of the magic numbers __________

16. If a radionuclide lies above and outside the band of stability, then the ejection of what particle will move it closer to this band?

(a) beta particle
(b) gamma ray photon
(c) positron
(d) a photon of gamma emission ________

New Terms

Write the definitions of the following terms, which were introduced in this section. If necessary, refer to the Glossary at the end of the text.

band of stability

odd-even rule

magic numbers

23.5 Transmutation

Review

When transmutation is caused by the bombardment of nuclei with high energy particles (e.g., ${}^{4}_{2}He$, ${}^{1}_{1}p$, or ${}^{2}_{1}d$, generally a compound nucleus first forms. It then sheds its excess energy by emitting a different particle or gamma radiation. Exactly what mode of decay is taken by the compound nucleus depends only on the energy it acquired by the initial bombardment and particle capture, not on the particle captured. Hundreds of isotopes, nearly all of them radioactive, and including all of the transuranium elements have been made this way.

Self-Test

17. To make a compound nucleus of ${}^{27}_{13}Al$ from each of the following bombarding particles, what must be the target isotope? Give its symbol.

    (a) proton ________

    (b) deuteron ________

    (c) alpha particle ________

18. What particle or photon must the compound nucleus, ${}^{27}_{13}Al^*$, eject to change into each nuclide? Give the name and symbol.

    (a) ${}^{23}_{11}Na$ ________

    (b) ${}^{26}_{12}Mg$ ________

    (c) ${}^{27}_{13}Al$ ________

19. What is the general name for elements 93-103?

    ____

    For elements 93-109? ____________________

New Terms

Write the definitions of the following terms, which were introduced in this section. If necessary, refer to the Glossary at the end of the text.

compound nucleus

transmutation

transuranium elements

23.6 Detecting and Measuring Radiation

Review

The abilities of various radiations to create ions (and radicals) in their wakes or to make phosphors scintillate account both for the hazards of radiations and their ease of detection. You should be able to describe in general terms how the Geiger counter, the cloud chamber, a scintillation counter, and a dosimeter work.

In learning the units for various measurements discussed in this section, notice that the *becquerel* (Bq) is the SI version of the *curie* (Ci), and that both describe the activity of a radioactive source, not the energy of its radiations. The becquerel and the curie are thus extensive quantities—they depend on both the mass of the source and the half-life or half-lives of whatever radionuclides are present.

The *gray* (Gy) is the SI version of the *rad*, and both refer to the energy absorbed by a quantity of matter because of the radiation it receives, not to the activity of the source and not even solely to the actual energy associated with the radiation. Thus the rad and the gray are also extensive quantities. They depend on the duration of the exposure as well as the energy of the radiation itself. (This energy—for example, in electron volts—is an intensive property. Each radionuclide's radiations have specific values of energy.)

The *rem* is always some fraction (it can be a very large fraction) of a rad (or a gray). The exact fraction depends on the kind of radiation, because the damage caused in a tissue varies with this factor even when different radiations have identical energies and duration of exposure. The rem is the best unit for describing potential harm to humans, because it is concerned with the effects on tissue.

Because of several radionuclides in the earth's crust as well as cosmic radiation, we are bathed constantly in a low level of *background radiation* averaging close to 360 mrem per year for each person in the United States—more depending on an individual's use of medical radiations. In all applications, workers can protect themselves to a considerable extent by using dense shielding materials (e.g., lead) and by getting at some distance from the source. For every doubling of the distance, the radiation intensity drops by a factor of four (*inverse square law*).

Self-Test

20. What is the name of a radiation detector that
    (a) uses photographic film or plates? ____________
    (b) employs supercooled vapors? ____________
    (c) contains a phosphor? ____________
    (d) lets radiation generate ions in a gas at low pressure? ____________
21. One rd = ____________ J/g
22. One Gy = 1____________ (supply units)
23. One Bq = ____________
24. One Ci = ____________ Bq
25. 1 Gy = ____________ rd
26. How are even very low doses in rems dangerous to humans?
    ____
27. What is the relationship of the rem to the rad, in general terms?
    ____
28. "Rad" stands for ____________
29. "Rem" stands for ____________
30. The exposure we all experience per year to background radiation is about

    (a) 3 rem (b) 36 rem (c) 360 mrem (d) 1600 μCi ____________

31. If the intensity of radiation is 100 units at a distance of 1.50 m from a source, how far away must one move to reduce the exposure by 1.00 unit?

(a) 0.0015 m (b) 15 m (c) 12.2 m (d) 1500 m ________________

New Terms

Write the definitions of the following terms, which were introduced in this section. If necessary, refer to the Glossary at the end of the text.

background radiation
becquerel (Bq)
curie (Ci)
gray (Gy)
inverse square law
ionizing radiations
rad (rd)
rem

23.7 Applications of Radioactivity

Review

In nearly all applications, gamma emitters are the best kinds of radionuclides. This radiation is the most penetrating of all, and therefore it is the easiest to detect, and it lets the scientist use very small quantities of the radionuclide. All of its radiation serves the purpose because little if any is blocked.

In *tracer analysis,* the ability of a body fluid to enter a particular tissue can be traced if the fluid contains a small concentration of a radionuclide (e.g., ${}^{99m}_{43}Tc$ as TcO_4^-).

When a sample is bombarded by neutrons (in *neutron activation analysis*), its various nuclei capture neutrons and become compound nuclei that then emit gamma radiation. Which frequency of gamma radiation comes out is determined by the kinds of atoms in the sample, and the intensities at these frequencies give measures of the concentrations of the atoms.

For *radiological dating,* pairs of radionuclides have to be identified and their relative concentrations in the sample measured. For dating very ancient rock formations, members of the pair might belong to the same radioactive disintegration series—for example, uranium-238 and lead-206—with the lighter one assumed to be produced solely by the decay of the heavier one at a rate of decay that has held constant over the millenia. For dating organic remains, the relative concentrations of carbon-14 and carbon-12 are used. As long as the living thing (e.g., a tree) lives, its level of carbon-14 is presumed to be constant. Once it dies, it no longer takes in carbon-14 and now the decay of this radionuclide at a known rate means that the age of any object made from the living thing (e.g., a wooden article) can be measured.

Self-Test

32. What method involving the use of radionuclides would be used in each situation?

    (a) Determine the existence and concentration of lead as an impurity in the fingernails of children who have eaten chips of lead-based paints.

    ____

    (b) Measure the age of the Laurentian shield of bedrock in southern Ontario province in Canada.

    ____

    (c) Measure how well blood circulates through a region of the lower leg suspected of having an early stage of gangrene with the hope of doing no more serious an amputation than absolutely necessary.

    ____

33. How does carbon-14 originate in the upper atmosphere? Write a nuclear equation.

    ____

34. In what chemical form is carbon-14 taken in by plants? Write a chemical formula. ______________________

35. For the carbon-14 method to work without any correction factors, what would have to be true about the ratio of carbon-14 to carbon-12 in all living things both today and in the past?

    ____

36. When carbon-14 dating methods were used on a sample of wood taken from a doorpost of an ancient archaeological site, it was found to have a specific activity of 382 Bq/g. How old was this wood sample according to calculations uncorrected for the factors that are known to cause some errors in the method? (Calculate to two significant figures.)

    ____

New Terms

Write the definitions of the following terms, which were introduced in this section. If necessary, refer to the Glossary at the end of the text.

neutron activation analysis
tracer analysis
radiological dating

23.8 Nuclear Fusion

Review

Nuclear fusion by the generation of bare nuclei and causing them to fuse requires that electrons be stripped from atoms and that the nuclei be forced close enough together and long enough for the strong force to overcome the electrostatic force. To minimize the electrostatic force, only the nuclei of hydrogen isotopes—particularly deuterium and tritium—are being studied as fusion fuel, because they bear only one unit of positive charge.

To create a plasma in which nuclei can fuse initially costs a large amount of energy, but fusion releases more than this cost. A net production of useful fusion energy will occur only when the product of the particle density of the plasma and the confinement time equals or exceeds the Lawson number, believed to be 3×10^{14} s/cm^3.

Inertial confinement is one approach to the containment of the plasma. The fuel—deuterium and tritium—is enclosed in a small hollow pellet which is bombarded by a laser. As the pellet explodes outward, there is an oppositely directed implosion that both heats the fuel and compresses it to achieve the necessary plasma density. The reaction is

$$^{3}_{1}\text{H} + ^{2}_{1}\text{H} \rightarrow ^{4}_{2}\text{He} + ^{1}_{0}\text{n}$$

The deuterium for this is made from water where D is 0.015% of all H atoms. The neutrons produced by this fusion deliver energy to and react with a surrounding lithium blanket to generate tritium.

$$^{6}_{3}\text{Li} + ^{1}_{0}\text{n} \rightarrow ^{4}_{2}\text{He} + ^{3}_{1}\text{H}$$

Magnetic confinement is the other approach to the containment of the plasma. The plasma is held within the confines of a magnetic field shaped for this purpose—a magnetic bottle.

Self-Test

37. What is the central scientific problem to the achievement of useful energy by fusion?

    ____

38. Since the strong force can hold nucleons together, what keeps it from operating at any plasma density once plasma has been formed?

    ____

39. What two variables are involved in the Lawson number?

    ____

40. What does the Lawson number tell us? ____________________

    ____

41. What are two reasons why isotopes of hydrogen have been picked over other isotopes for research on controlled fusion?

    ____

    ____

42. What two approaches are being studied to the confinement of plasma?

    ____

43. How is an X ray laser involved in inertial confinement fusion?

    ____

44. What is a tokamak? ______________________________

New Term

Write the definition of the following term, which was introduced in this section. If necessary, refer to the Glossary at the end of the text.

fusion

23.9 Nuclear Fission

Review

The capture of a neutron by a U-235 nucleus produces an unstable, compound nucleus that breaks apart—undergoes *fission*. The products are isotopes of intermediate atomic number, neutrons, and energy. If the neutrons can be slowed enough by moderators (e.g., graphic or heavy water or ordinary water), some may be captured by unchanged U-235 nuclei and so launch a chain reaction. (Plutonium-239 is also a fissile isotope.) The difference in binding energy between U-235 and the product isotopes is released largely as heat that can change water to steam and so drive electrical turbines.

To operate a reactor safely, control rods can be used to capture enough neutrons to make the multiplication factor equal 1. Now the reactor is said to be critical, because exactly as many neutrons are left at the end of the fission cycle as started the cycle.

Since the concentration of fissile isotope in the fuel elements of a nuclear reactor is small, a critical mass of the isotope is not possible and an atomic bomb explosion cannot occur. Should the coolant be lost, then the reactor could melt through its containment vessel. And heat could cause a steam explosion that would rupture the vessel (as occurred at Chernobyl, Russia).

Radioactive wastes include gases, liquids, and solids. Iodine-131, cesium-137, and strontium-90 are particularly problems when they get into the environment because the blood can carry them throughout the body where their radiations cause harm. Long-lived solid wastes must be kept out of human touch for centuries.

Self-Test

45. Why can nuclei capture neutrons much more easily than protons?

____

46. What is nuclear fission? __________________________

____

47. The isotopes initially formed from nuclear fusion have neutron-to-proton ratios that are too high or too low? ____________________

How do they adjust these ratios? ____________________

____

48. Which is generally higher, the sum of the binding energies of the nuclei produced by fission or the binding energy of the U-235 nucleus?

____

49. What makes the fission of U-235 self-sustaining? ________________

____

50. What is meant by "pressurized" in the pressurized water reactor?

____

51. What is the function of each of the following in a pressurized water nuclear reactor?

(a) The moderator.

____

(b) The cladding.

____

(c) The primary coolant loop.

____

(d) The secondary coolant loop.

____

(e) The control rods.

____

52. Why do each of the following radionuclide wastes pose human health problems?

(a) I-131

____

(b) Cs-137

____

(c) Sr-90

____

New Terms

Write the definitions of the following terms, which were introduced in this section. If necessary, refer to the Glossary at the end of the text.

fissile isotope

fission

nuclear chain reaction

23.10 The Breeder Reactor

Review

Under the right conditions, U-238 atoms can be changed to Pu-239 atoms, which are fissile. This change is called "breeding" because a non-fissile but fertile isotope (U-238) is changed to a fissile isotope. Since the world's supply of U-238 is very large compared to U-235, the prospects of successful breeder reactors opens up huge new supplies of atomic energy. It also increases the supply of Pu-239 which can be diverted to making atomic bombs.

Self-Test

53. What isotope is the source of Pu-239 in breeding? ________________

54. Why would breeding increase the world's supply of nuclear fuel?

____

____

New Term

Write the definition of the following term, which was introduced in this section. If necessary, refer to the Glossary at the end of the text.

breeder reactor

Answers to Self-Test Questions

1. 14.4 ng
2. Because it is the energy *lost* from the system when some mass changed to energy as the nucleons formed into a nucleus.
3. Some mass of the nucleons changed to energy when the nucleus formed.
4. High stability
5. (a) strong force (b) electrostatic force (c) electrostatic force (d) strong force (d) strong force
6. neutrons
7. (a) ${}^{279}_{111}X \rightarrow {}^{0}_{-1}e + {}^{279}_{112}Z + {}^{0}_{0}\gamma$
   (b) ${}^{279}_{111}X \rightarrow {}^{4}_{2}He + {}^{275}_{109}Z + {}^{0}_{0}\gamma$
   (c) ${}^{279}_{111}X \rightarrow {}^{0}_{1}e + {}^{279}_{110}Z$
   (d) ${}^{279}_{111}X \rightarrow {}^{1}_{0}n + {}^{278}_{111}Z$
   (e) ${}^{279}_{111}X + {}^{0}_{-1}e \rightarrow {}^{279}_{110}Z$ + X rays

8. (a) beta particle (b) X ray photon (c) beta particle (d) alpha particle (e) positron (f) neutron (g) positron (h) alpha particle (i) neutron (or gamma ray photon, only)
9. b
10 c
11. a
12. b
13. c
14. a
15. a
16. a
17. (a) $^{26}_{12}Mg$ (b) $^{25}_{12}Mg$ (c) $^{23}_{11}Na$
18. (a) alpha particle (b) proton $^{1}_{1}p$ (c) gamma ray photon $^{0}_{0}\gamma$
19. actinide elements; transuranium elements
20. (a) dosimeter (b) cloud chamber (c) scintillation counter (d) Geiger counter
21. 10^{-5}
22. J/kg
23. 1 disintegration/s
24. 3.7×10^{10}
25. 100
26. They generate unstable ions and radicals that initiate other chemical changes of danger to the individual.
27. The rem is a fraction of a rad, the fraction depending on how damaging a particular rad dose is in tissue.
28. radiation absorbed dose
29. radiation equivalent for man
30. c
31. b
32. (a) neutron activation analysis (b) radiological dating (c) tracer analysis
33. $^{1}_{0}n + ^{14}_{7}N \rightarrow ^{15}_{7}N^{*} \rightarrow ^{14}_{6}C + ^{1}_{1}p$
34. CO_2
35. a constant ratio of C-14 to C-12
36. 7.2×10^3 years
37. To get the fusing nuclei close enough for a long enough time so that the strong force can overcome the electrostatic force.
38. The electrostatic force keeps the nuclei too far apart to allow the strong force (which acts over a shorter distance) to work.
39. Particle density and confinement time.

40. A long confinement time would require a correspondingly greater plasma density, or a less particle density would require a long confinement time for the plasma.
41. Their nuclei bear only single plus charges, and their fusion gives the highest gain in binding energy per nucleon.
42. Inertial confinement and magnetic confinement.
43. Its energy causes the explosion of the fuel pellet and provides some heat, and the implosion of the fuel raises the plasma density.
44. A device for creating a magnetic "bottle" for the confinement of plasma under high temperature and plasma density.
45. Neutrons are electrically neutral and so are not repelled by nuclei as they approach.
46. The spontaneous breaking of an unstable nucleus roughly in half.
47. Too high. They eject neutrons.
48. The sum of the binding energies of the products.
49. It produces more neutrons that needed to cause further fission events.
50. The water in the primary coolant loop is under such high pressure that even at high temperatures it is in the liquid state.
51. (a) Convert high energy neutrons into slower (thermal)neutrons
    (b) Hold both the fuel in place and retain radioactive wastes.
    (c) Remove heat from the cladding elements as it is produced by fission.
    (d) Remove heat from the primary coolant loop and let this heat generate steam under pressure to drive the turbines.
    (e) Manage the flux of neutrons in the core so that the reactor will be critical during operation and go subcritical at shutdown.
52. (a) Concentrates in the thyroid gland where its radiation could cause thyroid cancer or other loss of thyroid function.
    (b) Transported by the blood wherever sodium goes.
    (c) Is attracted to bone tissue.
53. U-238
54. Nearly all of natural uranium now goes to waste, but breeding would convert all of it to nuclear fuel.

Tools you have learned

Remove this chart from the Study Guide and keep it handy when tackling homework problems.

Tool	Function
Einstein equation $\Delta E = \Delta m_0 c^2$	To calculate the amount of energy associated with a change in mass accompanying a nuclear reaction.
Nuclear equations	To describe nuclear reactions in a way that accounts for changes in mass numbers and atomic numbers
Inverse square law $\frac{I_1}{I_2} = \frac{d_2^2}{d_1^2}$	To calculate the intensity of radiation (I_2) at a distance (d_2) from its intensity (I_1) at a another distance (d_1).
Carbon-14 dating $t = (8.26 \times 10^3 \text{ yr}) \ln \frac{0.227}{s}$	To calculate the age, t, of a biological remain from its specific activity, s.

Summary of Important Equations

Einstein equation

$$\Delta E = \Delta m_0 c^2$$

Inverse square law (or radiation protection)

$$\frac{I_1}{I_2} = \frac{d_2^2}{d_1^2}$$

Carbon-14 dating

$$t = (8.26 \times 10^3 \text{ yr}) \ln \frac{0.227}{s}$$

Chapter **24**

ORGANIC COMPOUNDS AND POLYMERS

There are probably ten times as many organic compounds as any other kind, so this chapter can only serve as an introduction to a large field of chemistry. Yet the study of organic compounds assumes many of the features of the study of any other field. We learn in this chapter about classifying organic compounds into families defined by functional groups. We see how such groups confer common chemical properties to all members of the same family, but that the nonfunctional, hydrocarbon groups contribute much to physical properties. In our study of polymers, we learn that enormous molecular size alone affects properties in useful ways.

Learning Objectives

Throughout your study of this chapter, keep in mind the following objectives:

1 To learn about the major structural features of organic molecules and the importance of functional groups to the study of organic chemistry.

2 To learn what structural features characterize hydrocarbons—alkanes, alkenes, alkynes, and aromatic—and how the physical properties of substances whose molecules are dominated by the same features are hydrocarbon-like.

3 To learn the principles of formal (IUPAC) nomenclature—characteristic family name endings, rules for identifying "parents," rules for numbering parent chains or rings, and the names of hydrocarbon groups (alkyl groups).

4 To learn to predict physical states and general physical properties from structure, properties such as solubilities in water or nonpolar solvents.

5 To learn to "read" a molecular structure to tell if a substance is likely to undergo addition, substitution, oxidation, reduction (hydrogenation), or hydrolysis reactions.

6 To learn *how* certain reactions occur:

(a) how the carbon-carbon double bond undergoes an addition reaction;

(b) how an alcohol undergoes dehydration.

7 To learn to predict the products that form when

(a) 1° and 2° alcohols are oxidized

(b) alcohols undergo dehydration

(c) alcohols and carboxylic acids form esters

(d) ammonia or amines and carboxylic acids form amides

(e) amines neutralize strong acids

(f) aldehydes and ketones are hydrogenated

(g) esters or amides are hydrolyzed or saponified

(h) the benzene ring undergoes halogenation, nitration or sulfonation.

8 To learn the fundamental features of polymer structure and how molecular size influences properties.

9 To learn how to write the structure of an addition polymer from the structure of the monomer.

24.1 The Nature of Organic Chemistry

Review

The idea of the functional group is the main idea in this Section. Organic chemistry is organized around families defined by such groups, and we introduce many of the most important families here. We note that some groups bear similarities in structure and therefore in chemistry to simple inorganic species. Amines are ammonia-like in some of their reactions, for example. So, like ammonia, the amines react with acids. Substances whose molecules have oxygen or nitrogen atoms in them tend to be more polar and so more soluble in water than others of the same size

Besides functional groups, organic molecules contain hydrocarbon-like, nonfunctional groups. Be sure to understand the value of using one symbol, R, for all such groups, no matter how many carbons are present. The fact that they do not function, that they do not chemically change in most (if not all) of the reactions of a family, is what makes such a simplifying symbol possible. Alkyl groups, R, have few chemical reactions because they are essentially nonpolar and therefore unattractive to ionic or polar reactants.

Self-Test

1. Circle what is most likely the functional group in

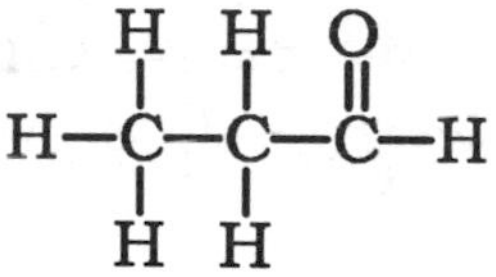

$$H-\overset{\displaystyle H}{\underset{\displaystyle H}{\overset{|}{\underset{|}{C}}}}-\overset{\displaystyle H}{\underset{\displaystyle H}{\overset{|}{\underset{|}{C}}}}-\overset{\displaystyle O}{\overset{\|}{C}}-H$$

2. Which molecule would be more polar, $CH_3—CH_3$ or $CH_3—NH_2$?

____

3. What forms in the following reaction?

   $CH_3—NH_2 + H—Br \rightarrow$ _______________

4. What forms in the following reaction?

   $R—NH_2 + H—Cl \rightarrow$ _______________

5. What significance does the functional group have for the study of organic chemistry? _______________

____

____

6. Study the members of the following pairs of compounds and decide if they represent *isomers*, or are *identical*, or *neither*.

   (a) $CH_3OCH_2CH_3$ and $CH_3CH_2CH_2OH$ _______________

   (b) 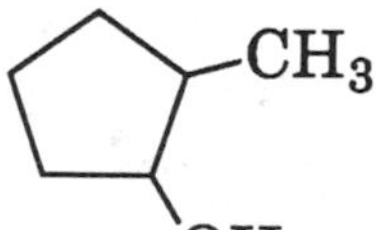 [cyclopentane ring with CH_3 and OH on adjacent carbons] and [six-membered ring containing O with CH_3 substituent] _______________

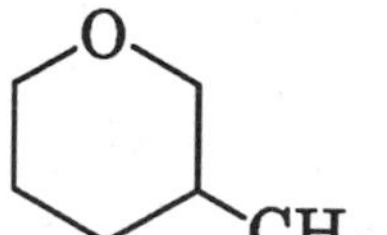

   (c) $CH_3CH_2CH_2OH$ and $HOCH_2CH_2CH_3$ _______________

7. Which compound in question 7 is a heterocyclic compound?

____

New Terms

Write the definitions of the following terms, which were introduced in this section. If necessary, refer to the Glossary at the end of the text.

branched chain
functional group
heterocyclic compound
organic chemistry
ring compound
straight chain compound

24.2 Hydrocarbons

Review

When the molecules of a compound have no double or triple bonds, it is a *saturated compound.* Otherwise, it is *unsaturated.* The molecules of *hydrocarbons* are made solely of C and H. The *saturated hydrocarbons*, have only single bonds and are called *alkanes* (or *cycloalkanes*). Whether *straight-chain*, *branched-chain*, or *ring*, the saturated hydrocarbons have very few chemical properties. All hydrocarbons are hydrophobic; they do not dissolve in water.

IUPAC Rules. Here's how to sort out the IUPAC rules for all families. Regardless of the family, the IUPAC rules all begin with the idea of a "parent" unit—a molecular portion which is named and whose carbon skeleton is numbered. Each family has its own rule for identifying the parent. For the alkanes, the "parent" is the longest continuous sequence of carbons. This parent is always, by itself, a straight-chain alkane. You should learn the names of these alkanes through C-10 (Table 24.2 in the text).

The ending of the name of a compound is the same for all members of a given family; each family has its own characteristic name ending. Thus "-ane" is the name ending for all alkanes and cycloalkanes.

To number the carbon atoms in a parent, there is a rule for each family for deciding which position is numbered 1. For the alkanes, the rule is to number from whichever end of the parent chain is nearest the first branch.

Hydrocarbon groups, called *alkyl groups,* have their own names, and you should learn the names and structures for those having one up to three carbons—the methyl, ethyl, propyl, and isopropyl groups.

The locations of the alkyl groups on the parent chain are identified by the numbers assigned to the carbons of the parent.

When you assemble all the parts of a name into one whole, remember that each alkyl group must be associated with a number; that multiplier prefixes, like di- and tri-, sometimes have to be added to the names of alkyl groups; that two numbers in a name are always separated by a comma; and that a hyphen is always used to separate a number from a word-part of a name. These are well illustrated in Example 24.1. The learning of the IUPAC rules, however, comes with your own practice.

The parent of an alkene must be the longest chain *that includes the double bond.* The parent is numbered from whichever end gets to the double bond first, regardless of the location of alkyl branches.

Many alkenes exhibit *geometric isomerism* (cis-trans isomerism) and have molecules that differ only in geometry at the double bond. Notice that this isomerism in alkenes is possible only when neither end of the double bond holds identical groups.

Alkene Chemistry. The carbon-carbon double bond has pi electrons, and this makes the double bond somewhat electron-rich and attractive to protons and other electron-poor species. So alkenes undergo *addition reactions* in which the reactant (or a catalyst) is able to donate H^+ to the double bond. Thus alkenes react with hydrogen chloride and with water, provided there is an acid-catalyst. When gaseous molecules of HCl add, H^+ from HCl goes to one end of the double bond using the pi electrons to make a new C—H bond. This leaves the other end, now electron poor, to accept the chloride ion. Overall, we have

$$>C=C< + H-Cl \longrightarrow H-\overset{|}{\underset{|}{C}}-\overset{|}{\underset{|}{C}}-Cl$$

Compare this addition with the addition of water and note that again one end of the double bond gets an H and the other end gets the rest of the adding molecule, OH in this case. Now the product is an alcohol.

$$>C=C< + H-O-H \xrightarrow[\text{catalyst}]{\text{acid}} H-\overset{|}{\underset{|}{C}}-\overset{|}{\underset{|}{C}}-O-H$$

Two halogens, Cl_2 and Br_2, also add to carbon-carbon double bonds. One Cl or Br goes to one end and the other Cl or Br goes to the other end of the double bond. H_2 adds similarly, but special catalysts are needed.

As a study goal, practice writing the structures of the products of the following alkenes with hydrogen chloride, water (in the presence of an acid catalyst), Cl_2, Br_2, and H_2 (assuming the special conditions).

$$H_2C=CH_2 \qquad CH_3CH=CHCH_3$$

Aromatic Hydrocarbons. Despite considerable formal unsaturation, *aromatic hydrocarbons* do not give addition reactions because the benzene ring strongly resists anything that breaks up its unique pi electron network. Instead, *substitution reactions* occur, which leave the benzene ring system intact. Practice writing the structures of what forms when benzene reacts with concentrated sulfuric acid, nitric acid, and the two halogens, Cl_2 and Br_2, when an iron(III) halide catalyst is present.

Self-Test

8. Write the IUPAC names of the following.

(a) $CH_3CH_2CH_2CH_2CH_3$

____

(b) 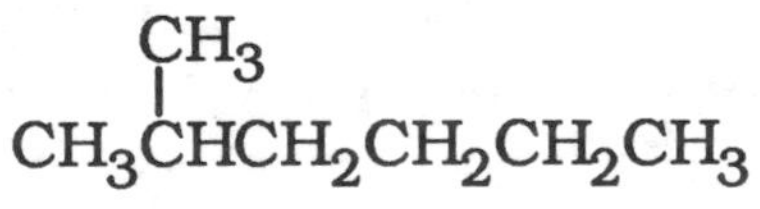
```
      CH3
      |
CH3CHCH2CH2CH2CH3
```

____

(c) 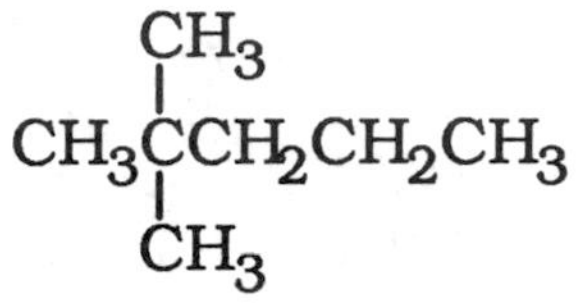
```
      CH3
      |
CH3CCH2CH2CH3
      |
      CH3
```

____

(d) 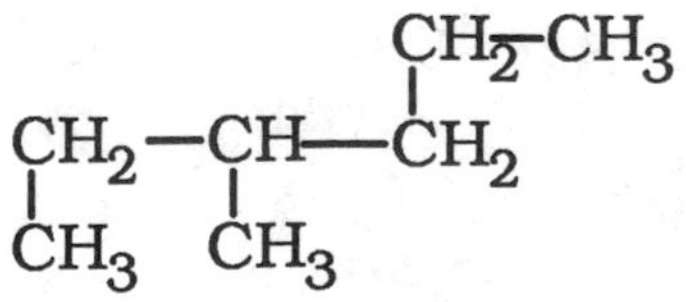
```
                CH2-CH3
                |
CH2-CH-----CH2
|       |
CH3  CH3
```

____

(e)

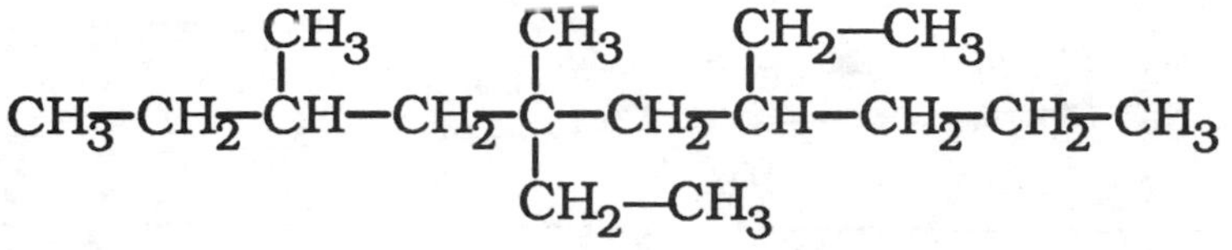

$$\begin{array}{c} CH_3 CH_3 CH_2-CH_3 \\ CH_3-CH_2-\overset{|}{C}H-CH_2-\overset{|}{\underset{|}{C}}-CH_2-\overset{|}{C}H-CH_2-CH_2-CH_3 \\ CH_2-CH_3 \end{array}$$

____

9. Write the structure of 2,3,3-trimethyl-4-ethylheptane.

10. Write the IUPAC names and structures of the isomers of C_5H_{12}.

11. Write the IUPAC name of $CH_3-CH_2-\overset{\overset{CH_2}{\|}}{C}-\underset{\underset{CH_3}{|}}{C}H-CH_2-CH_3$

____

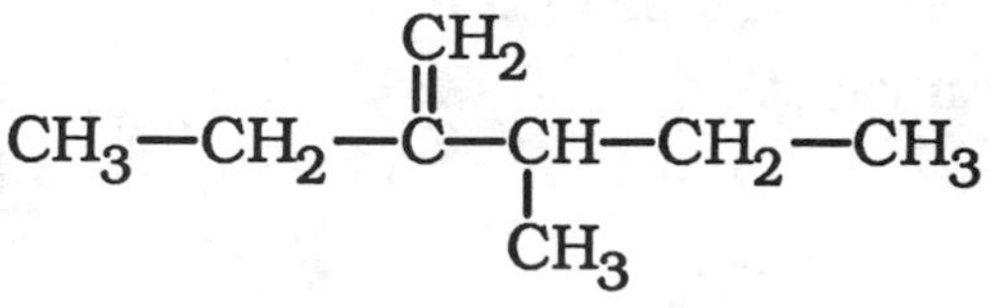

12. Write the structures of the *cis* and *trans* isomers of 4-methyl-2-pentene.

_______________ _______________

13. Write the condensed structural formulas for the products of the following reactions.
    (a) The addition of hydrogen to propene _______________
    (b) The addition of water to 3-hexene _______________
    (c) The addition of chlorine to 1-pentene _______________
    (d) The reaction of bromine in the presence of $FeBr_3$ with benzene.
    ____

14. Compare and contrast the chemical behavior toward hot concentrated sulfuric acid of cyclohexane, cyclohexene, and benzene.
    (a) Cyclohexane _______________
    (b) Cyclohexene _______________
    (c) Benzene _______________

New Terms

Write the definitions of the following terms, which were introduced in this section. If necessary, refer to the Glossary at the end of the text.

addition reaction	geometric isomers
alkanes	hydrocarbons
alkenes	saturated organic compounds
alkyl groups	substitution reactions

alkynes
geometric isomerism
unsaturated organic compounds

24.3 Alcohols and Ethers, Organic Derivatives of Water

Review

To be an alcohol, the molecule must have the O—H group (or it can be written H—O) attached to a *saturated* carbon atom, one with four *single* bonds. (Any molecule, of course, can have two or more such groups. In fact most organic compounds have more than one functional group, either alike or different.)

A molecule has the ether group if it carries O attached to two hydrocarbon groups, like alkyl groups, cycloalkyl groups, or benzene rings. The ether group has few chemical reactions, none in our study.

When you see an alcohol group, you think "This is a water-like group that confers on the molecule the following properties."

1. The OH group helps to make the substance more soluble in water and have a higher boiling point than the corresponding hydrocarbon.
2. The molecules of the substance can be oxidized—made to lose hydrogen—provided that the carbon holding the OH group also holds H.

   (a) Primary alcohols are oxidized to aldehydes and thence to carboxylic acids.

   (b) Secondary alcohols are oxidized to ketones.

   (c) Tertiary alcohols are not oxidized by simple loss of H_2.
3. The molecules of the alcohol can be made to eliminate H—OH and so acquire carbon-carbon double bonds.
4. The molecules of the alcohol will undergo the substitution (replacement) of OH by Cl, Br, or I by the action of the corresponding concentrated HX solution (where X = Cl, Br, or I).

   For the sake of completeness, anticipating the next section, add a fifth property.
5. Alcohols react with carboxylic acids to give esters.

Self-Test

15. Write the structure of 2,3-dimethyl-1-butanol.

16. Write the IUPAC name of the following compound.

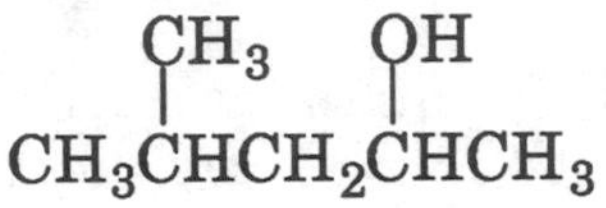

$$\begin{array}{c} \quad CH_3 \quad OH \\ CH_3\overset{|}{C}HCH_2\overset{|}{C}HCH_3 \end{array}$$ ____________________________

17. Write the structures of the products that could be made by the oxidation of each compound. If no oxidation can occur, state so.

 (a) 2-propanol

 ____

 (b) 1-propanol

 ____

 (c) 2-methyl-2-butanol

 ____

18. Examine the following structures and then answer the following questions.

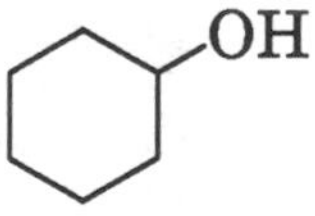

OH

A

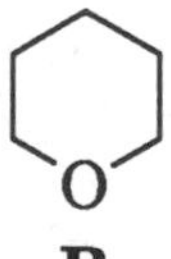

O

B

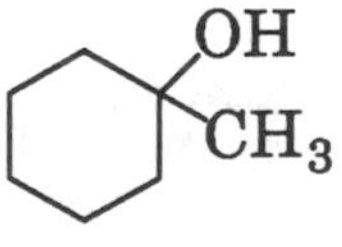

OH
CH_3

C

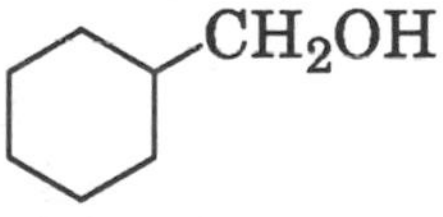

CH_2OH

D

 (a) Which compound(s) cannot be oxidized? ______

 (b) Which compound(s) cannot be dehydrated? ______

 (c) Which compound(s) can be oxidized to a ketone? ______

 (d) Which compound(s) can be oxidized to an aldehyde?______

(e) Which compound(s) can be oxidized to a carboxylic acid (with the same number of carbons)? ______

(f) Which compound(s) can be dehydrated (to alkenes)?______

19. Write the structure of the organic product in each situation.

(a) OH (cyclohexanol) $\xrightarrow[\text{heat}]{H_2SO_4}$ ______________________

(b) OH (cyclohexanol) + $Cr_2O_7^{2-}$ $\xrightarrow{H^+}$ ______________________

New Terms

Write the definitions of the following terms, which were introduced in this section. If necessary, refer to the Glossary at the end of the text.

alcohol	ether
elimination reaction	substitution reaction

24.4 Amines, Organic Derivatives of Ammonia

Review

Amines are alkyl derivatives of ammonia and so, like ammonia, are proton acceptors or Brønsted bases. Molecules of amines can also accept and donate hydrogen bonds. These are the essential characteristics of amines, but to anticipate a later section, amines can be converted to amides.

The protonated forms of amines, like the ammonium ion, are proton donors or Brønsted acids. They can neutralize strong base, like OH^-.

Self-Test

20. What forms when methylamine neutralizes hydrochloric acid? Write the structure.

21. If $CH_3CH_2NH_3^+Cl^-$ reacts with aqueous sodium hydroxide, what forms? Write the structure.

New Term

Write the definition of the following term, which was introduced in this section. If necessary, refer to the Glossary at the end of the text.

amine

24.5 Carbonyl Compounds

Review

The first task is to learn to recognize by name the functional groups that involve the carbonyl group when you see them in a complex structure. This is like being able to recognize a wiggley blue line on a map as representing a river, because functional groups are the "map signs" of organic structures. Then you learn what chemical and physical properties to associate with each "map sign."

When you learn functional groups, be sure not to tie their structures to particular positions on a page. The *sequences* of atoms are what count, not whether they are written left-to-right or top-to-bottom. For example, all of the following structures are *esters* because each has the "carbonyl-oxygen-carbon" sequence of atoms that defines the ester group..

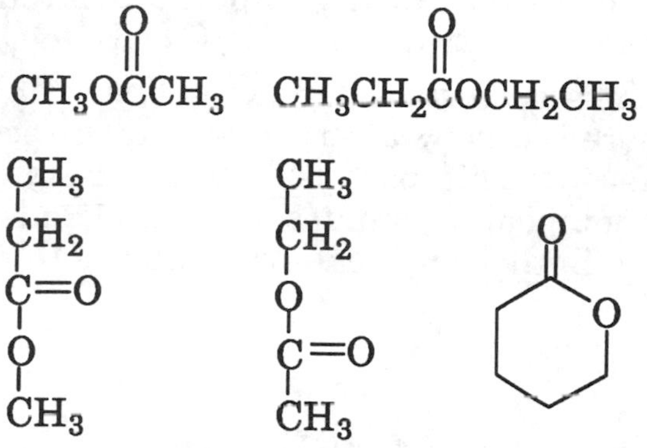

$$CH_3O\overset{O}{\overset{\|}{C}}CH_3 \qquad CH_3CH_2\overset{O}{\overset{\|}{C}}OCH_2CH_3$$

$$\begin{array}{c} CH_3 \\ | \\ CH_2 \\ | \\ C{=}O \\ | \\ O \\ | \\ CH_3 \end{array} \qquad \begin{array}{c} CH_3 \\ | \\ CH_2 \\ | \\ O \\ | \\ C{=}O \\ | \\ CH_3 \end{array}$$

Some important esters, e.g., those in vegetable oils or animal fats, have three ester groups per molecule.

It's important to get *aldehydes* and *ketones* straight, because they differ so much in ease of oxidation. In the *aldehyde group,*

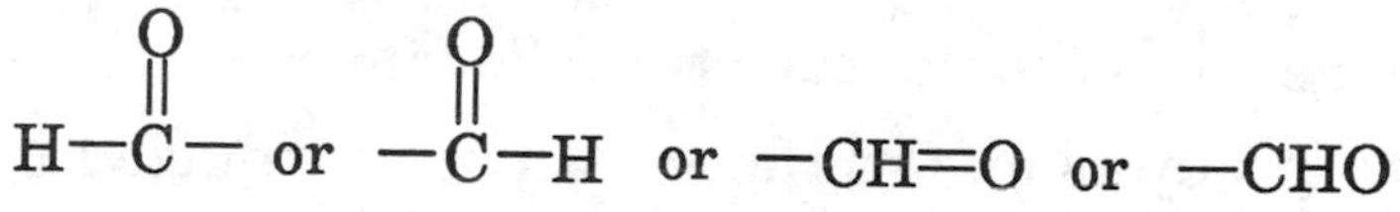

$$H{-}\overset{O}{\overset{\|}{C}}{-} \text{ or } {-}\overset{O}{\overset{\|}{C}}{-}H \text{ or } {-}CH{=}O \text{ or } {-}CHO$$

Representations of the aldehyde group

the carbonyl group holds at least one hydrogen atom. In addition it holds either another hydrogen atom or it is joined to a carbon atom. The carbon can be saturated or unsaturated. In ketones the carbonyl is flanked on both sides by carbon atoms.

Carboxylic acids all have the "carbonyl-oxygen-hydrogen" sequence.

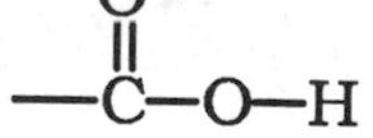

$${-}\overset{O}{\overset{\|}{C}}{-}O{-}H$$

Thus all of the following structures are of carboxylic acids,

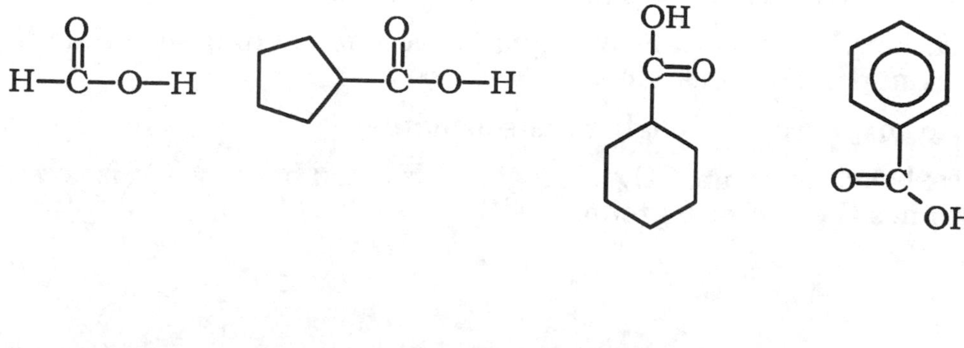

$$H{-}\overset{O}{\overset{\|}{C}}{-}O{-}H$$

You will often see the carboxylic acid group abbreviated as CO_2H or COOH, and sometimes it's written "backward" as HO_2C or HOOC.

Be sure to catch the structural difference between an *amine* and an *amide*. In the amide group, there is *always* a carbonyl group directly attached to the nitrogen atom. The difference is important because the properties are so different. Amines are basic compounds; amides are not. Amides are broken apart by water; amines are not. Both groups are polar and both can participate in hydrogen bonds.

$$\text{Amides:}\quad CH_3-\overset{\overset{\displaystyle O}{\|}}{C}-NH_2 \qquad CH_3-\overset{\overset{\displaystyle O}{\|}}{C}-NH-CH_3 \qquad CH_3-\overset{\overset{\displaystyle O}{\|}}{C}-\overset{\overset{\displaystyle CH_3}{|}}{N}-CH_2CH_3$$

$$\text{Amines:}\quad CH_3CH_2-NH_2 \qquad CH_3CH_2-NH-CH_3 \qquad CH_3CH_2-\overset{\overset{\displaystyle CH_3}{|}}{N}-CH_2CH_3$$

As for the chemical properties of carbonyl compounds, concentrate on the following characteristics.

Concerning *aldehydes*, you should learn the following.

1. The aldehyde group is one of the most easily oxidized of all functional groups, being changed by oxidation to the carboxyl group.
2. The aldehyde group adds hydrogen, catalytically, to give 1° alcohols.

♦♦♦

Concerning *ketones*, there are only two properties that we studied.

1. The keto group strongly resists oxidation.
2. The keto group adds hydrogen, catalytically, to give 2° alcohols.

♦♦♦

With respect to *carboxylic acids,* learn the following.

1. Carboxylic acids readily neutralize strong base, like OH^-, and form the corresponding carboxylate ions, RCO_2^-.
2. Carboxylic acids react with alcohols (acid-catalysis) to give esters.
3. The carboxyl group can be changed to the amide group by a reaction with either ammonia or an amine.
4. The carboxyl group strongly resists oxidation.
5. The carboxylate group, CO_2^-, is a good Brønsted base; when it accepts H^+, it becomes the carboxyl group, CO_2H, again.

♦♦♦

Esters undergo the following reactions.

1. The ester group is hydrolyzed (acid catalysis) to give the carboxylic acid and the alcohol from which the ester is made.
2. The ester group is saponified (action of aqueous alkali, like NaOH) to give the parent alcohol and the carboxylate ion of the parent carboxylic acid.

♦♦♦

Amides have the following properties.

1. The nitrogen of an amide is not a proton acceptor (not a Brønsted base), unlike the nitrogen of an amine (or ammonia).
2. Amides react with water to give the parent carboxylic acid and amine (or ammonia).

Self-Test

Questions 22-28 refer to the following structures. Some questions draw on knowledge from the preceding sections.

1. $CH_3CH_2\overset{O}{\overset{\|}{C}}{-}O{-}CH_3$
2. $HOCH_3$
3. $H{-}\overset{O}{\overset{\|}{C}}{-}CH_3$
4. O (cyclohexanone ring structure)
5. $HO{-}\overset{O}{\overset{\|}{C}}CH_2CH_3$
6. CH_3CH_2OH
7. $CH_3{-}\overset{O}{\overset{\|}{C}}{-}O{-}CH_2CH_3$
8. $CH_3\overset{O}{\overset{\|}{C}}{-}OH$
9. $HO{-}CH_2{-}\overset{O}{\overset{\|}{C}}{-}CH_3$
10. $CH_3{-}O{-}\overset{O}{\overset{\|}{C}}{-}CH_2{-}O{-}CH_3$
11. CH_3, O (benzene ring structure)

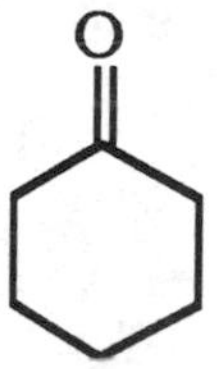

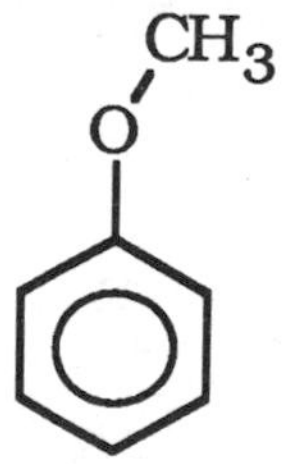

22. Which structures contain each of the following groups?

(a) ester group ____________________

(b) ether group ____________________

(c) alcohol group ______________

(d) carboxylic acid group ______________

(e) aldehyde group ______________

(f) ketone group ______________

23. Which compound is the most easily oxidized? ______________

24. Which compound(s) will rapidly neutralize sodium hydroxide at room temperature? ______________

25. Two of the compounds shown will react to give compound 1. Which are they? ______________

26. Compound 7 will react with water to give two of the compounds. Which two are they? ______________

27. Which compound has a benzene ring? ______________

28. Which compound will react with hydrogen to give compound 6? ______________

Questions 29-34 refer to the structures below. Some of the questions require knowledge of material given earlier.

1: $NH_2{-}CH_2{-}C(=O){-}CH_3$

2: $NH_2{-}CH_2CH_3$

3: $CH_3C(=O){-}NH_2$

4: NH_3

5: CH_3NH_2

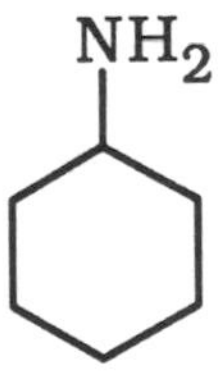

6: cyclohexane ring bearing NH_2

7: $CH_3C(=O){-}OH$

8: NH_2CH_3

9: CH_3NHCH_3

10: $CH_3NHC(=O)CH_3$

11: $CH_3NH_3^+$

29. The named groups given next are found in which structures?

(a) the amine group ______________

(b) an amino ketone ______________

(c) the amide group ______________

30. Which two structures are of the same compound? ______________

31. The hydrolysis of structure 10 gives which two compounds?

____

32. Which structures represent compounds that neutralize aqueous acids rapidly at room temperature?

____

33. Which structure has a carboxyl group? ______________

34. Which structure results when an amine neutralizes an acid?

____

24.6 Organic Polymers

Review

Polymers are *macromolecules* with repeating structural units provided by the *monomer* molecules in the *polymerization* reaction. The alkenes give *addition polymers*, like polyethylene and polypropylene, with extremely long molecules. The polyolefins and polyacrylates are addition polymers. Rubber is a polymer of a diene and an *elastomer*.

When two (or more) different monomers are polymerized, the product is a *copolymer*. The regularity of the structure of a copolymer can vary. The principal structural types of copolymers are alternate, block, random, graft and cross-linked.

Condensation polymers, like polyesters (from dicarboxylic acids and dialcohols) and polyamides (from dicarboxylic acids and diamines) are generally alternate.

Polymers have the chemical properties characteristic of their functional groups, but the reactions are usually much slower than those given by smaller members of the functional group families. Very long molecules with repeating sites capable of interchain attractions (like hydrogen bonds) lend themselves to make strong fibers.

Self-Test

35. Write the two structures of the polymers of each of the following monomers. One structure is to show three monomer units, and the second structure is to show the condensed representation.

(a) Ethylene

(b) Propylene

(c) 1-Chloroethene

36. If monomers A and B copolymerize, how might their structures be represented if they form

(a) an alternate copolymer

(b) a graft copolymer

37. Write the structure of the monomer from which the following addition polymer is made.

$$\left(CH_2 - \underset{CO_2H}{\overset{CH_3}{\underset{|}{\overset{|}{C}}}} \right)_n$$

____

38. What force of attraction is possible between molecules of nylon that partly accounts for the strength of nylon fibers?

____

New Terms

Write the definitions of the following terms, which were introduced in this section. If necessary, refer to the Glossary at the end of the text.

macromolecule
monomer
polymer
polymerization

Answers to Self-Test Questions

1. 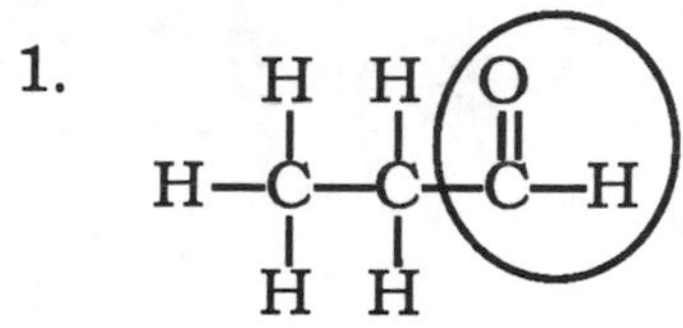

2. $CH_3—NH_2$
3. $CH_3—NH_3^+ + Br^-$
4. $R—NH_3^+ + Cl^-$
5. It greatly simplifies it. There are very few kinds of functional groups among the millions of organic compounds, and each kind displays mostly the same set of reactions.
6. (a) isomers, (b) isomers, (c) identical
7. The second compound of part (b).
8. (a) pentane, (b) 2-methylhexane, (c) 2,2-dimethylpentane, (d) 3-methylhexane, (e) 5,7-diethyl-3,5-dimethyldecane (or 3,5-dimethyl-5,7-diethyldecane)
9.

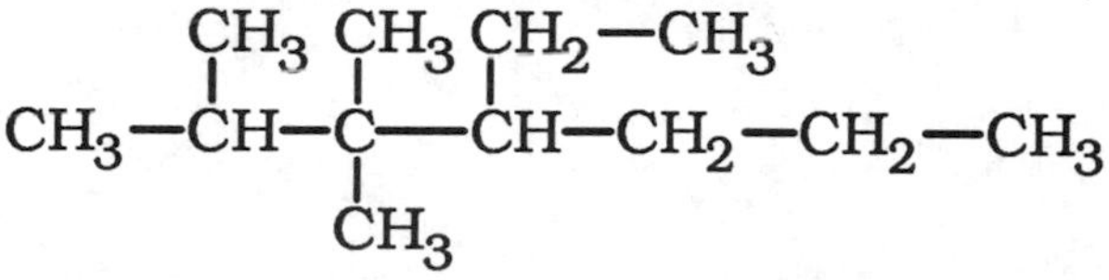

$$\begin{array}{l} \quad\quad\; CH_3\;\; CH_3\;\; CH_2-CH_3 \\ CH_3-CH-C-CH-CH_2-CH_2-CH_3 \\ \quad\quad\quad\quad\; CH_3 \end{array}$$

10. $CH_3CH_2CH_2CH_2CH_3$ pentane

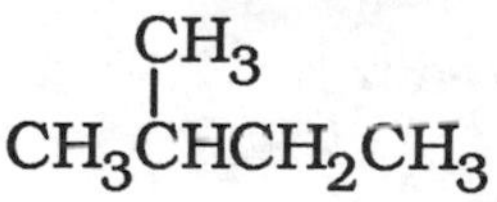

$$\begin{array}{l} \quad\;\; CH_3 \\ CH_3CHCH_2CH_3 \end{array}$$ 2-methylbutane

$$\begin{array}{l} \quad\;\; CH_3 \\ CH_3CCH_3 \\ \quad\;\; CH_3 \end{array}$$ 2,2-dimethylpropane

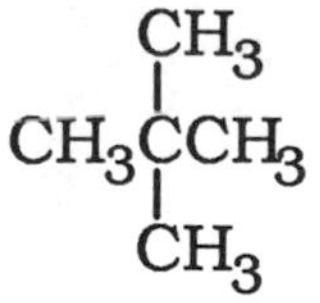

11. 2-ethyl-3-methyl-1-pentene (or 3-methyl-2-ethyl-1-pentene)
12.

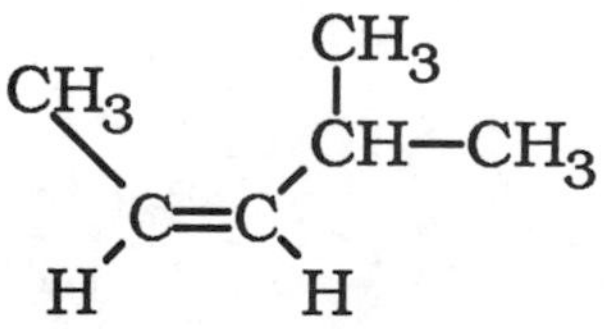

$$\begin{array}{l} \quad\quad\quad\quad CH_3 \\ CH_3 \quad\quad CH-CH_3 \\ \quad\quad C=C \\ H \quad\quad\quad H \end{array}$$

cis -isomer

$$\begin{array}{l} CH_3 \\ \quad\quad\quad\quad H \\ \quad\quad C=C \\ H \quad\quad\quad CH-CH_3 \\ \quad\quad\quad\quad\; CH_3 \end{array}$$

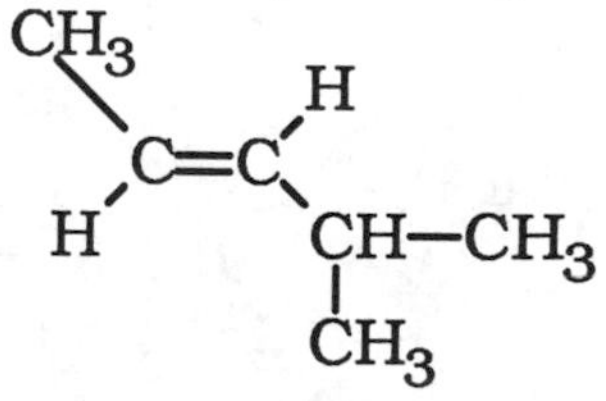

trans -isomer

13. (a) $CH_3CH_2CH_3$

(b) $CH_3CH_2CH_2\underset{\displaystyle OH}{\underset{|}{C}}HCH_2CH_3$

(c) $Cl{-}CH_2\underset{\displaystyle Cl}{\underset{|}{C}}HCH_2CH_2CH_3$

(d)  $C_6H_5{-}Br$ (+ HBr)

14. (a) No reaction

(b) 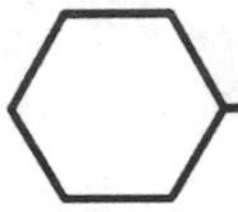 $C_6H_{11}{-}O{-}SO_3H$ forms [The product of an addition reaction]

(c)  $C_6H_5{-}SO_3H + H_2O$ forms [The product of a substitution reaction]

15. $CH_3\overset{\displaystyle CH_3}{\overset{|}{C}}H{-}\overset{\displaystyle CH_3}{\overset{|}{C}}HCH_2OH$

16. 4-methyl-2-pentanol

17. (a) $CH_3\overset{\displaystyle O}{\overset{\|}{C}}CH_3$ (b) CH_3CH_2CHO which is further oxidized to $CH_3CH_2CO_2H$ (c) no oxidation

18. (a) B,C, (b) B, (c) A, (d) D, (e) D, (f) A, C, D

19. (a)  cyclohexene (b) 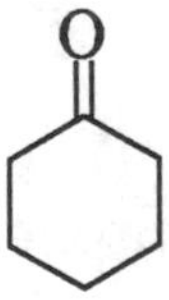 cyclohexanone

20 $CH_3NH_3^+Cl^-$
21. $CH_3CH_2NH_2$ (+ H_2O + NaCl)
22. (a) 1, 7, 10 (b) 10, 11 (c) 2, 6, 9 (d) 5, 8 (e) 3 (f) 4, 9
23. 3
24. 5, 8
25. 2, 5
26. 6, 8

27. 11
28. 3
29. (a) 1, 2, 5, 6, 8, 9 (Structure 4 is ammonia, not an amine.), (b) 1, (c) 3, 10
30. 5, 8
31. 5 (or 8) and 7
32. 1, 2, 4, 5, 6, 8, 9
33. 7
34. 11

35. (a) etc.$-CH_2CH_2CH_2CH_2CH_2CH_2-$etc. or $-\!(CH_2CH_2)\!_n-$

(b) etc.$-CH_2CH(CH_3)CH_2CH(CH_3)CH_2CH(CH_3)-$etc. or $-\!(CH_2CH(CH_3))\!_n-$

(c) etc.$-CH_2CH(Cl)CH_2CH(Cl)CH_2CH(Cl)-$etc. or $-\!(CH_2CH(Cl))\!_n-$

36. (a) ABABABABAB —etc.

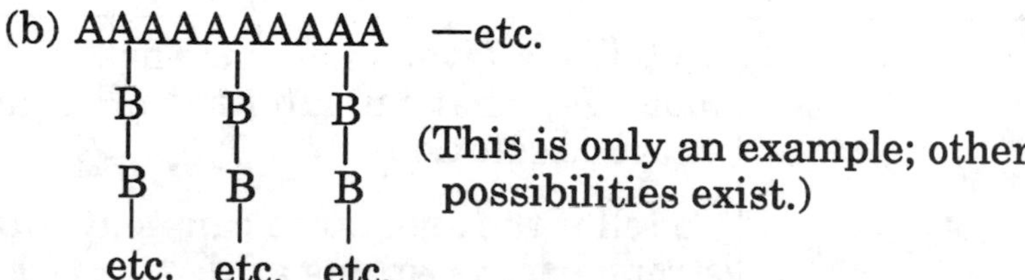

37. $CH_2{=}C(CH_3)CO_2H$

38. hydrogen bonds

Tools you have learned

Remove this chart from the Study Guide and keep it handy when tackling homework problems.

Tool	*Function*
Functional group structures	To recognize to which family (or families) a given structural formula belongs.
IUPAC rules of naming compounds	To construct names from structures. To write structures from names.
Oxidizable groups alkene double bond alcohol group (1° and 2°) aldehyde group	To tell if a given structure is vulnerable to attack by an oxidizing agent
pH-affecting functional groups Can lower the pH: carboxyl group protonated amine Can raise the pH: amino group carboxylate anion	To tell of a given substance has molecules that can alter the pH of an aqueous solution. To tell if the substance can neutralize strong base or strong acid.
Hydrolyzable groups esters and amides	To tell if a substance is vulnerable to attack by water (assuming the right catalyst or promoter).
Polymers and repeating units	To tell if a macromolecule is a polymer (with a repeating unit). To write the structure of the monomer(s)

Chapter 25

BIOCHEMICALS

All of living processes in nature have a molecular basis, and *biochemistry* describes them. The complex molecules of biochemistry have functional groups like those in simpler substances and so they have similar chemical properties. This chapter is meant to introduce you to the major kinds of biochemicals.

Learning Objectives

In this chapter, you should keep in mind the following goals.

1. To learn the names some of the kinds of biochemicals that are vital to living systems and the principal functional groups that characterize them.
2. To be able to look at a structural formula of a biochemical and assign it to its appropriate family.
3. To learn the structure of glucose, both its cyclic and its open-chain forms and learn how glucose is related to the nutritionally important disaccharides and polysaccharides.
4. To learn how to write the structure of a molecule—a triacylglycerol—typically found among animal fats or vegetable oils.
5. To learn the structural differences between the animal fats and vegetable oils and what "polyunsaturated" means.
6. Given the structure of a triacylglycerol, write an equation for its digestion, saponification, or hydrogenation.
7. To be able to describe the lipid bilayer arrangement of animal cell membranes.
8. Given the structures of any two α-amino acids, to write the structures of the possible dipeptides they could form.
9. To be able to write the sequence of atoms joined in the "backbone" of a polypeptide and explain how polypeptides can be alike in backbones but still different.
10. Given the structure of a simple polypeptide—e.g., a dipeptide or tripeptide—to be able to write an equation for its digestion.

11 To be able to distinguish between the terms "protein" and "polypeptide."

12 To be able to describe in general terms the causes of unique overall shapes for proteins and explain how such shapes are important at the molecular level of life.

13 To explain what enzymes are, what they do, and what the lock-and-key mechanism explains.

14 To learn the names and symbols of the substances directly involved in the chemistry of heredity; to explain what exons and introns are; and to describe in general terms what "replication" is.

15 To be able to describe what is meant by the "genetic code" and what are the different functions served by the types of nucleic acids—DNA, rRNA, ptRNA, mRNA, and tRNA—in the synthesis of polypeptides.

16 To be able to describe the technology of genetic engineering and the purposes served by it.

25.1 The Major Types of Biochemicals

Review

This very short section is meant only as an overview as well as to introduce you to the general kinds of biochemicals and how each serves in providing a living system with materials, energy, and information.

Self-Test

1. What is studied in the field of biochemistry?

   ____

2. What two kinds of substances are the chief sources of chemical energy for living systems?

   ____

3. The biochemicals most closely involved in providing information for living things are in what family?

   ____

4. What is the general name for the catalysts in cells? ________________

New Term

Write the definition of the following term, which was introduced in this section. If necessary, refer to the Glossary at the end of the text.

biochemistry

25.2 Carbohydrates

Review

The simplest *carbohydrates*—the *monosaccharides*—involve alcohol and aldehyde or ketone groups, so they partake of the properties of these systems. Their molecules are normally in cyclic forms that are in equilibrium with open-chain forms, and only in the open-chain forms are the aldehyde or keto groups present.

The *disaccharides* and *polysaccharides* give the monosaccharides when they react with water. These systems are made from the cyclic forms of the monosaccharides, being strung together by means of oxygen bridges. Water reacts at these bridges when disaccharides and polysaccharides are digested. We can write equations for the hydrolysis (digestion) of di- and polysaccharides as follows, and these equations will help you remember the important relationships.

Disaccharides:

$$\text{lactose} + H_2O \xrightarrow[\text{(hydrolysis)}]{\text{digestion}} \text{galactose} + \text{glucose}$$

$$\text{sucrose} + H_2O \xrightarrow[\text{(hydrolysis)}]{\text{digestion}} \text{glucose} + \text{fructose}$$

Polysaccharides:

$$\text{starch} + nH_2O \xrightarrow[\text{(hydrolysis)}]{\text{digestion}} n\text{glucose}$$

$$\text{cellulose} + nH_2O \xrightarrow[\text{(hydrolysis)}]{\text{acid catalysis}} n\text{glucose}$$

Starch is actually a mixture of two polysaccharides, amylose and amylopectin. Cellulose, the chief constituent of the cell walls of plants, is not digestible in humans, but its acid-catalyzed hydrolysis also gives glucose as the only product.

Self-Test

5. The functional group generally absent from carbohydrates is
   (a) alcohol (c) aldehyde
   (b) alkene (d) ketone ________
6. Animals store glucose units as
   (a) sucrose (c) starch
   (b) glycogen (d) cellulose ________
7. Sucrose digestion leads to
   (a) glucose and fructose (c) glucose only
   (b) galactose and glucose (d) malt sugar ________
8. The sugar in milk is
   (a) glucose (c) lactose
   (b) maltose (d) sucrose ________
9. A carbohydrate that makes up most of cotton is
   (a) cellulose (c) lactose
   (b) maltose (d) starch ________
10. The digestion of lactose gives
    (a) glucose (c) fructose and glucose
    (b) table sugar (d) galactose and glucose ________
11. Because of the many OH groups in glucose molecules, glucose is
    (a) insoluble in water (c) soluble in water
    (b) nonpolar (d) hypotonic ________
12. The digestion of starch is an example of
    (a) oxidation (c) neutralization
    (b) reduction (d) hydrolysis ________

25.3 Lipids

Review

The ester group is the key functional group in the *triacylglycerols*, members of the family of *lipids* that react with water during digestion to give long-chain carboxylic acids—fatty acids—and glycerol. The fatty acids often carry one or more alkene double bonds. It is not the presence of an ester group that defines the larger family of the lipids, however. To be a lipid, all that a natural product has to be is mostly hydrocarbon-like so that it tends to be far more soluble in nonpolar solvents than in water. Thus cholesterol, which has no ester group, is a lipid.

Self-Test

13. Which of the following compounds could *not* be obtained by the digestion of triacylglycerol?

(a) $CH_3CH_2CH_2CH_2CH_2CH_2CH_2CH_2CH_2CH_2CH_2\overset{O}{\overset{\|}{C}}OH$

(b) $CH_3CH_2CH_2CH_2CH_2CH_2CH_2CH_2CH_2CH_2CH_2CH_2\overset{O}{\overset{\|}{C}}OCH_3$

(c) $HOCH_2\underset{\underset{OH}{|}}{C}HCH_2OH$

(d)

$$CH_3CH_2CH_2CH_2CH_2CH_2CH_2CH_2CH{=}CHCH_2CH_2CH_2CH_2CH_2CH_2CH_2\overset{O}{\overset{\|}{C}}OH$$

____

14. Because the vegetable oils have several alkene groups per molecule, they are called

(a) polyunsaturated
(c) polymers
(b) polyenes
(d) polypeptides

____

15. When triacylglycerols are digested, the reaction is the hydrolysis of

    (a) glycerol (c) alkene groups

    (b) fatty acids (d) ester groups ________

16. The hydrocarbon-like portions of a phosphoglyceride are

    (a) cationic (c) hydrophilic

    (b) hydrophobic (d) anionic ________

17. The surfaces of the lipid bilayer are dominated by

    (a) hydrophilic groups (c) cholesterol

    (b) polyunsaturation (d) nonpolar tails ________

18. One of the services performed by proteins embedded in lipid bilayer membranes is

    (a) digestive enzyme function (b) enzyme manufacture

    (c) hormone synthesis (d) ion channels ________

New Terms

Write the definitions of the following terms, which were introduced in this section. If necessary, refer to the Glossary at the end of the text.

fatty acids triacylglycerols

lipids

25.4 Proteins

Review

All *proteins* consist of molecules of one or more *polypeptides*, and many proteins also include another organic molecule or a metal ion. Each polypeptide is a polymer of several (usually hundreds of) *α-amino acids*. The specific amino acids used, the number of times each is employed, and the order in which they are joined are the three factors that give each polypeptide its uniqueness. All but one of the some 20 amino acids are chiral, so all proteins are also chiral.

When amino acids link together, water splits out from the carboxyl group of one and the α-amino group of the next one to give a peptide bond. Polypeptides coil over much of their lengths into helices that are stabilized by hydrogen bonds. These helices usually undergo further kinking and folding. Thus each polypeptide has its own unique shape as well as unique amino acid sequence, and if this shape is lost, the protein no longer can function biologically.

Enzymes are proteins that catalyze reactions in cells. An enzyme molecule can accept only those substrate molecules that can fit to it.

Self-Test

19. The side chain in serine is —CH_2OH. Therefore, serine's structure is

(a)

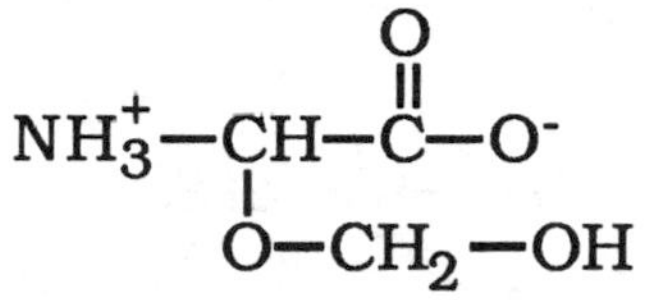

$$\overset{+}{N}H_3-CH(O-CH_2-OH)-\overset{O}{\overset{\|}{C}}-O^-$$

(b)

$$\overset{+}{N}H_3-CH(CH_2-OH)-\overset{O}{\overset{\|}{C}}-O^-$$

(c)

$$HO-CH_2-\overset{+}{N}H_2-CH(OH)-\overset{O}{\overset{\|}{C}}-O^-$$

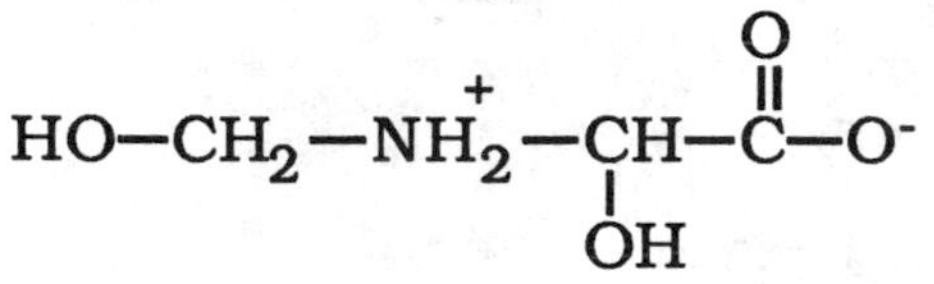

(d)

$$\overset{+}{N}H_3-CH_2-\overset{O}{\overset{\|}{C}}-O-CH_2-OH$$

____

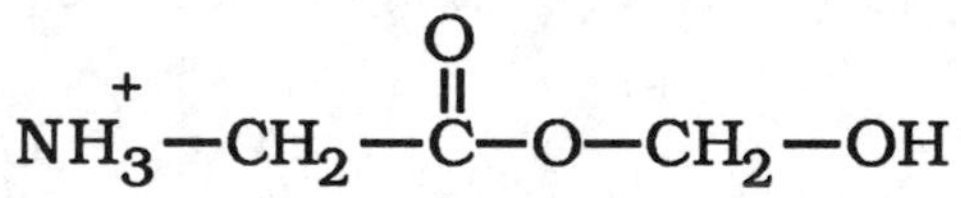

20. Which arrow points to the peptide bond?

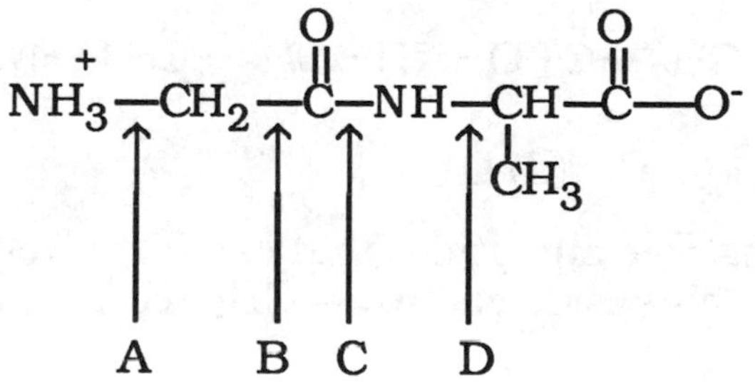

$$\overset{+}{N}H_3-CH_2-\overset{O}{\overset{\|}{C}}-NH-CH(CH_3)-\overset{O}{\overset{\|}{C}}-O^-$$

A (arrow to the NH_3–CH_2 bond) B (arrow to the CH_2–C bond) C (arrow to the C–NH bond) D (arrow to the NH–CH bond)

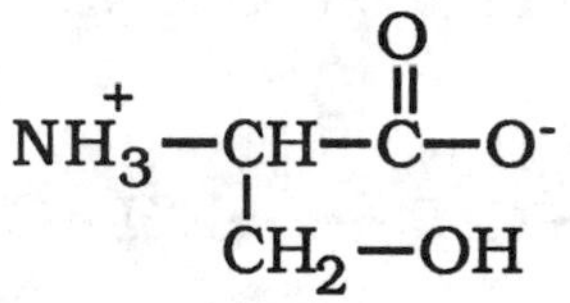

(a) A (b) B (c) C (d) D

____

21. The side chain in the following compound:

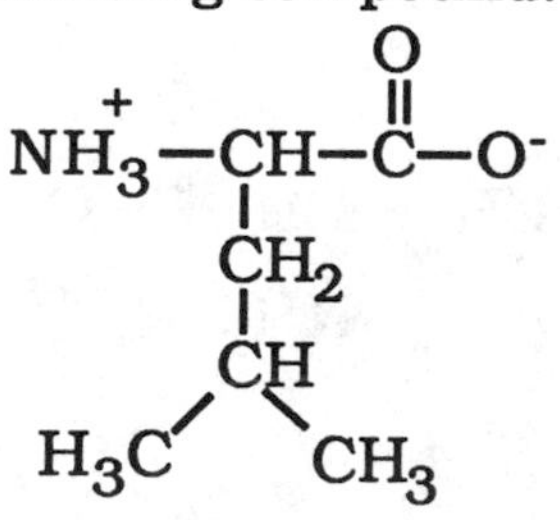

$$\overset{+}{N}H_3-CH(CH_2-CH(CH_3)_2)-\overset{O}{\overset{\|}{C}}-O^-$$

is

(a) 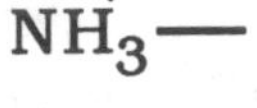 $\overset{+}{N}H_3$—   (b) 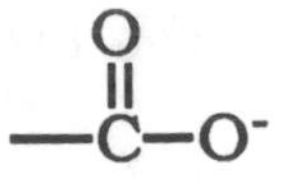 $—\overset{O}{\overset{\|}{C}}—O^-$

(c) an alkyl group   (d) 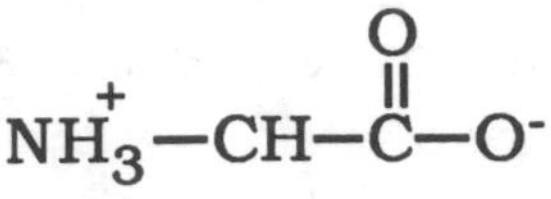 $\overset{+}{N}H_3—CH—\overset{O}{\overset{\|}{C}}—O^-$

____

22. Which structure best represents the nature of the main chain or "backbone" in polypeptides?

A $\overset{+}{N}H_3—CH_2—\overset{O}{\overset{\|}{C}}—NH—CH_2—\overset{O}{\overset{\|}{C}}—NH—CH_2—\overset{O}{\overset{\|}{C}}—etc.$

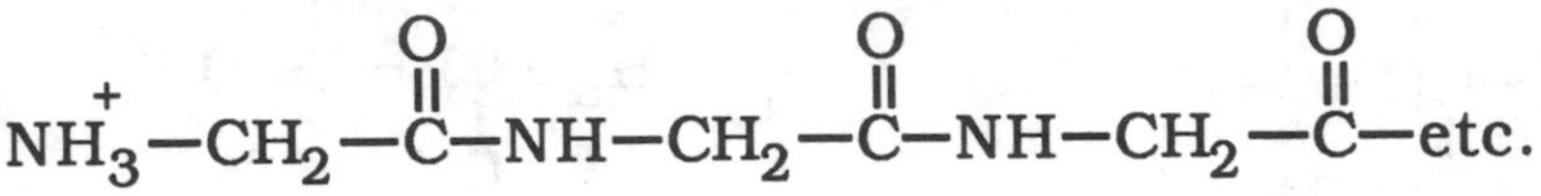

B $\overset{+}{N}H_3—CH_2—\overset{O}{\overset{\|}{C}}—O—\overset{O}{\overset{\|}{C}}—CH_2—NH—NH—CH_2—\overset{O}{\overset{\|}{C}}—etc.$

C $\overset{+}{N}H_3—CH_2—\overset{O}{\overset{\|}{C}}—\underset{NH—}{\underset{|}{C}H}—\overset{O}{\overset{\|}{C}}—\underset{NH—}{\underset{|}{C}H}—\overset{O}{\overset{\|}{C}}—etc.$

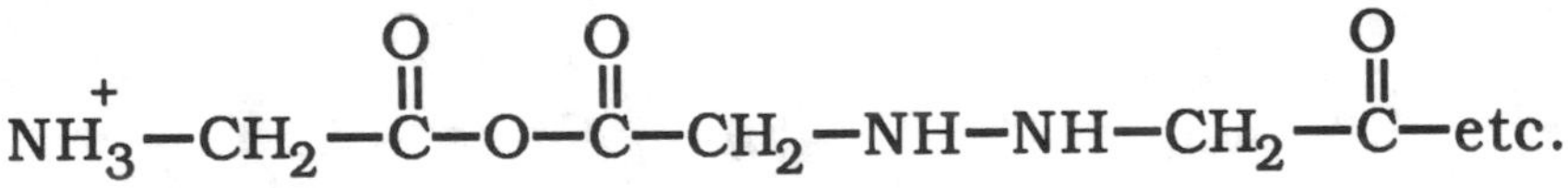

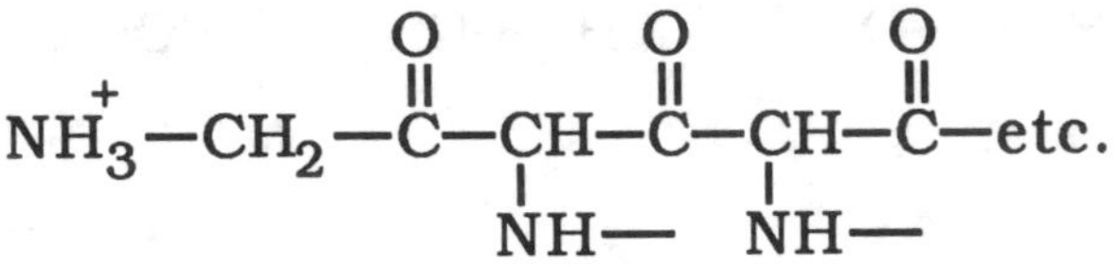

D $\overset{+}{N}H_3—CH_2—\overset{O}{\overset{\|}{C}}—O—NH—CH_2—\overset{O}{\overset{\|}{C}}—O—NH—CH_2—\overset{O}{\overset{\|}{C}}—O—etc.$

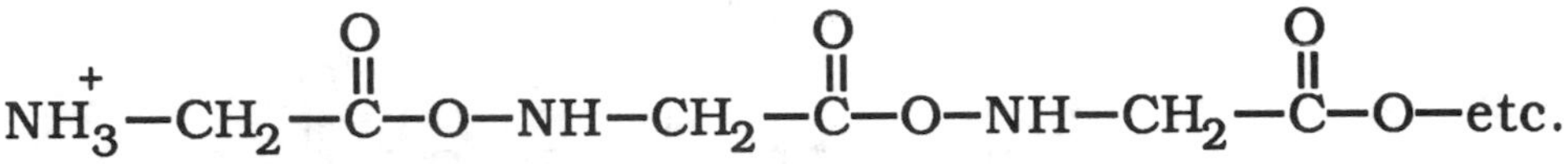

(a) A (b) B (c) C (d) D __________

23. One of the possible dipeptides that can form from alanine (side chain = CH_3) and cysteine (side chain = $—CH_2SH$) is

(a)
$\overset{+}{N}H_3—\underset{SH}{\underset{|}{C}H}—\overset{O}{\overset{\|}{C}}—NH—\underset{CH_3}{\underset{|}{C}H}—\overset{O}{\overset{\|}{C}}—O^-$

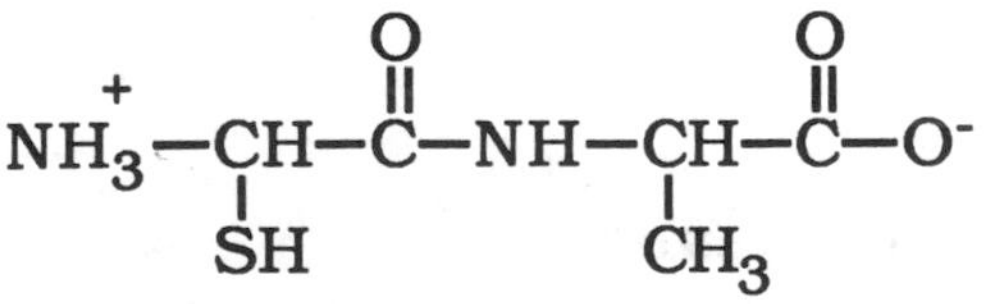

(b)
$\overset{+}{N}H_3—\underset{CH_3}{\underset{|}{C}H}—\overset{O}{\overset{\|}{C}}—S—CH_2—\underset{NH_2}{\underset{|}{C}H}—\overset{O}{\overset{\|}{C}}—O^-$

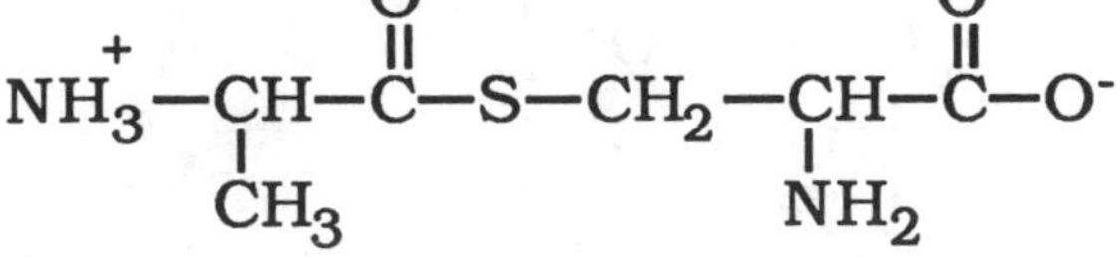

(c) $\overset{+}{N}H_3-CH_2-CH_2-\overset{O}{\overset{\|}{C}}-NH-\underset{CH_2SH}{\underset{|}{CH}}-\overset{O}{\overset{\|}{C}}-O^-$

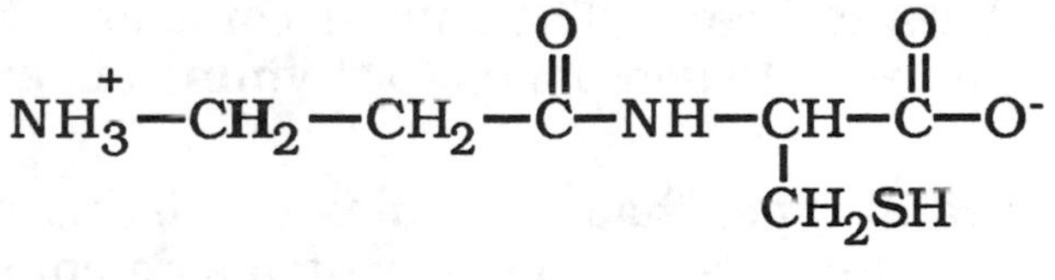

(d) $\overset{+}{N}H_3-\underset{CH_2SH}{\underset{|}{CH}}-\overset{O}{\overset{\|}{C}}-NH-\underset{CH_3}{\underset{|}{CH}}-\overset{O}{\overset{\|}{C}}-O^-$

____

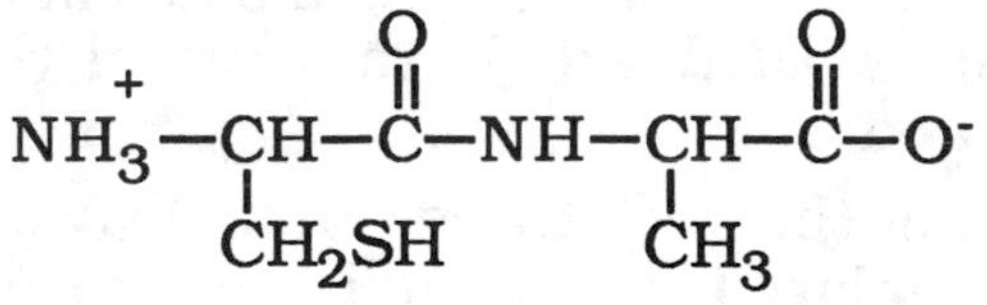

24. In protein chemistry, the work "helix" refers to
    (a) a building unit of a protein
    (b) a folded helix
    (c) a hydrogen bond unit
    (d) a coiled polypeptide __________
25. Enzymes are
    (a) catalysts (c) substrates
    (b) B-vitamins (d) denatured proteins __________

New Terms

Write the definitions of the following terms, which were introduced in this section. If necessary, refer to the Glossary at the end of the text.

α-amino acid	peptide bond
enzyme	polypeptide
lock-and-key mechanism	protein

25.5 Nucleic Acids and Heredity

Review

DNA either directs the synthesis of more of itself—*replication*—or it directs the apparatus for making polypeptide molecules having particular sequences of their amino acid side chains. DNA is one of the two kinds of *nucleic acids,* and the backbone of DNA is an alternating sequence of deoxyribose-phosphate units, each one bearing a side-chain amine or base. The kind, number, sequence, and hydrogen-bonding abilities of these amines—there are four of them—determines the properties of the *gene* units of individual DNA *double he-*

lices, according to the Crick-Watson theory. The amines come as matched base pairs, with adenine (A) pairing by hydrogen bonds to thymine (T) and guanine (G) pairing to cytosine (C).

RNA, which occurs in several types, has molecules consisting of alternating sequence of ribose-phosphate units, each one bearing a side-chain amine or base. They are the same bases as in DNA except that uracil (U) replaces thymine (T). Base pairing occurs A and U.

Just prior to cell division, each of the two strands in a DNA double helix uses the pairing requirements of its amines to guide replication.

Between cell divisions, the sequence of amines in DNA directs the synthesis of primary transcript RNA, ptRNA, in which the bases are complementary to those of the parent DNA. This process—*transcription*—transfers the genetic message to RNA. Both the *intron* and *exon* segments of DNA are transcribed into ptRNA, but only the exon units of DNA make up parts of a gene. The base sequences of ptRNA that came from introns are next deleted, and the sequences that came from the exons are joined to give, after this processing, a molecule of messenger RNA, mRNA. Each adjacent series of three bases on mRNA is a *codon*. It will eventually direct a particular amino acid unit into place during the synthesis of a polypeptide. Thus the *genetic code* is the match-up between codons on mRNA and the amino acids used to make polypeptides.

The mRNA next becomes associated with a cluster of ribosomal RNA molecules (rRNA) and proteins at a particle called a ribosome. Here the synthesis of a polypeptide is directed by the mRNA. Transfer RNA molecules, tRNA, bear amino acid units to this synthesis site. An adjacent series of three bases, called an *anticodon*, on tRNA can fit by hydrogen bonds only to its matching codon on mRNA. This overall process—*translation*—uses the transcribed genetic message (on mRNA) to give the polypeptide structure.

Genetic defects can arise at any stage, but those that are inherited occur as incomplete or faulty sequences of bases on DNA.

Viruses consist of nucleic acids—some with DNA and others with RNA—combined with proteins.

In *genetic engineering,* DNA material corresponding to some desired protein is inserted into a cell where it then proceeds to make the protein for which it is coded. When a cell's DNA—it might be a cell of some bacterium or a yeast—is altered by new DNA, the resulting DNA is called *recombinant DNA*. The technique has been used to manufacture human insulin, human growth hormone, and other proteins.

Self-Test

26. The one-gene—one-enzyme relationship involves a master code that consists of a distinctive sequence of ____________________

along a backbone in mRNA. These consist of three consecutive

______________ joined to ______________________________________ units on the mRNA backbone.

27. The force of attraction responsible for the pairing of amines in the

__________ double helix is the ______________________________

and (use the code letters) __________ pairs with T and ____________ pairs with G.

28. The product of replication is another ______________________

29. The sequences of bases in DNA that together make up a whole gene are called ________________, and the sequences that separate these from each other and are not associated with genes are called ____________.

30. The type of RNA made at the direction of DNA is

____

31. The type of RNA whose base sequence is complementary just to exons is called ______________________________________

32. The overall series of events from DNA to the RNA that directs polypeptide synthesis is called ______________________________

33. After translation has occurred, the product is a ____________________

34. The RNA that carries amino acid units is called ___________________

35. A ______________ is a particle made of nucleic acid and proteins that is capable of causing an infection.

36. When a ______________ in a bacterial cell is made to accept a DNA molecule unrelated to the normal inventory of the cell, it then carries a DNA referred to as ________________, and when such a bacteria is made to manufacture some desired proteins, the overall operation is called____________________________________

New Terms

Write the definitions of the following terms, which were introduced in this section. If necessary, refer to the Glossary at the end of the text.

anticodon
DNA
DNA double helix
exon
genetic code
genetic engineering
intron
recombinant DNA
replication
RNA
transcription
translation

Answers to Self-Test Questions

1. The organic compounds present in living cells or that have been made from them.
2. carbohydrates and lipids
3. nucleic acids
4. enzymes
5. b
6. b
7. a
8. c
9. a
10. d
11. c
12. d
13. b
14. a
15. d
16. b
17. a
18. d
19. b
20. c
21. c
22. a
23. d
24. d
25. a
26. codons; bases; ribose
27. DNA; hydrogen bond; A; C
28. DNA double helix
29. exons; introns

30. primary transcript RNA (ptRNA)
31. messenger RNA (mRNA)
32. transcription
33. polypeptide
34. transfer RNA (tRNA)
35. virus
36. plasmid; recombinant DNA; genetic engineering

Tools you have learned

Remove this chart from the Study Guide and keep it handy when tackling homework problems.

Tool	*Function*
Structure types for biochemicals glucose structure (cyclic) triacylglycerol structure polypeptide "backbone" nucleic acid "backbone"	To enable one to use the structure of an unclassified biochemical and place it in its proper family.
Protein uniqueness the sequence of side chains on the polypeptide backbone	To explain how polypeptides are alike and how they are different.
Nucleic acid uniqueness the sequence of bases on the sugar-phosphate backbone	To explain how nucleic acids are alike and how they are different. To explain how genetic information is stored in nucleic acids

NOTES

NOTES

NOTES

NOTES

NOTES

NOTES

NOTES

NOTES

NOTES

NOTES

NOTES

NOTES

NOTES